AF664997

INSTRUMENTATION IN EARTHQUAKE SEISMOLOGY

MODERN APPROACHES IN GEOPHYSICS

VOLUME 22

The titles published in this series are listed at the end of this volume.

INSTRUMENTATION IN EARTHQUAKE SEISMOLOGY

By

JENS HAVSKOV

University of Bergen,
Norway

and

GERARDO ALGUACIL

University of Granada,
Spain

Springer

A C.I.P. Catalogue record for this book is available from the Library of Congress.

ISBN 978-94-017-5113-1 ISBN 978-1-4020-2969-1 (eBook)
DOI 10.1007/978-1-4020-2969-1

Additional material to this book can be downloaded from http://extras.springer.com

Printed on acid-free paper

CD-Rom includes colour figures and with permission SEISAN and SEISLOG software downloaded from http://www.geo.uib.no/seismologi/software/software.html

Table of contents

Preface

This book is primarily intended for seismologists and technicians working with seismological instruments. Seismologists tend to take their data for granted, hoping that the black boxes of seismographs and processing software will take care of all the nasty problems, much like driving a car and not worrying too much of how it works. But cars stop or malfunction and so do seismographs. Thus a basic understanding of seismic instrumentation is essential, even for the seismologist who is never going to turn a knob on an instrument.

Instrumentation is not just a topic for seismologists, since most equipment is in fact installed and maintained by non-seismologists, so this group of professionals has just as much a need for information on instrumentation.

Early versions of the book have been used for a lab course in seismic instrumentation given at the University of Bergen since 1999 and also in a few other countries. The experience from these courses has helped to improve the book, and we think that it is suitable for students in seismology.

There have been numerous publications on instrumental topics in seismology, with very special emphasis on particular subjects, but few -if any- general textbooks. Some overview was made by Lee and Stewart (1981), which, on the instrumental side, mainly dealt with micro-earthquake networks. The old Manual of Seismological Observatory Practice (MSOP) (Wilmore, 1979) dealt with all the classical analog seismographs, but is mostly outdated now. The New Manual of Seismological Observatory Practice (NMSOP, Bormann (Ed.), 2002, www.gfz-potsdam.de/bib/nmsop_formular.html)) is the most up to date book on seismic instruments and one of the authors of this book (JH) has also participated in making NMSOP. So, why another book? The NMSOP deals with a lot of issues in addition to seismic instruments, and we felt there would be a need for a book which further expands on the instrumental topics, much more than was possible within NMSOP, and put it all together in one volume. The intention with this book is that it should be a practical tool with only the amount of theory needed to understand the basic principles and that answers to most practical problems should be found here.

The book particularly deals with seismic sensors, their response functions and how to make calibration and correct for the instrument. More details of the signal processing are the matter of Scherbaum's book 'Of Poles and Zeros' (Scherbaum, 2001), published in the same series as this one. We consider Scherbaum's a companion book to ours, and have taken care to have a minimum of overlap.

The book tries to make an overview of some of the current equipment on the market as well as references to where it can be obtained. We are well aware that this can only be a snapshot in time due to the fast changes currently taking place and by the time the book is published, new equipment will be available and not mentioned. On the other hand, there is equipment mentioned that has been sold for more than 20 years, so even outdated equipment will serve to illustrate how the technology is used. And the basic principles behind nearly all equipment sold today have been known for some time and are less likely to change rapidly.

Mentioning particular equipment does not mean we endorse that equipment and we have not tried to include all equipment available.

Most of the signal processing examples are made with the earthquake processing software SEISAN (Havskov and Ottemöller, 2000) (www.geo.uib.no) and there are also examples using the data acquisition software SEISLOG (Utheim et al, 2001) (www.geo.uib.no). These software systems are included on the CD with the book as well as color versions of some of the figures.

Many people have given corrections and comments, in particular students who had to read the book. We are grateful to all of them. Einar Mæland has revised sections 6.2 and 6.3 in detail and Ole Meyer has checked the appendix. Several companies have provided technical details for their equipment and reviewed relevant text: Tony Russell (Earth Data), Nathan Pierce (Güralp Systems), Ogie Kuraica (Kinemetrics), Dieter Stoll (Lennartz), Paul Passmore (RefTek), Christoph Kundig (GeoSig) and Robert Leugoud (PMD Eentec). Christoph Kundig has in addition reviewed the whole book. Peter Bormann has done a great job of finding a lot of inconsistencies. Miguel Abril revised the appendix and some chapters. Mathilde Bøtter Sørensen did a very complete final review. The bulk of the book was written during one of the authors (JH) sabbatical stay at the University of Granada with support from the University of Bergen.

CHAPTER 1

Introduction

Seismology would be a very different science without instruments. The real big advances in seismology happened from around 1900 and onwards and were mainly due to advancement in making more sensitive seismographs and devising timing systems, so that earthquakes could be located. Later, the importance of accurate measurement of the true ground motion became evident for studying seismic wave attenuation, and the Richter magnitude scale depends on being able to calculate the ground displacement from our recorded seismogram (Figure 1.1).

The ability to do earthquake location and calculate magnitude immediately brings us into two basic requirement of instrumentation: Keeping accurate time and determining the frequency dependent relation between the measurement and the real ground motion. In order to get there, we need to know a bit more.

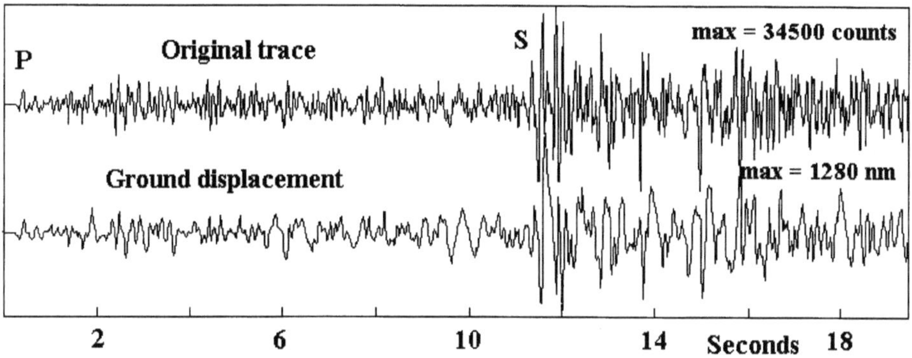

Figure 1.1 The top trace shows the original digitally recorded signal from a magnitude 3 earthquake recorded at a distance of 120 km. The maximum amplitude is just a number (called counts). The bottom trace shows the signal converted to true ground displacement in nm from which the magnitude can be calculated. The distance to the earthquake is proportional to the arrival time difference between the S-wave and P-wave, so having 3 stations makes it possible to locate the earthquake. The seismometer is a 1 Hz sensor with velocity output.

So what are the main topics of instrumental seismology? It all starts with being able to measure the ground motion, and this is the most important topic. Then it follows recording and/or transmission to a central site.

We usually talk about ground displacement (formally measured in m) since this is what seismologists like to use. Like a fault has a displacement of 2 m during an earthquake, we can talk of a ground displacement of 2 m, although we normally are measuring much smaller motions. On the other hand, engineers seem to think that acceleration

(m/s^2) is the most natural unit, since it is directly related to force and the peak ground acceleration is an often quoted measure.

The range of amplitudes measured in seismology is very large. The natural background noise, highly frequency dependent, sets the limit for the smallest amplitudes we can measure, which is typically 1 nm displacement at 1 Hz (Chapter 3), while the largest displacement is in the order of 1 m. This is a dynamic range of 10^9. The band of frequencies we are interested in also has a large range, from 10^{-5} to 1000 Hz (Table 1). The challenge is therefore to construct seismic instruments, both sensors and recorders, which cover at least part of this large frequency and dynamic range.

Table 1.1 Typical frequencies generated by different seismic sources.

Frequency (Hz)	Type of measurements
0.00001-0.0001	Earth tides
0.0001-0.001	Earth free oscillations, earthquakes
0.001-0.01	Surface waves, earthquakes
0.01-0.1	Surface waves, P and S waves, earthquakes with M > 6
0.1-10	P and S waves, earthquakes with M> 2
10-1000	P and S waves, earthquakes, M< 2

Earlier, analog instrument were usually made to record one type of ground motion like velocity. Traditionally, seismologists prefer recording weak motion displacement or velocity, for easy interpretation of seismic phases, while engineers use strong motion acceleration, whose peak values are directly related to seismic load on structures. Today it makes less of a difference, since due to advancement in sensor and recording systems, the weak motion instruments can measure rather strong motions and the strong motion sensors are almost as sensitive as the weak motion sensors. The digital recording furthermore makes it easy to convert from acceleration to velocity etc., see Figure 1.2.

1.1 Sensor

Since the measurements are done in a moving reference frame (the earth's surface), almost all seismic sensors are based on the inertia of a suspended mass, which will tend to remain stationary in response to external motion. The relative motion between the suspended mass and the ground will then be a function of the ground's motion (Figure 1.3). The swinging system will have a resonance frequency f_0

$$f_0 = \frac{1}{2\pi}\sqrt{k/m} \tag{1.1}$$

where k is the spring constant and m the mass. If the ground displacement frequency is near the resonance frequency, we get a larger relative motion (depending on damping) and, as it turns out, below the resonance frequency, the relative displacement, due to ground displacement, decreases (Chapter 2).

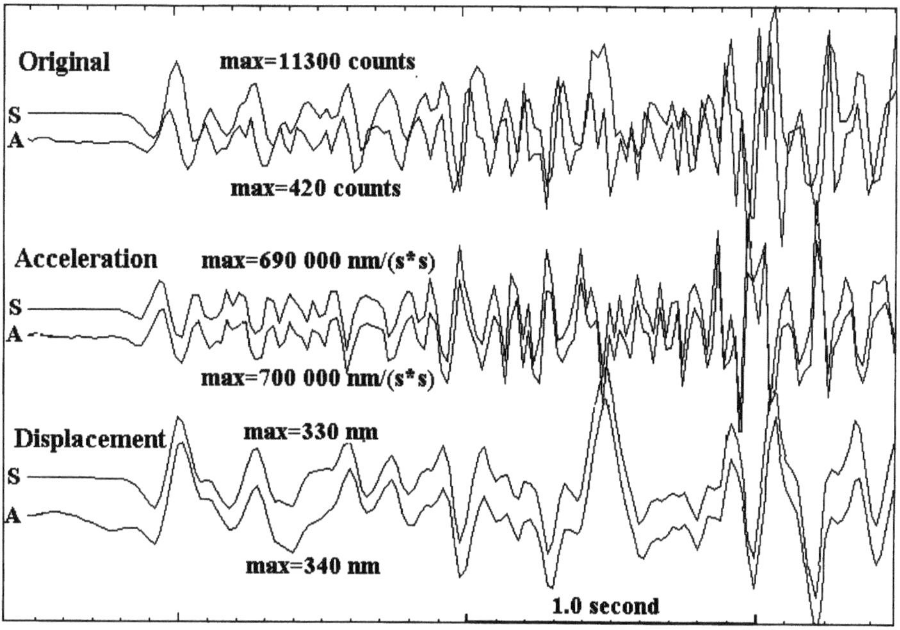

Figure 1.2 Acceleration and displacement. The seismogram in the figure is the first few seconds of the P-wave of the signal seen in Figure 1.1. On the site there is also an accelerometer installed (A) next to the seismometer (S). The top traces show the original records in counts. The signal from the seismometer is similar to the accelerometer signal, but of lower dominant frequencies and the amplitudes are different. The middle traces show the two signals converted to accelerations and the bottom traces, converted to displacement (frequency band 1-20 Hz). The signals are now very similar and of the same amplitude. This example clearly demonstrates that, with modern instruments and processing techniques, we can use both accelerometers and (velocity-sensitive) seismometers and get the same result.

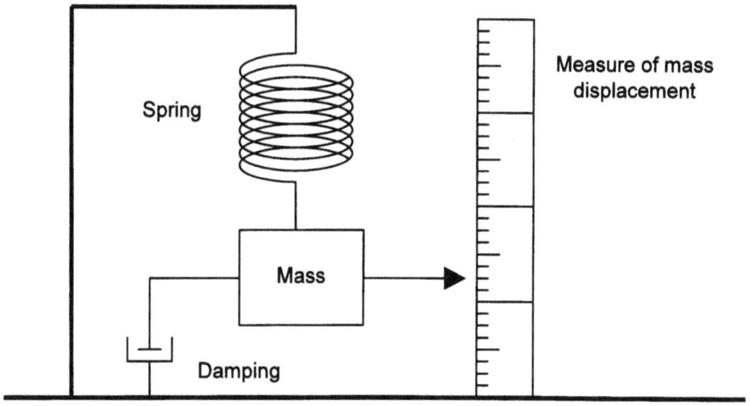

Figure 1.3 Principle behind the inertial seismometer. The damping of the motion can be mechanical, but is usually electro-magnetic.

The sensor is moving with the ground and there is not any fixed undisturbed reference available. So displacement or velocity cannot be measured directly. According to the inertia principle, we can only observe the motion if it has a non zero acceleration. So if we put a seismic sensor in a train, we can only measure when the train is accelerating or braking. As seismologist, we want to measure displacement, but this is not possible to do directly. We have to measure the ground acceleration and integrate twice. Displacement at very low frequencies produces very low accelerations since

$$\ddot{u} \propto f^2 u \tag{1.2}$$

where u is the ground displacement and f the frequency. It is therefore understandable why it is so difficult to produce seismometers that are sensitive to low frequency motion. So, while it is quite straight forward to make a sensor which records equally well a given acceleration level at all frequencies, even DC, it is much more difficult to measure low frequency displacements. Much of the advance in recent years in seismic instruments has been the ability to build sensors with a better low frequency sensitivity. Earlier, sensors sensitive to low frequency were made with sophisticated mechanical systems, which by all sorts of tricks were able to have a low resonance frequency. It was, however, not possible to make sensors with a stable resonance frequency much lower than 0.03 Hz. Today, purely mechanical sensors are only constructed to have resonance frequencies down to about 1.0 Hz (short period sensors), while sensors that can measure lower frequencies are based on the Force Balance Accelerometer (FBA) Principle of measuring acceleration directly.

The sensor mass displacement is linearly proportional to the external acceleration, even at zero frequency since this just corresponds to a permanent change in the external force. As we shall see in Chapter 2, this linearity does not hold for frequencies above the resonance frequency, where the mass displacement due to ground acceleration will drop proportional to frequency squared. But, since it is easy to construct a swinging system with a high resonance frequency (small mass and/or stiff spring, (1.1)), we can easily make an accelerometer by just measuring the mass displacement of a spring suspended mass for frequencies below the resonance frequency. So we have our mechanical accelerometer, at least in theory. The problem is now how to measure this mass displacement, particularly at low frequencies, where the acceleration is small. The popular velocity transducer (gives out a voltage proportional to the mass velocity relative to the frame) is not a good choice since the velocity goes down with frequency and is zero at zero frequency. So, even if there is a permanent acceleration, we cannot measure it. Special mass displacement transducers have been introduced, but it is hard, if not impossible, to make them accurate for the large dynamic range we need in seismology. However, for a small displacement measurement, they can be made very sensitive and accurate. This is what is used in the so-called Force Balance Accelerometer Principle (Figure 1.4).

The Force Balanced Accelerometer (FBA) has a feedback coil, which can exert a force equal and opposite to the inertia force due to the largest acceleration we want to measure. The displacement transducer sends a current to this force coil through a resistor R in a negative feedback loop. The polarity of the current is such that it opposes any motion of the mass, and it will try to prevent the mass from moving at all with respect to the frame. A small permanent acceleration on the mass will therefore result in a small permanent current and a large acceleration will need a large current. The current is in fact proportional to the ground acceleration, so the voltage over the resistor gives a direct measure of acceleration.

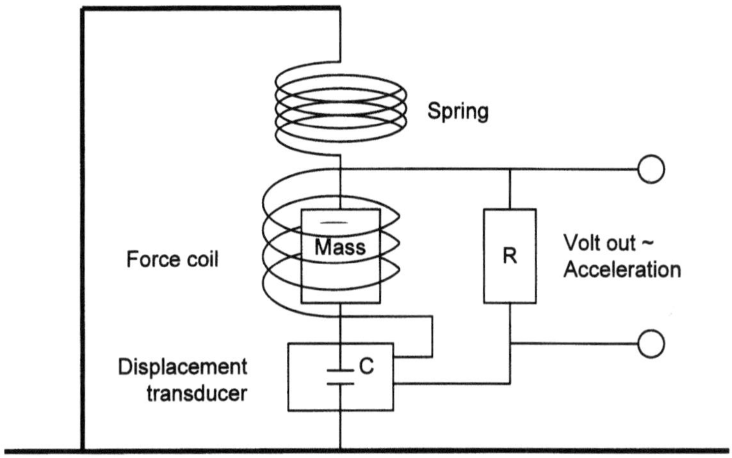

Figure 1.4 Simplified principle behind Force Balanced Accelerometer. The displacement transducer normally uses a capacitor *C*, whose capacitance varies with the displacement of the mass. A current, proportional to the displacement transducer output, will force the mass to remain stationary relative to the frame.

The FBA principle is now the heart of nearly all modern strong motion and broadband sensors (sensors recording in a large frequency band like 0.01 to 50 Hz). By connecting an integrating circuit, either in the feedback loop, or after, the sensor can give out a voltage proportional to velocity. However, due to the inertial principle, there must be a low frequency limit (velocity is zero at zero frequency!) and this limit is set by the mechanical-electrical qualities of the sensor. Currently, the best broadband sensors have a limit of about 0.0025 Hz, much lower than it was ever possible with purely mechanical sensors. The FBA principle also has the advantage of making linear sensors with a high dynamic range, since the mass almost does not move. Currently, the best sensors have a dynamic range of 10^6 - 10^8.

So, in summary, mechanical sensors are the most used when the natural frequency is above 1 Hz, while the FBA type sensors are used for accelerometers and broadband sensors. This means that a lot of sensors now are 'black boxes' of highly complex mechanical and electronic components of which the user can do very little. However, as we shall see in the calibration sections, Chapter 10, it is still possible to do a series of simple tests to check the functionality of the instrument.

1.2 Recorder

The next challenge is to be able to actually record the signals from the seismometer without loosing quality. Analog recording is still used (Figure 1.6) since it gives a very fast and easy overview of the seismicity. However, the dynamic range is very limited (at most 200 in contrast to 10^8 for the best recorders). Analog recorders are very expensive to buy and operate, so they are rarely installed anymore. In addition, few digital seismic recorders give out an analog signal required of the analog recorders, so an additional digital to analog converter must be installed.

The development of high dynamic range and broadband sensors would not have improved seismology a lot unless the recording technique had followed. Fortunately, the last 20 years have seen a fast development in digital recording, which almost has been able to keep up with the improvement in sensor development. Converting the analog signal to a digital signal means converting the continuously varying signal to discrete samples of the analog signal, see Figure 1.5.

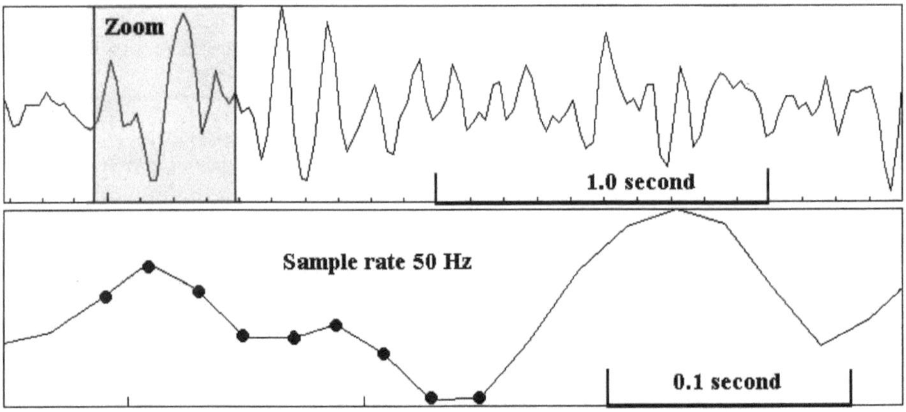

Figure 1.5 Example of a digitized earthquake signal. The top trace shows the signal with a low time resolution and the bottom trace, the zoom window with a high resolution. In the zoom window, the discrete samples are clearly seen. The sampled points are indicated with black dots.

The digitizer will thus convert a voltage to a number, just like we do with a digital voltmeter. The difference is that we have to do it many times a second on a varying amplitude signal. From the figure, we see that we are loosing information between each two samples, so that intermediate changes cannot be seen after digitization. We can also see two important aspects of the digitizer: The sample interval (step in time direction) and the resolution (step in amplitude direction), which corresponds to one step in the numerical output from the digitizer. If more high frequency content is needed, it is just a question of increasing sample rate (except that the dynamic range usually deteriorates with a higher sample rate) and we have to sample at least at twice the rate of the frequency of the signal we want to record. The step in the amplitude direction depends on the digitizer resolution and is the smallest step it can resolve. The best digitizers can usually resolve 100 nV, which then ideally should correspond to the number 1. So, what about the dynamic range? Assuming that the number 1 out of the digitizer corresponds to the real input signal (and not digitizer noise), then the largest number corresponds to the dynamic range. Digitizers are usually classified as 12 bit, 16 bit or 24 bit which has to do with the number of discrete values the digitizer uses. A 12 bit converter has 2^{12} levels and a 24 bit converter 2^{24} levels or $\pm 2^{11}$ and $\pm 2^{23}$ respectively. This corresponds to dynamic ranges of 2048 and 8.4 10^6, respectively. Now most digitizers are based on 24 bit analog-to-digital converters (ADC), however few have a true 24 bit performance, since shorting the input gives out counts significantly different from zero. So the digitizers do not quite match the best sensors, although there are a few digitizers approaching and above the true 24 bit performance. The standard quality 24 bit digitizers usually only reach 21-22 bit dynamic range.

Digitizer and recorder are often one unit although they physically are two separate parts. Recording is now mostly based on hard disks, while some units use solid-state memory. The trend is to use more and more powerful computers. While many units on the market today use microcontrollers in order to save power, the trend is to use complete single board PC's due to the low power consumption of these new generation PC's. The advantage of using real PC's with 'normal' operating systems is that all standard software, e.g. communication, can be used instead of the manufacturers' special fix for his special recorder.

Since large hard disks can be used for recording, recorder memory is no longer a problem and continuous recording is now easy to do. However, nearly all recorders also have a facility for additionally storing only the time segment containing real events and the system is said to have a trigger. This means that a program continuously monitors the incoming signal levels to decide when a real event occurs.

1.3 Stations and networks

So now we have a complete system and it is just putting it in the field, or isn't it? If we are going to be able to record a signal level of 1 nm, we cannot just put the sensor anywhere. Apart from the earth's natural background noise, which well may be larger than 1 nm at 1 Hz (Chapter 3), we must take other noise disturbances into account like traffic, wind induced noise etc. So before installing a station, a noise survey must be made (Chapter 7). In addition, broadband sensors require special attention due to the large sensibility to low frequency disturbances created mainly by temperature induced tilt. So while it may be simple and cheap to install a short period seismometer, a broadband installation might be time consuming and expensive. In some cases, it may even be necessary to install the sensor in a borehole to get away from the noise near the surface. Another method of improving the signal to noise ratio, is to use a seismic array (Chapter 9).

One station does not make a network, which is what we normally want for locating earthquakes. Setting up a network is mainly a question of communication. This field is in rapid development and will probably completely change our concept of how to make seismic networks in the future. Earlier, a network was often a tightly linked system like the classical microearthquake network of analog radio telemetric stations a few hundred kilometers away sending analog data from the field stations to a central digital recorder (Figure 1.6). While these types of networks still are being built using digital transmission, the trend is for field stations to become independent computer nodes that, by software, can be linked together in *virtual seismic networks*, so called to distinguish them from the more classical physical networks. So, in the virtual network, the communication system is a design part separated from the seismic stations themselves. The communication can be based on radio, satellite or public data communication channels.

Since data from many different stations are used together, all signals must be accurately timed. Today, this is not a problem since the GPS (satellite based Global Positioning System) system delivers very accurate time and GPS receivers are cheaply available.

Figure 1.6 The central recording site at the Andalusia Seismic Network. The network consists of analog seismic station sending analog signals to the central recording site. Here the signals are digitized using the PC and events triggering the system are stored on disk. In addition, data are recorded in analog form on drum recorders (helicorder). In this photo we see an example where a seismic swarm is recorded, making it possible to get an easy overview of the recent activity.

1.4 Arrays

Ground motion measured by a seismic station is a sample of one point of a spatial wave-field that propagates in time. A set of stations may be deployed in an area in such a way that the whole wave-field in this area is sampled and the data processing of all waveforms is performed jointly. We call this set of seismic stations a *seismic array*. Linear arrays have been used for seismic surveys for long, but the term is now applied to 2-D or even 3-D arrangements for studying the seismic source, wave propagation, shallow structure, etc.

The use of arrays makes it possible, for example, to "direct" the sensitivity of the network to a source point, with an improved signal-to-noise ratio (SNR) over the individual stations. Its applications have been extended in recent years from global seismology and nuclear test ban verification to local and regional seismology, including the seismo-volcanic sources, to be used as a powerful tool for seismologists.

1.5 Instrument correction and calibration

We are now recording a lot of nice signals and some seismologist is bound to start using it; after all, that is why we set up seismic stations. The signal will then be displayed on screen and the maximum amplitude will be e.g. 23838. This is not of much help if we want to know what the true ground motion is, so we now have to figure out what the relation is between the true ground motion and the numbers out of the computer. This involves two steps: (1) Getting the true parameters for each unit of the sensor and the recorder, which might involve doing laboratory work, but at least consulting manufacturer manuals and (2) Being able to calculate the systems response function $T(f)$, which simply is defined as

$$T(f) = \frac{Z(f)}{U(f)} \tag{1.3}$$

where f is the frequency, Z in the recorded signal and U the true ground signal. Knowing the response function, we can then calculate the true ground signal. Unfortunately, the story does not end here since the function $T(f)$ can be specified in many different ways depending on data format and seismic processing systems, and it is not uncommon to get different results. So working with response information can be quite confusing. Figures 1.1 and 1.2 show examples of signals corrected for instrument response.

By now, the seismologists should have complete correct data available for a well calibrated seismic station installed at an ideal location with a perfect operating seismic network, if all recommendations in this book have been followed!! But it is not time to relax yet. When analyzing the data, it might become evident that not all is well. The hope is that by knowing more about seismic instruments, problems should be spotted soon, and it should be relatively easy to find out where the problem is. Help is then at hand to do some calibrations and tests (Chapter 10). Repairing the equipment is another story and best left to the true experts.

CHAPTER 2

Seismic sensors

A seismic sensor is an instrument to measure the ground motion when it is shaken by a perturbation. This motion is dynamic and the seismic sensor or seismometer also has to give a dynamic physical variable related to this motion. In the earlier seismometers, this variable was the displacement of a stylus representing the amplified ground motion. In present day instruments, the physical output that a seismometer yields is always a voltage.

Ideally, we can consider a seismometer as a black box whose input is ground motion (represented by a kinematic variable: displacement, velocity or acceleration) and its output is voltage -or pen displacement, in older equipment-.

Our objective is to measure the ground motion at a point with respect to this same point undisturbed. Unfortunately it is not so easy to measure this motion and the seismic sensor is the single most critical element (and can be the most expensive) of a seismograph (seismometer and recording unit). The main difficulties are:

1: The measurement is done in a moving reference frame, in other words, the sensor is moving with the ground and there is not any fixed undisturbed reference available. So displacement cannot be measured directly. According to the inertia principle, we can only observe the motion if it has an acceleration. Seismic waves cause transient motions and this implies that there must be acceleration. Velocity and displacement may be estimated, but inertial seismometers cannot detect any continuous component[1] of them.

2: The amplitude and frequency range of seismic signals is very large. The smallest motion of interest is limited by the ground noise. From Figure 3.3, it is seen that the smallest motion might be as small as or smaller than 0.1 nm. What is the largest motion? Considering that a fault can have a displacement of 10 m during an earthquake, this value could be considered the largest motion. This represents a dynamic range of $(10/10^{-10}) = 10^{11}$. This is a very large range and it will probably never be possible to make one sensor covering it. Similarly, the frequency band starts as low as 0.00001 Hz (earth tides) and could go to 1000 Hz. These values are of course the extremes, but a good quality all round seismic station for local and global studies should at least cover the frequency band 0.01 to 100 Hz and ground motions from 1 nm to 10 mm.

---

[1] Actually, the seismometer can measure only velocities or displacements associated to non-zero values of acceleration. As it will be shown, the response for ground velocity and for ground displacement approaches zero as frequency decreases. This is an essential characteristic of inertial seismometers and cannot be overcome. Other types of sensors measure strain (strain-meters) and in this case they may respond to static relative displacements, since they are not based on inertia.

It is not possible to make one single instrument covering this range of values and instruments with different gain and frequency response are used for different ranges of frequency and amplitude. Sensors are labeled e.g. short period (SP), long period (LP) or strong motion. Today, it is possible to make instruments with a relatively large dynamic and frequency range (so called broadband instruments (BB)) If the sensor has a period of more then 200 s, it is called a very broadband sensor and sometimes labeled VBB. Since this abbreviation also is used for velocity broad band sensor (which covers most BB sensors), the abbreviation BB will be used for all broad band sensors in this text. The tendency goes in the direction of increasing both the dynamic and frequency range. This means that where earlier two or three sensors were used, one sensor can today often do the same job, although still not covering the complete range. So the development goes towards the black box concept.

2.1 The standard inertia seismometer

Since the measurements are done in a moving reference frame (the earth's surface), almost all seismic sensors are based on the inertia of a suspended mass, which will tend to remain stationary in response to external motion. The relative motion between the suspended mass and the ground will then be a function of the ground's motion.

Figure 2.1 shows a simple seismometer that will detect vertical ground motion. It consists of a mass suspended from a spring. The motion of the mass is damped using a 'dash pot', so that the mass will not swing excessively near the resonance frequency of the system. A ruler is mounted on the side to measure the motion of the mass relative to

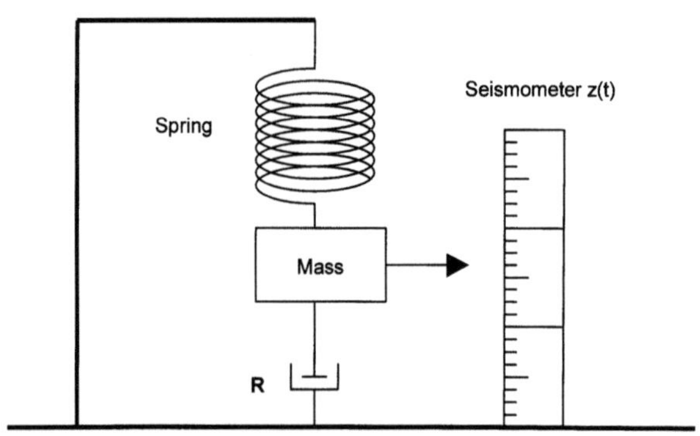

Figure 2.1 A mechanical inertial seismometer. R is a dash pot.

the ground. If the system is at rest, what would happen if a sudden (high frequency) impulse like the first swing of a P-wave occurred? Intuitively one would expect that the mass initially remains stationary while the ground moves up. Thus the displacement of the ground can be obtained directly as the relative displacement between the mass and the earth as read on the ruler. It is also seen that the ground moves up impulsively, the mass moves down relative to the frame, represented by the ruler, so there is a phase shift of π in the measure of ground displacement. Similarly if the ground moves with a very fast sinusoidal motion, one would expect the mass to remain stationary and thus the ground sinusoidal motion could be measured directly. The amplitude of the measurement would also be the ground's amplitude and the seismometer would have a gain of 1. The seismometer thus measures relative displacement directly at high frequencies and we can say the seismometer response function (motion of mass relative to ground motion) is flat at high frequencies. The phase shift is also π in this case.

What about very low frequencies? With the ground moving very slowly, the mass would have time to follow the ground motion, in other words, there would be little relative motion and less phase shift. Thus the gain would be low. Finally, what would happen at the resonance frequency? If the damping is low, the mass could get a new push at the exact right time, like pushing a swing at the right time, so the mass would move with a larger and larger amplitude, thus the gain would be larger than 1. For this to happen, the push from the ground must occur when the mass is at an extreme position (top or bottom) and there must be a phase shift of -π/2, see Figure 2.2.

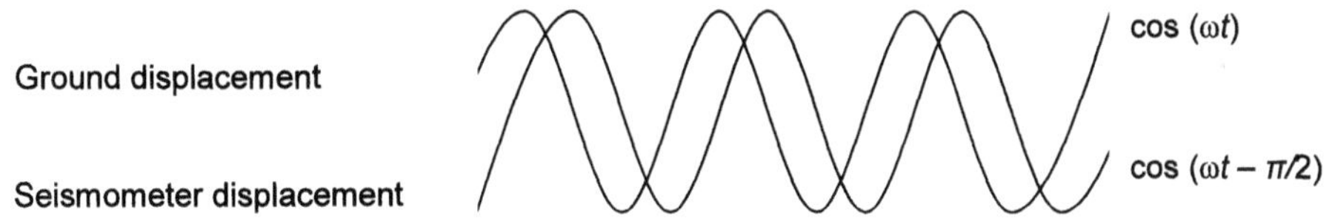

Figure 2.2 Phase shift between the ground motion and the seismometer motion at the resonance frequency. For simplicity, the amplitude is shown to be the same for both signals. Note the sign convention for phase shift.

The sign is negative since the maximum amplitude of the mass movement will be delayed relative to the ground motion, see Figure 2.2. We now have a pretty good idea of how the response of the seismometer will look like and Figure 2.3 shows a schematic representation using a sensor with a resonance frequency of 1 Hz. Note that the phase shifts given are relative to the coordinate system as defined in Figure 2.1. For real seismographs, the convention is that a positive ground motion (vertical up) gives a positive signal, so the measuring scale will have to be inverted, such that for high frequencies and/or an impulsive input, an upwards ground motion gives a positive signal.

Figure 2.3 is of course not an exact representation, but it shows, by comparison to the exact figure (Figure 2.4) that it is possible to get a quite good idea of the frequency response of a mechanical seismometer without using any mathematics. In the following, the exact relationship will be derived.

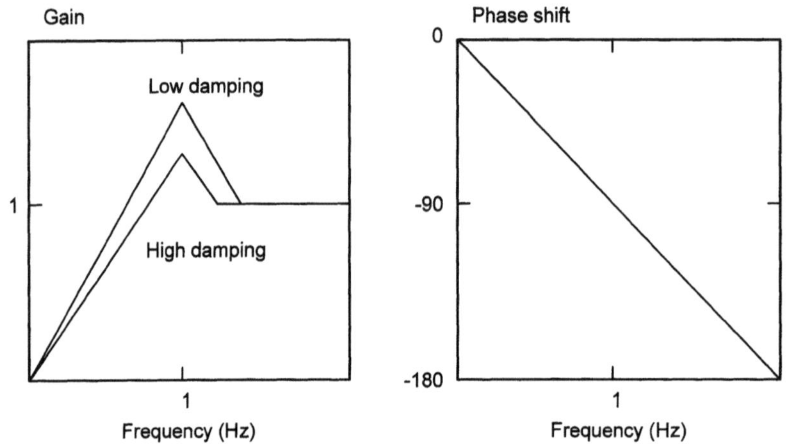

Figure 2.3 The amplitude and phase response of a mechanical seismometer. The curves are derived qualitatively. The free period of the seismometer is 1.0 s. The phase shift is in degrees. For the gain, a curve for low and high damping is shown. For the exact curves, see Figure 2.4.

2.2 Seismometer frequency response

Let $u(t)$ be the ground's vertical motion and $z(t)$ the displacement of the mass relative to the ground, both positive upwards. There are two real forces acting on the mass m: The force of the deformed spring and the damping.

Spring force. $-kz$, negative since the spring opposes the mass displacement, k is the spring constant. The resonance angular frequency of the mass–spring system is $\omega_0 = \sqrt{k/m}$, where $\omega_0 = 2\pi/T_0$ and T_0 is the corresponding natural period.

Damping force. $-d\dot{z}$, where d is the friction constant. Thus the damping force is proportional to the mass times the velocity and is negative since it also opposes the motion.

The acceleration of the mass relative to an inertial reference frame, will be the sum of the acceleration $\ddot{z}$ with respect to the frame (or the ground) and the ground acceleration $\ddot{u}$.

Since the sum of forces must be equal to the mass times the acceleration, we have:

$$-kz - d\dot{z} = m\ddot{z} + m\ddot{u} \tag{2.1}$$

For practical reasons, it is convenient to use ω_0 and the seismometer damping constant, $h = \frac{d}{2m\omega_0}$ instead of k and d, since both parameters are directly related to measurable quantities. (2.1) can then be written:

$$\ddot{z} + 2h\omega_0\dot{z} + \omega_0^2 z = -\ddot{u} \tag{2.2}$$

This equation shows that the acceleration of the ground can be obtained by measuring the relative displacement of the mass, z, and its time derivatives. Before solving (2.2), let us evaluate it.

If the frequency is high, the acceleration will be high compared to the velocity and displacement and the term $\ddot{z}$ will dominate. The equation can then be written approximately as

$$\ddot{z} = -\ddot{u} \tag{2.3}$$

This shows that the motion of the mass is nearly the same as the motion of the ground with reversed sign or a phase shift of π. This is the same result as obtained qualitatively above. If the frequency is low, the z-term will dominate and the equation can be approximated as

$$\omega_0^2 z = -\ddot{u} \tag{2.4}$$

Therefore, for small frequencies, the relative displacement of the mass is directly proportional to the negative ground acceleration, $z = -\ddot{u}/\omega_0^2$ and the sensitivity of the sensor to low frequency ground acceleration is inversely proportional to the squared natural frequency of the sensor. We can simply say that the seismometer works as a spring scale. This is also the principle behind the gravity meter. In the qualitative evaluation above, it seemed that there was no phase shift, so how can the minus sign be explained? This is simply due to the phase shift between displacement and acceleration. If the displacement is written $u(t) = \cos(\omega t)$, then by double differentiating, the acceleration is obtained as $\ddot{u}(t) = -\omega^2 \cos(\omega t) = -\omega^2 u(t)$. The fact that the seismometer measures acceleration linearly at low frequencies is used for constructing accelerometers.

If the damping is very high, (2.2) is

$$2h\omega_0\dot{z} = -\ddot{u} \tag{2.5}$$

and the seismometer velocity is proportional to the ground acceleration, or, integrating once, the seismometer displacement is proportional to ground velocity. This property is also used in special applications, see later, under active seismometers.

In the general case, there is no simple relationship between the sensor motion and the ground motion, and (2.2) will have to be solved, so that the input and output signals can be related. Ideally, we would like to know what the output signal is for any arbitrary input signal, however that is not so easy to solve directly. Since an arbitrary signal can be described as a superposition of harmonics (Fourier series, see Chapter 6), the simplest way to solve (2.2) is to assume an input of a harmonic ground motion and solve for the solution in the frequency domain. Lets write the ground motion as

$$u(t) = U(\omega)e^{i\omega t} \tag{2.6}$$

where $U(\omega)$ is the complex amplitude and ω is the angular frequency. Note that, unless otherwise specified, we mean angular frequency when saying frequency. (2.6) is written in complex form for simplicity of solving the equations and the real part represents the actual ground motion, see also Chapter 6. Since a seismometer is assumed to represent a linear system, the seismometer motion is also a harmonic motion with the same frequency and amplitude is $Z(\omega)$

$$z(t) = Z(\omega)e^{i\omega t} \tag{2.7}$$

We then have

$$\begin{aligned} \ddot{u} &= -\omega^2 U(\omega)e^{i\omega t} \\ \dot{z} &= i\omega Z(\omega)e^{i\omega t} \\ \ddot{z} &= -\omega^2 Z(\omega)e^{i\omega t} \end{aligned} \tag{2.8}$$

Inserting in (2.2) and dividing by the common factor $e^{i\omega t}$, we can calculate the relationship between the output and input as $T(\omega)=Z(\omega)/U(\omega)$, the so called displacement frequency response function:

$$T_d(\omega) = \frac{Z(\omega)}{U(\omega)} = \frac{\omega^2}{\omega_0^2 - \omega^2 + 2\omega\omega_0 hi} = \frac{\omega^2((\omega_0^2 - \omega^2) - 2ih\omega\omega_0)}{(\omega_0^2 - \omega^2)^2 + 4\omega^2\omega_0^2 h^2} \tag{2.9}$$

From this expression, the amplitude displacement response $A_d(\omega)$ and phase displacement response $\Phi_d(\omega)$ can be calculated as the modulus and phase of the complex amplitude response:

$$A_d(\omega) = |T_d(\omega)| = \frac{\omega^2}{\sqrt{(\omega_0^2 - \omega^2)^2 + 4h^2\omega^2\omega_0^2}} \tag{2.10}$$

$$\Phi_d(\omega) = a\tan\left(\frac{\mathrm{Im}(T_d(\omega))}{\mathrm{Re}(T_d(\omega))}\right) = a\tan\left(\frac{-2h\omega\omega_0}{\omega_0^2 - \omega^2}\right) \tag{2.11}$$

and $T_d(\omega)$ can be written in polar form as

$$T_d(\omega) = A_d(\omega)e^{i\Phi_d(\omega)} \quad (2.12)$$

From (2.10), we can again see what happens in the extreme cases. For high frequencies, we get

$$A_d(\omega) \rightarrow 1 \quad (2.13)$$

This is a constant gain of one and the sensor behaves as a pure displacement sensor. For low frequencies,

$$A_d(\omega) \rightarrow \frac{\omega^2}{\omega_0^2} \quad (2.14)$$

which is proportional to acceleration. For a high damping,

$$A_d(\omega) \approx \frac{\omega}{2h\omega_0} \quad (2.15)$$

and we have a pure velocity sensor, however the gain is low since h is high.

Figure 2.4 shows the amplitude and phase response of a sensor with a natural period of 1 s and damping from 0.25 to 4. As it can be seen, a low damping ($h < 1$) results in a peak in the response function. If $h=1$, the seismometer mass will return to it rest position in the least possible time, without overshooting, the response curve has no peak and the seismometer is said to be critically damped. From the shape of the curve and (2.10), we can see that the seismometer can be considered a second order high pass filter for ground displacement. Seismometers perform optimally at damping close to critical. The most common value to use is $h=1/\sqrt{2} = 0.707$. Why exactly this value and not 0.6 or 0.8? In practice it does not make much difference, but the value 0.707 is a convenient value to use when describing the amplitude response function. Inserting $h=0.707$ in (2.10), the value for $A_d(\omega_0)=0.707$. This is the amplitude value used to define the corner frequency of a filter or the -3 dB point. So using $h=0.707$ means we can describe our response as a second order high pass filter (actually the same as a Butterworth filter, see Chapter 6) with a corner frequency of ω_0.

When the damping increases above 1, the sensitivity decreases, as described in (2.10), and the response approaches that of a velocity sensor. From Figure 2.4, we see that for $h=4$, the response approaches a straight line indicating a pure velocity response.

2.3 Seismometer frequency response, alternative solution

In the literature, the complex frequency displacement response function is often found as

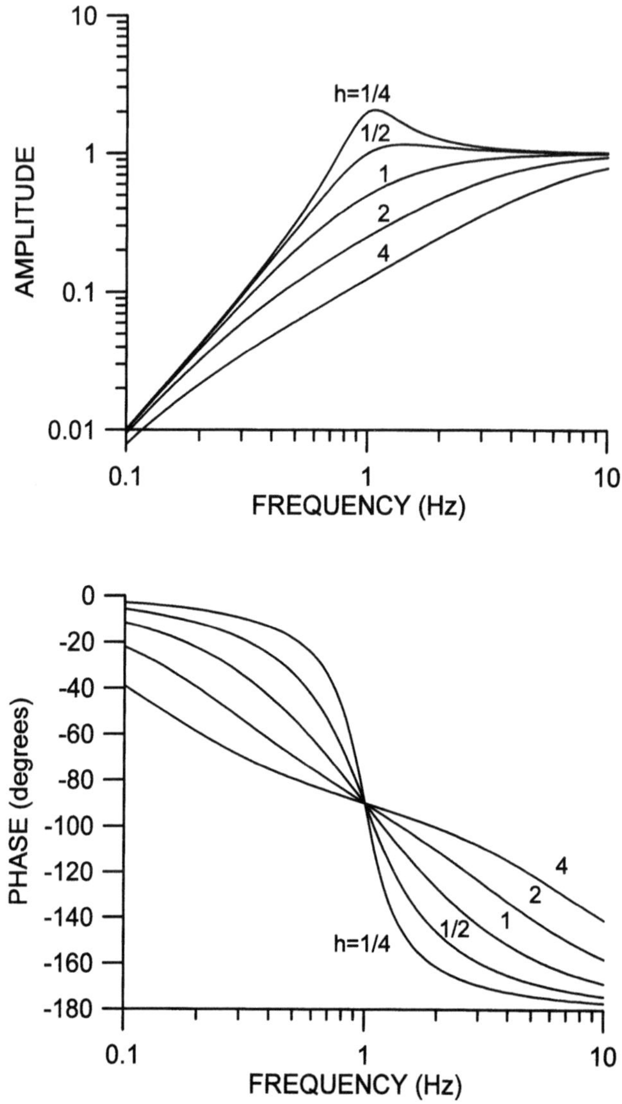

Figure 2.4 The amplitude and phase response functions for a seismometer with a natural frequency of 1 Hz. Curves for various level of damping h are shown.

$$T_d(\omega) = \frac{\omega^2}{\omega_0^2 - \omega^2 - i2\omega\omega_0 h} \tag{2.16}$$

The amplitude response is the same as above (2.10), but the phase response is

$$\Phi(\omega) = a\tan\left(\frac{2h\omega\omega_0}{\omega_0^2 - \omega^2}\right) \tag{2.17}$$

From Figure 2.4 and (2.11) it was seen that at the resonance frequency, the motion of the seismometer mass is delayed $\pi/2$ or the phase shift is $-\pi/2$ (Figure 2.2).

From (2.17) the phase shift is $\pi/2$. Obviously both delays cannot be correct. If the phase shift is measured, a negative phase shift is observed if it is defined as in Figure 2.2. So, where does the positive phase shift come from? The difference comes from the way (2.2) is solved. The trial solution leading to (2.9) was of the form $f_1(\omega) = e^{i\omega t}$. If instead, a trial solution of the form $f_2(\omega) = e^{-i\omega t}$ is used, the alternative frequency response function (2.16) is found. The real signals for both trial functions are

$$\begin{aligned} \mathrm{Re}(f_1) &= \cos(\omega t) \\ \mathrm{Re}(f_2) &= \cos(-\omega t) = \mathrm{Re}(f_1) \end{aligned} \tag{2.18}$$

So the same physical ground motion is represented by both functions, although we might consider that f_2 is the same as f_1 with a negative time. However, if the same phase shift Φ is added to the two trial functions to simulate what happens to the output signals, they become different.

$$\begin{aligned} \mathrm{Re}(f_1) &= \cos(\omega t + \Phi) \\ \mathrm{Re}(f_2) &= \cos(-\omega t + \Phi) \neq \mathrm{Re}(f_1) \end{aligned} \tag{2.19}$$

Obviously, we have to have the same physical solution independently of which trial solution is used so this explains why the phase shift of the solution using f_2 is the same as for f_1 but with opposite sign, since we then get

$$\begin{aligned} \mathrm{Re}(f_1) &= \cos(\omega t + \Phi) \\ \mathrm{Re}(f_2) &= \cos(-\omega t - \Phi) = \mathrm{Re}(f_1) \end{aligned} \tag{2.20}$$

Unfortunately, both forms of the frequency response function are used in the literature without clarifying the difference. This might seem of academic interest; however, when correcting for the instrument response by spectral transformation (see Chapter 6), the correct instrument response must be used depending on how the Fourier transform is defined. The definition used in this book and the most common to use is:

$$x(t) = \frac{1}{2\pi} \int_{-\infty}^{\infty} X(\omega) e^{i\omega t} d\omega \tag{2.21}$$

$$X(\omega) = \int_{-\infty}^{\infty} x(t) e^{-i\omega t} dt \tag{2.22}$$

where $x(t)$ is the time signal and $X(\omega)$ the Fourier transform. If the Fourier transform as (2.22) is used, then (2.9) must be used. If (2.16) is used, the correction for phase will be wrong. If the Fourier transform has been defined with opposite sign,

$$X(\omega) = \int_{-\infty}^{\infty} x(t) e^{i\omega t} dt \tag{2.23}$$

the alternative definition (2.16) must be used. In many cases there is little difference in the corrected signal except for a time shift, but if significant energy is present near the seismometer natural frequency, some differences can be observed, see example in Figure 2.5.

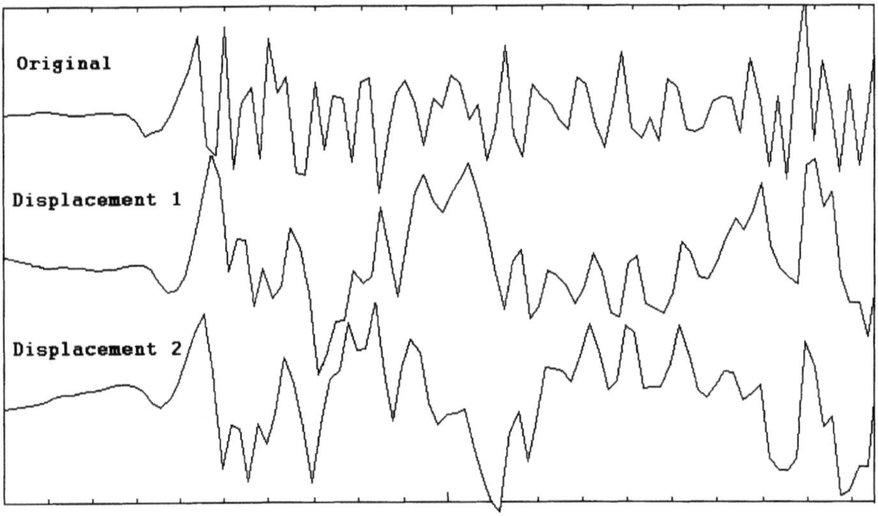

Figure 2. 5 Effect of using the wrong instrument correction. The top trace shows the P-phase of a local earthquake. The distance between the tics is 0.1 s. Displacement 1 trace shows the displacement in the frequency band 1-20 Hz using the 'correct' equation while the trace Displacement 2 shows how the displacement trace looks with the "wrong" correction for phase.

2.4 The velocity transducer

Nearly all traditional seismometers use a velocity transducer to measure the motion of the mass (example in Figure 2.6). The principle is to have a moving coil within a magnetic field. This can be implemented by having a fixed coil and a magnet that moves with the mass or a fixed magnet and the coil moving with the mass. The output from the coil is proportional to the velocity of the mass relative to the frame and we therefore call this kind of electromagnetic seismometer a velocity transducer. Two new constants are brought into the system:

Generator constant G. This constant relates the velocity of the mass to the output of the coil. It has a unit of V/ms^{-1}. Typical values are in the range 30 to 500 V/ms^{-1}.

Generator coil resistance R_g. The resistance of the generator coil (also called signal coil) in Ohms (the coil is built as hundreds or thousands of turns of thin wire winding).

The signal coil makes it possible to damp the seismometer in a very simple way by loading the signal coil with a resistor. When current is generated by the signal coil, it will oppose to the mass motion with a proportional magnetic force (see 2.4.1).

The frequency response function for the velocity transducer is different than for the mechanical sensor.

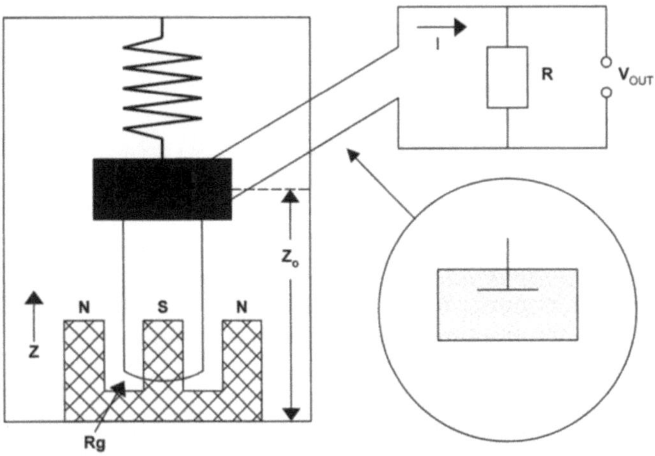

Figure 2. 6 A model of an electromagnetic sensor. The coil resistance is R_g, the damping resistor is R and the voltage output is V_{out}. The dashpot damping has been replaced by the damping from the coil moving in the magnetic field (redrawn and modified from Scherbaum, 2001, with kind permission of Kluwer Academic Publishers).

With the velocity transducer, the observed output signal is a voltage proportional to mass-frame velocity $\dot{Z}(\omega) = i\omega Z(\omega)$ and *G,* instead of $Z(\omega)$. The displacement response for the velocity sensor is then

$$T_d^v(\omega) = \frac{Z(\omega)}{U(\omega)} G = \frac{i\omega\omega^2 G}{\omega_0^2 - \omega^2 + i2\omega\omega_0 h} = \frac{i\omega^3 G}{\omega_0^2 - \omega^2 + i2\omega\omega_0 h} \tag{2.24}$$

and it is seen that the only difference compared to the mechanical sensor is the factors G and $i\omega$. For the response curve (2.24), the unit is $(\mathrm{ms}^{-1}/\mathrm{m})(\mathrm{V}/\mathrm{ms}^{-1})=\mathrm{V}/\mathrm{m}$. It is assumed in (2.24) that a positive velocity gives a positive voltage, see also 2.4.1.

In practice, the sensor output is always connected to an external resistor R_e (for damping control and because amplifiers have a finite input impedance). This forms a voltage divider. Thus the effective generator constant (or loaded generator constant) G_e becomes

$$G_e = G \cdot \frac{R_e}{R_e + R_g} \tag{2.25}$$

and this must be used in (2.24) instead of G.

2.4.1 ELECTROMAGNETIC DAMPING

Lets recall (2.1):

$$m\ddot{z} + d\dot{z} + kz = -m\ddot{u} \tag{2.26}$$

The term $d\dot{z}$ represents a damping force opposed and proportional to the velocity of the mass relative to the frame (or to the ground). This force may be due to the friction of a solid moving within a fluid, usually oil in the case of old mechanical seismometers. In a sensor with electromagnetic transducer, as was explained above, a voltage E is induced in the coil proportional to velocity $E = G\dot{z}$, where G is the generator constant. If the signal coil is shunted with an external resistance, let R_T be the total circuit resistance, then a current $I=E/R_T$ (neglecting self-induction) will flow through the circuit. This will cause a force on the mass proportional to this current and in the sense opposed to its motion, as is given by Lenz law. Due to energy conservation, it turns out (after a few derivations, e.g. Aki & Richards, 1980), that the constant of proportionality, the motor constant, is the same as the generator constant G, hence this damping force F_d will be

$$F_d = GI = G\frac{E}{R_T} = \frac{G^2}{R_T}\dot{z} \tag{2.27}$$

So the value of constant d in (2.26) is G^2/R_T, and dividing by the mass,

$$\frac{d}{m} = \frac{G^2}{R_T m} = 2h_e\omega_0 \tag{2.28}$$

where h_e represents the electromagnetic damping coefficient. Additionally, in many sensors of this type, the gap between the mass and the frame or the case is very small, and this causes air viscous flow damping to contribute to the total h in a small, but non-negligible, proportion. Thus, in general, the total damping is the sum of electromagnetic and mechanic contributions,

$$h = \frac{G^2}{2R_T m\omega_0} + h_m \tag{2.29}$$

and the critical damping resistance (total resistance for damping h = 1) *CDR* is given by

$$\frac{G^2}{2 \cdot CDR \cdot m\omega_0} + h_m = 1 \tag{2.30}$$

If h_m is small,

$$CDR \approx \frac{G^2}{2m\omega_0} \tag{2.31}$$

2.4.2 POLARITY CONVENTION

The convention is that a ground motion up, North or East should give a positive signal out of the seismometer. As mentioned in the introduction, the mechanical seismometer inverts the signal, so, in order to get a positive signal, a factor of –1 must be used. Formally, the velocity transducer also inverts the signal, since the current generated will oppose the motion of the mass and G is sometimes used with a negative sign, however, that is of course just a question of flipping the wires and seismometers are set up so that the polarity convention holds. This means that the final form of the frequency response, to use in practice, is (2.24) multiplied by –1 and the displacement response becomes

$$T_d^V(\omega) = \frac{-i\omega^3 G}{\omega_0^2 - \omega^2 + i2\omega\omega_0 h} \tag{2.32}$$

The minus sign is sometimes forgotten, which gives rise to a wrong phase shift of π. This also implies that all equations given for response functions, except (2.26), must be multiplied by –1 if the polarity convention should hold.

2.5 Instrument response curves, different representation

Usually, the standard displacement gain curve as shown in Figure 2.4 is used. Sometimes we also want to know what the velocity and acceleration response functions look like for both the mechanical seismometer and the velocity seismometer. Recall that the displacement response function (2.9) was

$$T_d(\omega) = \frac{Z(\omega)}{U(\omega)} = \frac{\omega^2}{\omega_0^2 - \omega^2 + i2\omega\omega_0 h} \tag{2.33}$$

If we now replace U(ω) with $\dot{U}(\omega) = i\omega U(\omega)$, the velocity response function becomes

$$T_v(\omega) = \frac{Z(\omega)}{\dot{U}(\omega)} = \frac{-i\omega}{\omega_0^2 - \omega^2 + i2\omega\omega_0 h} \tag{2.34}$$

And, likewise, the acceleration response function will be

$$T_a(\omega) = \frac{Z(\omega)}{\ddot{U}(\omega)} = \frac{-1}{\omega_0^2 - \omega^2 + i2\omega\omega_0 h} \tag{2.35}$$

Similarly, for the velocity sensor we get the response functions for velocity

$$T_v^v(\omega) = \frac{\omega^2}{\omega_0^2 - \omega^2 + i2\omega\omega_0 h} \tag{2.36}$$

and for acceleration

$$T_a^v(\omega) = \frac{-i\omega}{\omega_0^2 - \omega^2 + i2\omega\omega_0 h} \tag{2.37}$$

If the sensor is an accelerometer, the response curves are the same as for the velocity sensor multiplied by $i\omega$, like the velocity sensor curve was just the mechanical sensor curve multiplied by $i\omega$.

It is thus easy to convert from one type of curve to another since it is just a question of multiplying or dividing with ω (or $i\omega$ for the complex response). It can be a bit confusing to remember which is which and Figure 2.7 gives a schematic overview of the different representations of standard sensors.

Strictly speaking, none of the sensors are linear, in the sense that an arbitrary waveform of ground motion be exactly reproduced at scale, for any kind of response. The mechanical sensor has a flat displacement response above ω_0, but few such instruments are used, the most famous is probably the Wood Anderson seismometer (see Figure 2.12). The velocity sensor has a flat response for velocity for frequencies above ω_0, so by making a BB sensor with corner frequency at e.g. 0.01 Hz, we have a sensor with a wide frequency band linear to velocity. However, the sensor is still used to measure signals at frequencies below ω_0, so it is only free of distortion in a limited frequency band.

The accelerometer is the most broadband instrument available since it has a response flat for acceleration in the whole seismic frequency band. This is achieved by making the accelerometer with a ω_0 larger than the largest frequency of interest. Note that the shape of the accelerometer response is the same as that for the mechanical seismometer, only ω_0 has changed. So why shall we bother with BB sensors? Figure 2.7 partly gives the answer. The accelerometer is insensitive to displacement at low frequencies and too sensitive at high frequencies. So for a very sensitive accelerometer, it is difficult to cover the whole amplitude displacement range (see also discussion on active sensors).

A seismic sensor can also be considered as a filter. For example, the velocity sensor is a second order high pass filter for velocity, while it is a first order band pass filter for acceleration. Considering the sensors as filters, we can see why different sensors will make the same seismic signal appear different. In the examples above, a displacement signal in the frequency band 1-10 Hz will be recorded undisturbed by the mechanical sensor, be enhanced proportional to frequency for the velocity sensor and enhanced proportional to frequency squared for the accelerometer so the signal will appear more high frequency on the accelerometer than on the mechanical seismometer. This is of course not surprising since the mechanical sensor is recording pure displacement and the accelerometer pure acceleration and we have to differentiate two times in going from displacement to acceleration. Figure 2.8 shows an example of the same signal as displacement, velocity and acceleration.

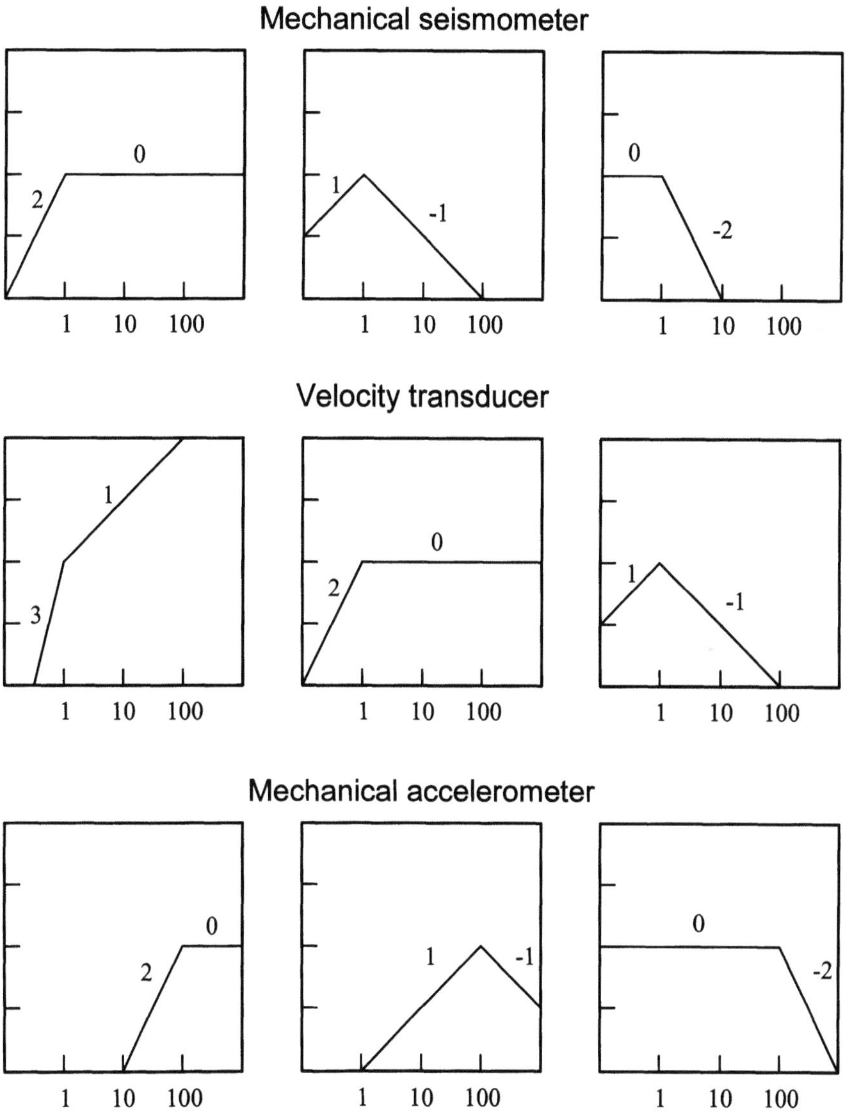

Figure 2.7 Schematically, amplitude frequency response of a mechanical 1 Hz sensor (top), a mechanical 1 Hz sensor with a velocity transducer (middle) and an accelerometer (100 Hz) (bottom). From left to right, the figures show the sensor response for ground displacement, velocity and acceleration, respectively. The axes are logarithmic and the horizontal axes show frequency in Hz. The asymptotic slope for each segment is indicated. Note how one curve translates into another by just adding or subtracting one unit of the slope corresponding to a multiplication or division by frequency.

It might well be asked if it matters whether we record displacement, velocity or acceleration, apart from what we are used to look at. In any case, many systems do not record pure displacement, velocity or acceleration except in a particular passband, (as seen from Figure 2.7), although this pass band can be quite wide, like for BB sensors and accelerometers. Seismologists are used to look at velocity, since that is what most sensors record, and there is a reluctance to e.g. use accelerometers instead of the standard SP sensor, as the signal look different. However, that is only a question of appropriate filtering after recording to make the signal look 'nice'. So, the question really boils down to whether the recording system has a sufficient frequency and dynamic range to be able to do appropriate processing after recording. For that reason, the velocity output offers a compromise between a system being too sensitive at either frequency extreme.

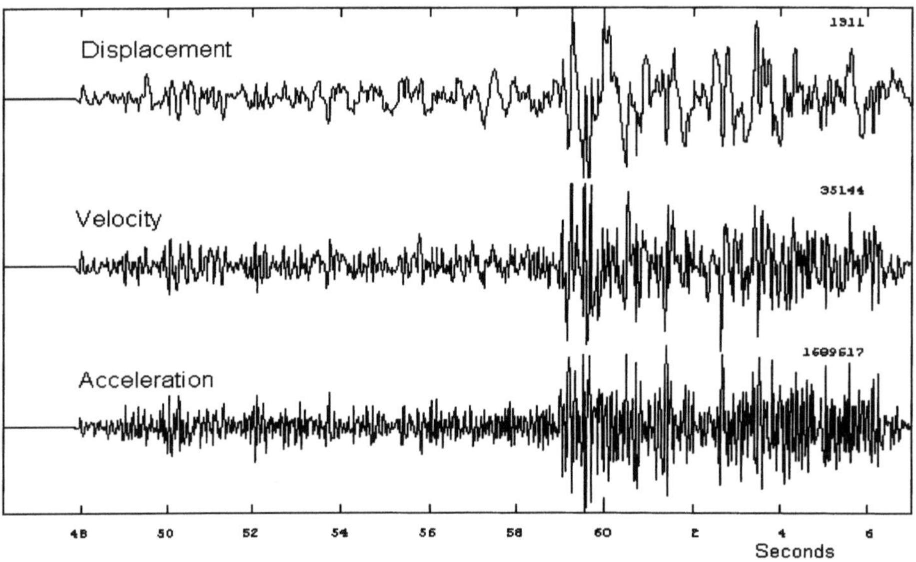

Figure 2.8 A signal from a magnitude 3 earthquake recorded with a 1 Hz seismometer. The 3 traces show displacement, velocity and acceleration, respectively, in the frequency band 0.3-20 Hz. The numbers on top of the traces at the right hand side are the maximum amplitudes in nm, nm/s and nm/s^2, respectively.

2.6 Sensor response to transient signals

Inertial seismometers, regardless of whether they are designed for a flat response to displacement, velocity or acceleration within a given band, are excited only by ground acceleration. This is a consequence of the lack of a reference "at rest" and the physical nature of the reference system, the sensor frame, in which we measure the mass motion.

In (2.1), the only external force is the ground acceleration multiplied by the sensor mass (the inertia force, since the whole sensor system moves with acceleration). If this term is null, we have just the equation for a free swinging damped oscillator. It is thus the ground acceleration that is the single cause for the forced motion of the mass, regardless of the response of the sensor, which can be flat to ground displacement, velocity or acceleration in a given frequency band.

To illustrate this, let us show the response to a transient ground displacement. Figure 2.9a represents a ramp lasting about two seconds. Both ends of the ramp have been smoothed to avoid unrealistic accelerations (this waveform may be, e.g., a simplified model of a near-field observation of an earthquake). Figure 2.9 shows the output transients of several types of sensors for this ground motion.

A somewhat less distorted record would be obtained for displacement and velocity with a longer period (or BB) sensor. Nevertheless, it should be kept in mind that it is just the ground acceleration that excites the motion of the mass. Consequently, the response of an inertial sensor for ground displacement and velocity essentially cannot record the DC component, therefore the gain of such sensors for both displacement and velocity tends to zero as frequency decreases (Figure 2.7).

2.7 Damping constant

The damping constant is obviously a very important parameter for a traditional seismic sensor and we will look a bit on how the damping constant affects the signal and how to measure the damping constant. If we give the mass an initial displacement z_0, it will swing forever if damping is zero (not possible in practice), gradually stop swinging with a low damping and exactly not swing (no overshoot) if damping is one (Figure 2.10). So there is obviously a relationship between the amount of over swing and the damping.

(2.1) can be solved for an initial constant displacement z_0 and $h < 1$, with no external force. The solution is (e.g., Scherbaum, 2001)

$$z(t)=\frac{z_0}{\cos\Phi}e^{-h\omega_0 t}\cos(\omega_1 t-\Phi) \tag{2.38}$$

where $\Phi = \arcsin(h)$ is the phase shift and ω_1 is the apparent frequency of the damped system:

$$\omega_1 = \sqrt{\omega_0^2(1-h^2)} \tag{2.39}$$

For an electromagnetic velocity transducer, the voltage output will be $V(t) = G_e\dot{z}(t)$, where G_e is the effective or loaded generator constant. By differentiating the equation of $z(t)$ and using the formula of sin of sum of angles and the value of Φ, the voltage output may be shown to be (Mac Arthur, 1985):

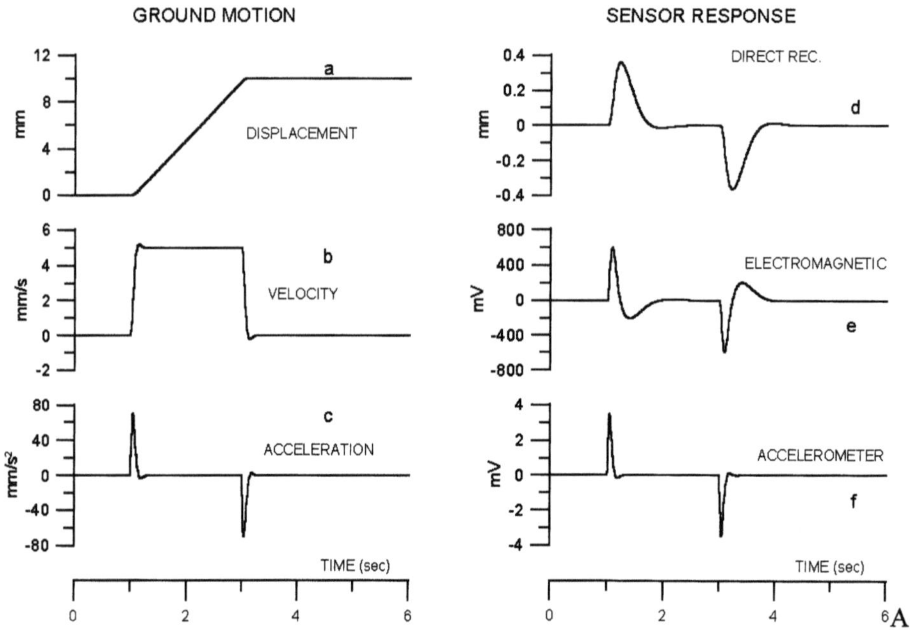

Figure 2.9 Transient response of sensors. Figure a represents the ramp signal, Figure b and c represent the ground velocity and acceleration respectively. Figure d shows the mass motion of a sensor with free period 1 sec (i.e. what we can consider the output of a direct-recording sensor like a pen connected to the mass). Figure e is the voltage output of this same sensor equipped with an electromagnetic (velocity) transducer with G_e=200V/m·s^{-1}. Trace f is the output of an accelerometer built as a sensor with a natural frequency of 20 Hz and a displacement transducer for measuring the mass motion with a sensitivity of 1000V/m.

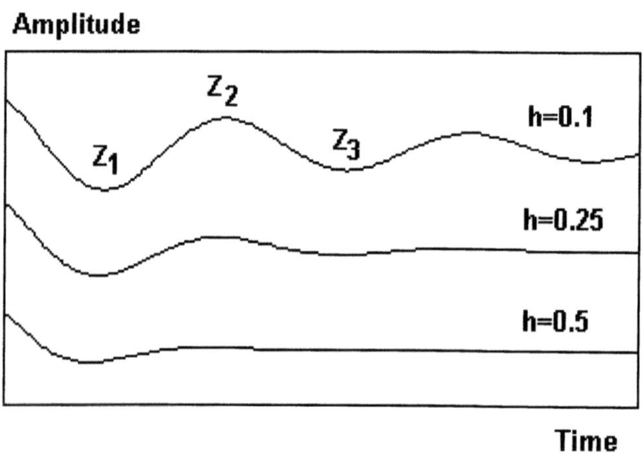

Figure 2.10 Effect of damping on the signal. The sensor has been given some initial positive displacement and then left swinging freely. The damping constant is *h*. The extreme amplitude values are z_1, z_2 and z_3. It is seen that for a damping of 0.5, there is little over swing.

$$V(t) = \frac{G_e z_0 k}{\omega_1 m} e^{-h\omega_0 t} \sin(\omega_1 t) \tag{2.40}$$

where k is the spring constant and m the seismometer mass. It is seen that a damped system will swing with a lower frequency than an undamped system. So, when measuring the free period of a seismometer, this must be taken into account. From (2.38), it follows that the damped system will swing with a frequency of ω_1 and with an amplitude decaying exponentially and that the higher the damping, the faster the swinging will die out. From (2.38), it can be shown that the amplitude ratio of two consecutive extreme values z_1/z_2 can be written as

$$\ln(z_1 / z_2) = \frac{\pi h}{\sqrt{1-h^2}} \tag{2.41}$$

Thus the actual damping of a given seismometer can be estimated by measuring the amplitudes of two consecutive extreme values. If the amplitude changes very little, the ratio between extreme *1* and extreme *n+1* (positive or negative amplitude) would be (seen from (2.41))

$$\ln(z_1 / z_n) = \frac{\pi h}{\sqrt{1-h^2}} n \tag{2.42}$$

This simple way of measuring damping will be further explored in Chapter 10.

The damping consists of two parts, the mechanical damping due to the air friction of the sensor motion h_m (also called open circuit damping) and the damping due to the electrical current through the coil h_e so $h=h_m + h_e$. The mechanical damping cannot be changed and range from a very low value of 0.01 to 0.3 (see Table 2.3 for typical values of damping) while the electrical damping can be regulated with the size of the external resistor to obtain a desired total damping. The electrical damping can be calculated as (see 2.4.1)

$$h_e = \frac{G^2}{2M\omega_0 R_T} \tag{2.43}$$

where M is the seismometer mass and R_T is the total resistance of the generator coil and the external damping resistor. Seismometer specifications often give the critical damping resistance *CDR*, which is the total resistance *CDR*= R_T required to get a damping of 1.0. From (2.43), it is seen that if we know the electrical damping h_{e1} for one value of R_T, R_{T1}, the required resistance R_{T2} for another required electrical damping h_{e2} can be calculated as

$$R_{T2} = R_{T1} \frac{h_{e1}}{h_{e2}} \tag{2.44}$$

If the mechanical damping is low, (2.44) can be written in terms of *CDR* as

$$R_{T2} = CDR / h_{e2} \tag{2.45}$$

and thus, the desired total resistance for any required damping can easily be calculated from *CDR*.

As an example, let us look at a classical 1 Hz sensor, the Geotech S13 (Table 2.3). The coil resistance is 3600 ohm and *CDR* = 6300 ohm. Since the open circuit damping is low, it can be ignored. The total resistance to get a damping of 0.7 would then from (2.45) be $R = 6300 / 0.7 = 9000$ ohm and the damping resistor to connect would have a value of 9000-3600 = 5400 ohm.

2.8 Construction of seismic sensors

The mass-spring system of the vertical seismometer serves as a very useful model for understanding the basics of seismometry. However, in practical design, this system is too simple, since the mass can move in all directions as well as rotate. So, nearly all seismometers have some mechanical device which will restrict the motion to one translational axis. Figure 2.11 shows how this can be done in principle for a vertical seismometer.

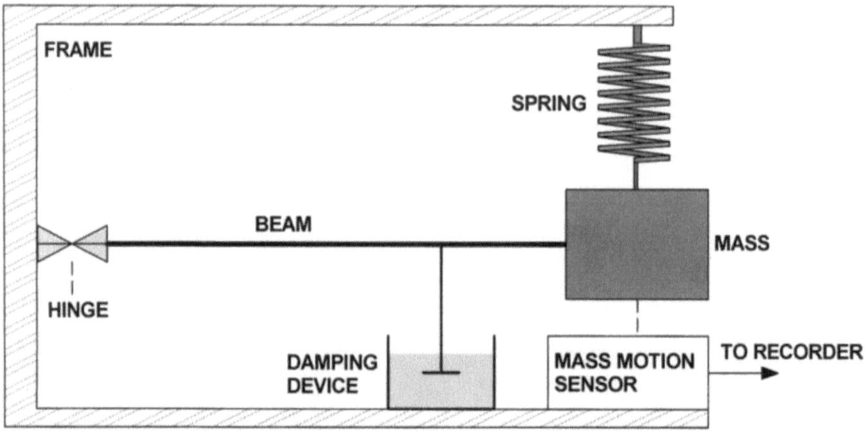

Figure 2.11 A vertical mass-spring seismometer where the horizontal motion has been restricted by a horizontal hinged rod. Usually, the hinge is a thin flexible leaf, to avoid friction.

It can be seen that due to the hinged mass, the sensor is restricted to move vertically. The mass does not only move in the z-direction but in a circular motion with the tangent to the vertical direction. However, for small displacements, the motion is sufficiently linear. The above pendulum arrangement is in principle the most common way to restrict motion and can also be used for horizontal seismometers. Pendulums are also sensitive to angular motion in seismic waves, which normally is so small that it has no practical importance.

The simplest pendulum sensitive to horizontal motion is a mass suspended with a string. For a small size mass m compared to the length L of the string, the natural frequency is

$$\omega_0 = \sqrt{g/L} \tag{2.46}$$

where g is the gravitational constant. For small translational ground motions, the equation of motion is identical to (2.1) with z replaced with the angle of rotation. Note that ω_0 is independent of the mass.

2.8.1 THE WOOD ANDERSON SHORT PERIOD TORSION SENSOR

This instrument was developed around 1925 and has played a key role in defining the Richter magnitude scale, the basis of all magnitude scales used today. This pendulum type instrument consists of a metal cylinder (Figure 2.12) tautly attached to an eccentric vertical suspension wire. The cylinder will rotate around the wire and a mirror, attached to the mass, is used for photographic recording of the displacement. This is one of the very few seismographs recording displacement photographically. The natural period was 0.8 sec and the nominal gain was 2800. It has later been determined that the gain in reality was only 2050 (Hutton and Boore, 1987), which is important, since local magnitude scales made today refer to the gain of the Wood Anderson seismograph. Some instruments are still in operation.

2.8.2 LONG PERIOD SENSORS

In order to obtain high sensitivity to low frequency seismic waves, we need a low natural frequency. With a standard mass-spring sensor, where

$$\omega_0 = \sqrt{k/m}\ , \tag{2.47}$$

a large mass combined with a soft spring will give a low frequency, but this arrangement is limited by the mechanical properties. For a pendulum, we see that to get a period of 20 s, a length of 100 m is needed, not very practical either. For both the pendulum and the spring system ((2.46) and (2.47)) it is seen that, if the restoring force is small, the natural frequency will be low. If gravity had been smaller, the natural frequency would have been smaller and the same is true if k had been smaller. The solution is to use astatic suspensions, where the restoring force is very small and, theoretically, any natural frequency can be obtained.

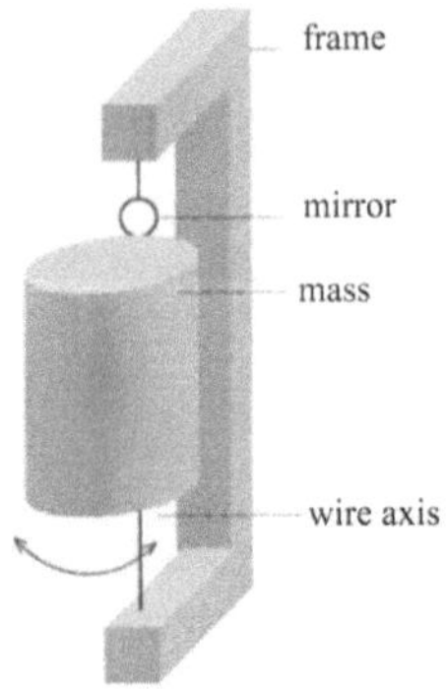

Figure 2.12 The Wood Anderson torsion seismometer, schematically. The mass (a few grams) is suspended on an eccentric taut wire and can rotate slightly. The height of the instrument is 36 cm.

2.8.3 GARDEN-GATE

The simplest *astatic* suspension for horizontal seismometers is the "garden-gate" pendulum (Figure 2.13). The mass moves in a nearly horizontal plane around a nearly vertical axis. The restoring force is now $g \sin(\alpha)$ where α is the angle between the vertical and the rotation axis, so the natural frequency becomes

$$\omega_0 = \sqrt{g \sin(\alpha) / L} \qquad (2.48)$$

where L is the vertical distance from the mass to the point where the rotation axis intersects the vertical above the mass (see Figure 2.13).

To obtain a natural frequency of 0.05 Hz with a pendulum length of 20 cm will require a tilt of 0.1 degree. This is about the lowest period that has been obtained in practice with these instruments. Making the angle smaller makes the instrument very sensitive to small tilt changes. The "garden-gate" was one of the earliest designs for long period horizontal seismometers. It is still in use but no longer produced. Until the new broadband sensors were installed in the GSN (Global Seismic Network), this kind of sensors was used in the WWSSN network (World Wide Standard Seismic Network). Figure 2.14 shows an example of a well know sensor.

This long period horizontal seismometer (S-5007-H) was typically used in the WWSSN system. The period was claimed to be adjustable from 6 to 60 s, but in practice it is hard to use a period longer then 30 s and 25 s is a stable value. The generator constant is 89 V/ms^{-1}. The instrument requires a very stable temperature environment and lots of space

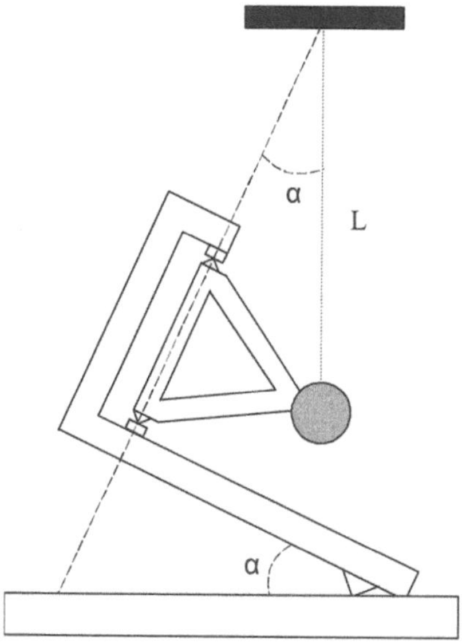

Figure 2.13 Principle of the garden gate pendulum. The tilt angle is exaggerated. A string pendulum must have length L to have the same period, so at a small angle α, L becomes very large. Redrawn from Figure 5.8 by E. Wielandt, in NMSOP, Vol. 1, Bormann (Ed.), 2002; copyright granted by IASPEI.

Figure 2.14 A typical horizontal long period seismometer made by Sprengnether. The two black cylinders seen at each side are the magnets and the coils are moving inside with the mass in the center. Leveling is done with the 3 screws at the corners of the base plate. Once the horizontal leveling is done (front screws), the back screw is used to adjust the period. The base plate side length is 63 cm and the weight 45 kg.

compared to a BB instrument with the same frequency band. Several of these instruments are still in use.

These garden gate LP sensors have a maximum motion of 10 mm with a linearity of 5%. Considering that the smallest displacement they can measure is at around 1nm, a rough estimate of the dynamic range is $10 mm/10^{-6}$ mm = 10^7 or 140 dB, not bad for such an old sensor. This can only be achieved with digital recording, while the old analog recording has a dynamic range of maybe 50 dB. Thus, if connected to a modern digitizer, they can make a good BB sensor.

2.8.4 INVERTED PENDULUM

The inverted pendulum (Figure 2.15) was also an early design. The mass was held in place by springs or a strong hinge. The most famous example is the Wiechert pendulum built around 1905. It had a mass of 1 ton, a free period of 10 s and a mechanical amplification of 240, which is low by today's standards. However, many such instruments were built and kept in operation worldwide for a long time period and thereby provided a large number of standard records, much like the WWSSN. Several instruments are still in operation and it is probably the seismograph, which has had the longest continuous operational history. The records obtained in recent years are used to compare pairs of earthquakes recorded at the same location many years apart.

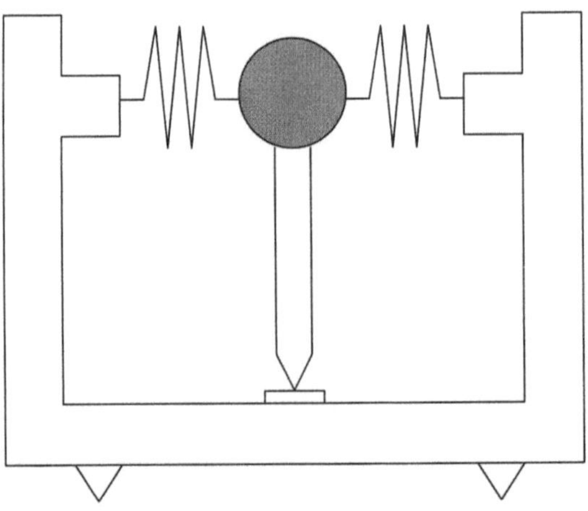

Figure 2.15 The inverted pendulum seismometer. Redrawn from Figure 5.7 by E. Wielandt in NMSOP, Vol. 1, Bormann (Ed.), 2002; copyright granted by IASPEI.

2.8.5 LACOSTE

The *astatic* spring geometry for vertical seismometers was invented by LaCoste (1934). The principle of the sensor is shown in Figure 2.16. The sensor uses a "zero length spring" which is designed such that the force $F=k \cdot L$, where L is the *total* length of the spring. Normal springs used in seismometers do not behave as zero length springs, since

$F=k \cdot \Delta L$, where ΔL is the change in length relative to the unstressed length of the spring. However, it is possible to construct a "zero length spring" by e.g. twisting the wire as it wound into a spring. The physical setup is to have a mass on a beam and supported by the spring so that the mass is free to pivot around the lower left hand point.

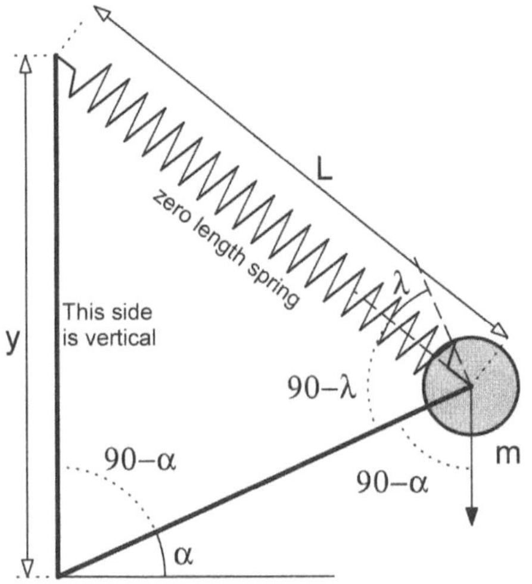

Figure 2.16 The principle behind the LaCoste suspension. The mass m is sitting on a hinge, which has an angle α with the horizontal and suspended by a spring of length L.

This system can be constructed to have an infinite period, which means that the restoring force must be zero at any mass position. This means that if the mass is at equilibrium at one angle, it will also be at equilibrium at another angle, which is similar to what was obtained for the garden gate horizontal seismometer. Qualitatively, what happens is that if e.g. the mass moves up, the spring force lessens, but it turns out the force from gravity in direction of mass motion is reduced by the exact same amount, due to the change in angle as will be shown below.

The gravity force F_g acting on the mass in direction of rotation can be written as

$$F_g = mg\cos(\alpha) \tag{2.49}$$

while the spring restoring force F_s acting in the opposite sense is

$$F_s = kL\cos(\lambda) \tag{2.50}$$

To replace λ with α, we can use the law of sines

$$\frac{L}{\sin(90-\alpha)} = \frac{y}{\sin(90-\lambda)} \quad \text{or} \quad \cos(\lambda) = \frac{y}{L}\cos(\alpha) \tag{2.51}$$

Equating F_s and F_g and including the expression for $cos(\lambda)$, we get

$$ky = mg \text{ or } y = mg/k \quad (2.52)$$

As long as this condition holds, the restoring force is zero independent of the angle. As with the garden gate seismometer, this will not work in practice and, by inclining the vertical axis slightly, any desired period within mechanical stabilization limits can be achieved. In practice, it is difficult to use free periods larger than 30 s.

The astatic leaf-spring suspension is comparable to the LaCoste suspension, but much simpler to make (Figure 2.17). The delicate equilibrium of forces in astatic suspensions makes them sensitive to external disturbances, so they are difficult to operate without a stabilizing feedback system (see section on active sensors, 2.10-2.15).

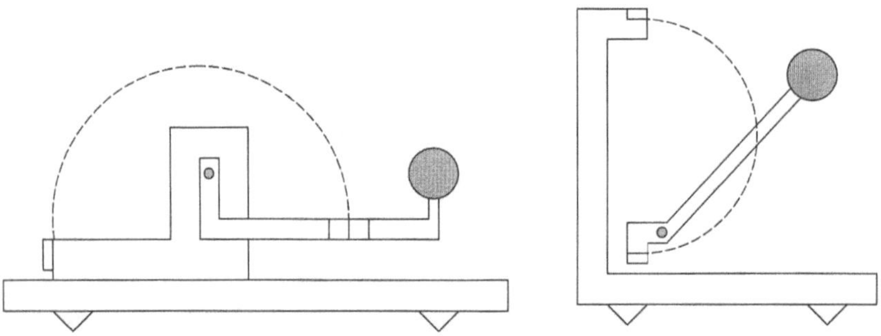

Figure 2.17 Leaf spring astatic suspensions. The figure to the left shows a vertical seismometer and to the right an oblique axis seismometer. Redrawn from Figure 5.10 by E. Wielandt in NMSOP, Vol. 1, Bormann (Ed.), 2002; copyright granted by IASPEI.

The LaCoste pendulums can be made to operate as vertical seismometers, as has been the main goal, or as sensors with an oblique axis (Figure 2.16, 2.44). Figure 2.18 shows the Sprengnether LP seismometer.

This long period vertical seismometer (S-5007-V) was typically used in the WWSSN system. The period was claimed to be adjustable from 6 to 40 s, but in practice it is hard to use a period longer than 30 s and 25 s is a stable value. The generator constant is 100 V/ms^{-1}. The instrument requires a very stable temperature environment and lots of space like the companion horizontal seismometer.

2.9 Sensor calibration coil

Most sensors with electromagnetic transducers include an auxiliary coil coaxial with the signal coil, which allows us to introduce a small force on the suspended mass (in earlier hinged pendulum seismometers, it was parallel but not coaxial to the signal coil). Manufacturer specifications include in this case the motor constant K of this coil. For a

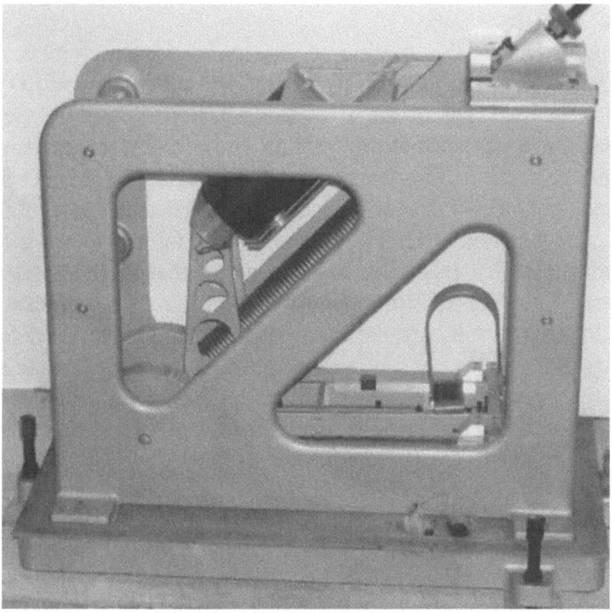

Figure 2.18 A typical vertical long period seismometer made by Sprengnether. The cylinder seen at top center is the magnet and the coil is moving with the mass seen bottom left. Leveling is done with the 3 screws at the corners of the base plate. Once the horizontal leveling is done (back screws), the front screw is used to adjust the period. The base plate side length is 60 cm and the weight 50 kg.

given current $I(t)$ injected into the calibration coil, the calibration force F_c exerted on the mass is $F_c(t)=KI(t)$, which is equivalent to the effect of a ground acceleration of

$$\ddot{u}(t) = \frac{KI(t)}{m} \tag{2.53}$$

The sign depends on the polarity of the coil connection.

The signals $I(t)$ most used for calibration are steps and sinusoidal functions. Once the motor constant and the mass of the calibration coil are known, it is possible in principle to determine all the other constants, as explained in Chapter 10.

The calibration coil should be used with caution, however, since some sensors have a spurious coupling with the signal coil, which, at higher frequencies (> 5 Hz), induces a signal, not related to mass motion, into the signal coil. This effect will be discussed further in Chapter 10. It is possible to introduce a transient signal for calibration using the signal coil, thus overcoming this problem.

In feedback sensors such as BB or FBA, the calibration signal is introduced in a sum point of the feedback loop, making an additional calibration coil unnecessary.

2.10 Active sensors

In the preceding sections, the traditional mechanical seismometers were described including some of their limitations, particularly with respect to low frequency sensitivity. These types of sensors are no longer produced for natural frequencies below 0.2 Hz or for sensors measuring strong motion and are now mostly built according to the 'force-balance' principle. This means that the external force on the sensor mass is compensated (or 'balanced') by an electronically generated force in the opposite direction, such that the mass remains stationary. Or, more exactly, nearly stationary, since some small motion is needed in order to detect that the mass tries to move. The force is generated by a current through a coil (similar to a calibration coil), so the current needed to balance the external force is therefore proportional to that force. By measuring the current, we have a measure proportional to the external acceleration and the sensor directly measures acceleration. Since the mass hardly moves, the linearity of the system is very good and because the amount of feedback can be varied, it is possible to have sensors that can work with very strong motions. Figure 2.19 shows a schematic drawing of a force balance accelerometer (FBA).

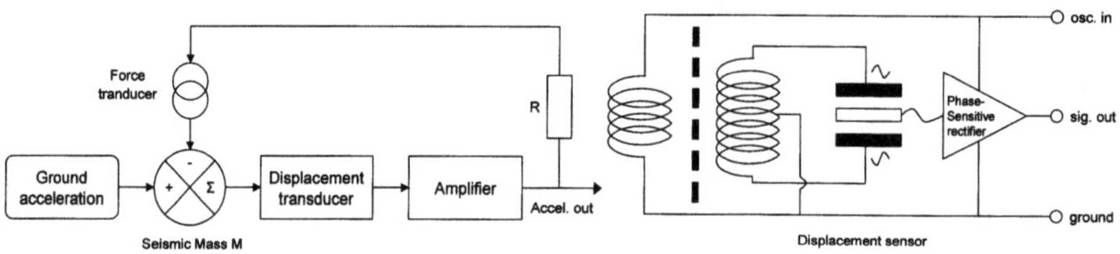

Figure 2.19 Left: Feedback circuit of a force balance accelerometer (FBA). The ground acceleration is balanced by the force generated by the feedback current through the force transducer. The displacement is sensed with the displacement transducer seen in detail right. In this example, the sensor mass moves with the center plate of the capacitor. Redrawn from Figures 5.15 and 5.16 by E. Wielandt in NMSOP, Vol. 1, Bormann, (Ed.), 2002; copyright granted by IASPEI.

The new key element in the sensor is the displacement transducer. A velocity transducer cannot be used since it has too low sensitivity at low frequencies. Most displacement transducers are of the variable capacitive half bridges type (Figure 2.19), which consists of a 3 plate capacitors where either the central plate or the outer plates move with the mass. Identical AC signals of opposite phase (e.g. 5 kHz) are put across the plates. The output is null as long as the center plate is in the middle. If it moves from the center, a small AC signal is output, which has an amplitude proportional to the capacitance difference between the two capacitors. The deviation sign is detected by the phase of this voltage relative to the input AC signal, so a phase-sensitive demodulator is used. The output signal is amplified and fed back to the force transducer through a resistor *R*. The current will be exactly large enough to balance the force on the mass, which therefore remains almost stationary. This can also be called a servo system (see also 2.14 and 2.15). The voltage over the resistor *R* is therefore proportional to the acceleration and we have an accelerometer, the so-called Force Balance Accelerometer. However, the same principle is used in almost all broadband seismometers as well.

Since a servo system will have delays, the system will have a limited upper frequency response, and some additional phase compensation is required in the loop to assure stability to e.g. avoid undesired oscillations. This is usually done either by including a phase-lead circuit (advancing the phase, Usher et al. (1977)) in the feedback or an integrator following the transducer. The natural frequency of the complete system is electrically increased, since the servo system will act as a stiff spring (see 2.15).

The feedback systems described here have several advantages over the traditional sensors. Since the mass hardly moves and is rigidly fixed to the frame (hinged), the linearity and cross axis sensitivity is much improved compared to traditional sensors. This is also the case for sensitivity for rotation and spurious modes of resonance. According to Pearce (personal communication), linearity can be improved by up to a factor of 1000.

2.11 Accelerometers

Accelerometers or strong motion sensors by definition measure large accelerations, typically up to 1-2 g. Earlier types were based on a mechanical sensor with a high natural frequency to ensure that the output was proportional to acceleration, and recording optically on film. Accelerometers have been of the FBA type for more then 30 years and have seen a steady improvement in performance. Initially, accelerometers were mostly considered devices for measuring 'strong motion' and only of interest for engineers since no weak motion was recorded, even if e.g. a 0.25 g instrument was used and the dynamic range was typically 80-100 dB. Now FBA's are available with a dynamic range of up to 155 dB and are advertised as weak motion sensors as much as accelerometers. What does 155 dB mean? For a 0.25 g instrument, the smallest detectable acceleration is then $0.25g/(10^{155/20}) = 5 \cdot 10^{-9}$ g $\approx 5 \cdot 10^{-8}$ m/s^{-2}. At 1 Hz, this corresponds to $5 \cdot 10^{-8}/4\pi^2 = 1$ nm displacement and 100 nm at 0.1 Hz. Comparing to Figure 3.3 in Chapter 3, it is seen that this is a fairly low noise level and the FBA would be quite acceptable as a seismometer. A better way of evaluating the noise level is by plotting the instrument sensitivity in comparison with the Peterson curves, see Figure 3.4. The improvement in sensitivity and dynamic range is almost completely due to the use of capacitive transducers, intrinsically noiseless, and to the feedback principle.

The generator constant G in V/g of an accelerometer can be calculated very simply. The output voltage over the resistor R is V for a given acceleration a, the motor constant of the feedback coil is σ (N/m) and the current through the coil is then V/R. Since the force generated by the coil $\sigma \cdot V/R$ is the same as the ground force on the mass, G can be calculated as

$$\sigma \cdot V/R = m \cdot a \tag{2.54}$$

or since $G = V/a$

$$G = mR/\sigma \tag{2.55}$$

Additional amplifiers can be added to the accelerometer to change the value of *G*, but essentially the generator constant is determined by only 3 passive components of which the mass by definition is error free, the resistor is a nearly ideal component and the force transducer can be made very precise because the motion is so small. So the FBA can be made very precise.

The FBA has a flat response down to DC, while, as discussed above, there is an upper frequency limit. The FBA therefore responds to static changes in the acceleration and e.g. a permanent tilt will give a permanent output. This can be used to calibrate the FBA, see Chapter 10.

2.12 Velocity broadband sensors

Traditional networks would have weak motion sensors for LP and SP to cover the spectrum of interest. The introduction of the velocity broadband sensor (hereafter just referred to as broadband sensor (BB)) covers the whole frequency band with the same sensitivity as the individual SP and LP sensors and therefore eliminates the need for 2 sensors. Accelerometers cover a large frequency band, so, why not just use an accelerometer? Sensors for weak motion have traditionally had larger gain than accelerometers, but, for broadband recording, a high gain output proportional to acceleration is unfavorable. Broadband in seismology means from long periods (about 0.01 Hz) to frequencies of some tens of Hz. Since the sensitivity to velocity increases proportional to frequency (Figure 2.7), a very sensitive accelerometer is easily saturated by noise at high frequency. At low frequencies, a permanent offset is easily generated (tilt or temperature) and might be saturating the system. Output signals at low frequency would be very low compared to levels at high frequency, requiring a very large dynamic range of the recording system. By changing the accelerometer to have a band limited output proportional to velocity, we can essentially avoid the above mentioned problems.

There are many possible ways of implementing a velocity broadband response by feedback, but not all are equally efficient in terms of dynamic range. One state-of-the art technique used by some sensors, consist of a judiciously chosen multiple feedback (see 2.15.2) in which the active components, noisy and able to get saturated, are reduced to minimum. The multiple feedback has, besides a force proportional to the mass displacement, a second contribution that is added to the feedback proportional to the time derivative of displacement. This is achieved simply by a large capacitor (Figure 2.20). The current through it is C times the derivative of the voltage. The feedback current is still proportional to acceleration (mass must still be kept in place) while the output voltage across the capacitor is a time integral of the current and therefore proportional to ground velocity. There is a low frequency limit under which the proportional feedback dominates over the differential one. The total response can be seen in Figure 2.21. In practice, a third feedback path is added with a current proportional to the integral of the BB output, to shape the response like that of a long period passive seismometer with velocity transducer. This provides also another output useful to monitor the mass free position (see 2.15.2).

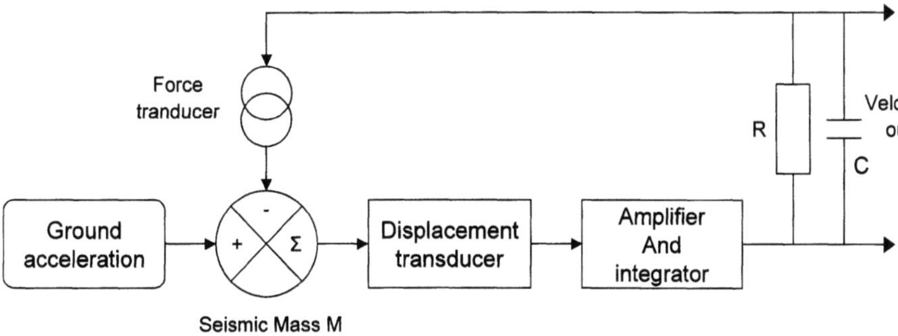

Figure 2.20 Velocity broadband seismometer. The ground acceleration is balanced by the force generated by the feedback current through the force transducer. The displacement is sensed with the displacement transducer (Figure 2.19). The signal is partly integrated before feedback and a capacitor is added to generate a velocity output. Partly redrawn from Figure 5.16 by E. Wielandt in NMSOP, Vol. 1, Bormann (Ed.), 2002; copyright granted by IASPEI.

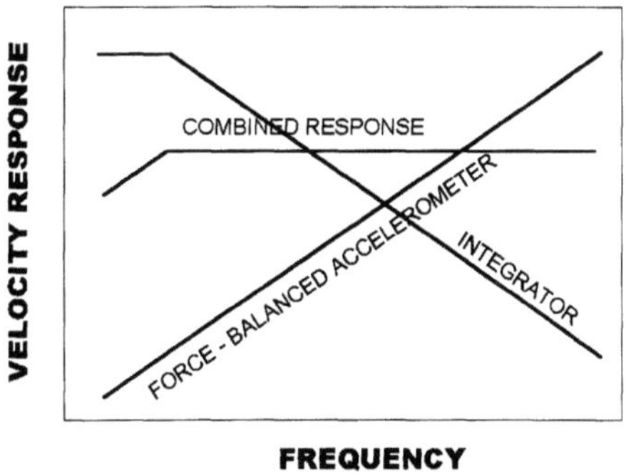

Figure 2.21 Simplified response elements of a velocity broadband sensor showing log-log velocity response of gain versus frequency. The idealized response of each element is shown (all slopes are 1 or −1), as well as the combined response.

The velocity response for a BB sensor is mainly determined by 4 passive components (*R*, *C*, mass and feedback coil) and can be considered to behave like a conventional velocity sensor. It can actually be described by the usual parameters: Free period, damping and generator constant and will often behave closer to the theoretical seismometer velocity response equation than passive velocity sensors.

At some high frequency the feedback current becomes too small (gain less than one due to the filter). This is the upper frequency limit of the feedback system and the response is no longer flat to velocity. At high frequencies, suspensions will also have parasitic resonances, so all BB sensors will have some upper frequency limit depending on both electronic and mechanical properties.

Besides capacitive transducers, other types of displacement sensitive devices have been used in some BB sensors, such as linear voltage differential transformers (LVDT). They have a wider displacement range but their sensitivity is lower than capacitive displacement sensors. This imposes a mechanical design of longer free period (Wielandt and Streckeisen, 1982).

BB sensors are available with a large range of properties and prices. Generally, the better the signal to noise ratio and the lower the corner frequency, the higher the cost. Examples will be given later and in Table 2.4

2.13 Extending the frequency response, inverse filtering and feed back

The simplest way of extending the frequency response is to use the 'inverse filter method'. This simply consists of an amplifier that selectively enhances the low frequency part of the signals, such that the flat part of the response curve is extended downwards in frequency. This requires a rather special filter with a response as shown in Figure 2.22.

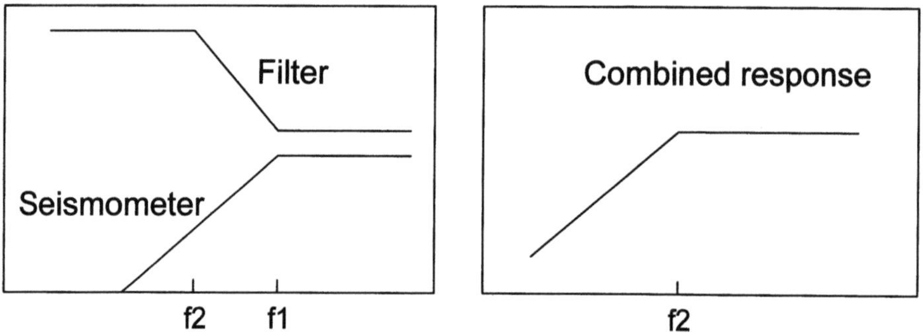

Figure 2.22 Enhancing frequency response by inverse filtering. The response functions are shown as log gain versus log frequency. The seismometer has corner frequency *f1* and the special filter corner frequencies *f1* and *f2* (left). To the right is seen the combined response.

The response after filtering looks like a seismometer with corner frequency *f2*. This method of extending the frequency response is limited by the signal output of the sensor and does not, like the FBA principle, improve dynamic range and linearity, see also discussion on sensor self noise below.

An alternative way is to use feedback. The FBA sensors are specially made for feedback with a feedback coil and a displacement transducer. But it is also possible to use feedback with a standard velocity transducer using the signal coil for both signal pickup and feedback. Figure 2.23 shows the principle schematically.

The principle is that part of the output is fed back to the signal coil, such that it opposes the motion of the mass by exerting a force proportional to the mass velocity. So, the mass will try to remain stationary like for a FBA sensor, with the same advantages of

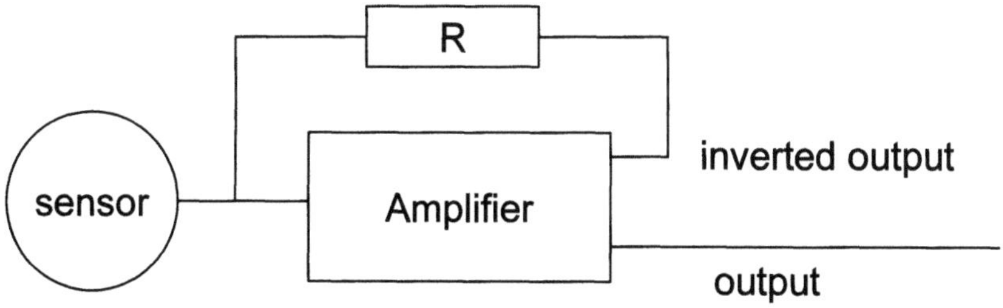

Figure 2.23 Velocity sensor with negative feedback. *R* is the feedback resistor. Actually, the external *load* that the sensor *sees* is equivalent to a negative resistance.

linearity. The strong negative feedback is equivalent to a very high damping since the feedback force is proportional to the mass velocity, and, as we have seen from (2.15), if the damping is high, then the output from the seismometer is proportional to acceleration (velocity for a mechanical sensor) within a limited band centered at the resonance frequency. So, this feedback sensor can simply be described by the standard velocity sensor response function (2.34). A natural question is if we could not just use a small damping resistor to obtain the same effect? Looking at (2.45) and using the example of the S13 sensor, a damping of e.g. 4 would require a total damping resistance of 6300/4= 1575 ohms. However, since the coil resistance is 3600 ohm, this is not possible unless the damping resistance is negative, which is how this special case is treated mathematically. So we can say that a strong negative velocity feedback is equivalent to a negative external damping resistance.

If a standard velocity response is desired, some filtering (response equalization) like for the BB sensor is required. Filtering is usually also required for stability reasons.

The advantage of this design is that quality sensors with extended frequency response can be constructed with inexpensive sensors like exploration type 4.5 Hz geophones and this type of sensors are sold commercially. Due to limitations in mechanical and electrical parameters, the sensor corner frequency cannot be extended more than a factor of 5, so a 4.5 Hz sensor becomes a 1.0 Hz sensor (see example later), and linearity and cross axis sensitivity is not as good as for the FBA type instruments.

2.14 Theoretical aspects of active sensors

The following sections will deal with more detailed theoretical calculation of response function, etc., for the various types of active sensor described in 2.10 to 2.13.

2.14.1 GENERAL RESPONSE OF A FEEDBACK SYSTEM

Let us consider now the way in which the feedback modifies the response of a system in general, by using the concept of the complex response function. Let T be the response of a system, whose input is $X(\omega)$ and whose output is $Y(\omega)$. By definition, the response is $T(\omega)=Y(\omega)/X(\omega)$. Now the output $Y(\omega)$ is connected as the input of another system, whose response is $\beta(\omega)$. The output of this one is fed back and subtracted from the complete system input $U(\omega)$. We suppose that connecting the output of a system to the input of another one does not affect its individual response (Figure 2.24).

The new input of T is now $X = U - \beta \cdot Y$, but since its output has to be $Y = T \cdot X$, or

$$Y = T \cdot (U - \beta \cdot Y) \tag{2.56}$$

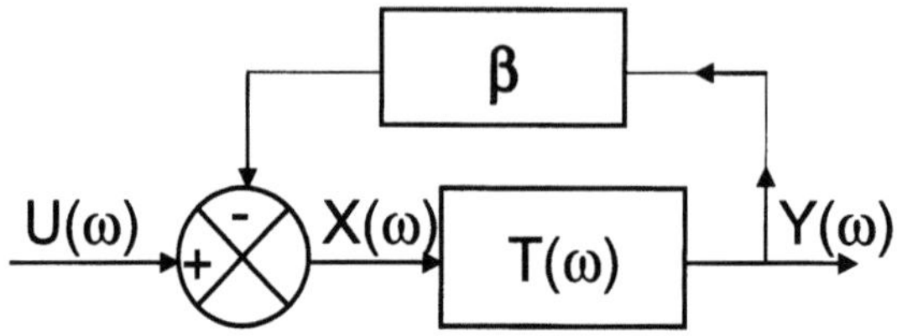

Figure 2.24 Feedback system

Thus, the response T_{af} of the whole system will be

$$T_{af}(\omega) = \frac{Y(\omega)}{U(\omega)} = \frac{T(\omega)}{1 + \beta(\omega) \cdot T(\omega)} \tag{2.57}$$

This is known as the Miller theorem. If the product of the open-loop response T and the loop gain is $\beta T >> 1$, then the closed-loop response T_{af} is completely controlled by the feedback loop:

$$T_{af} \approx \frac{1}{\beta} \tag{2.58}$$

The main point is that the shape of T is not significant, provided that the loop gain β·T is much higher than unity.

In practice, it is often required to include an integrator in the main path or a phase-lead system in the feedback to achieve a stable response. In the case of feedback active sensors, T is represented in most cases by the response of the mechanical system and a displacement transducer. The closed-loop overall response is mainly shaped by the feedback loop.

2.14.2 THE BASIC FORCE-BALANCED ACCELEROMETER

There is a number of schemes for implementing an active sensor with broad-band response to ground acceleration.. The force-balance principle will now be described in more detail. Let us consider first the block diagram of the Figure 2.25.

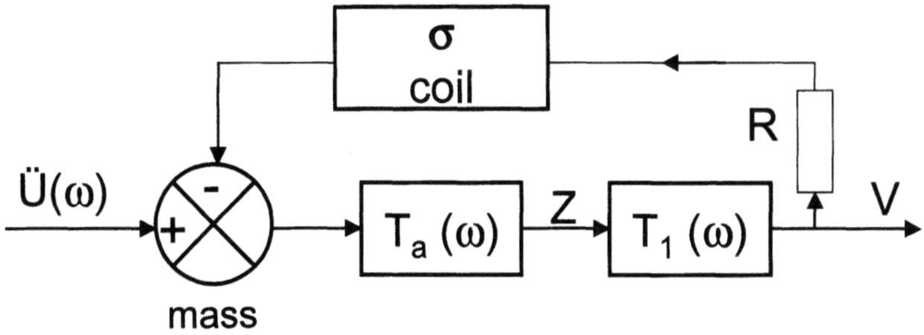

Figure 2.25 Block schematics of a FBA system.

A mechanical sensor is excited by ground acceleration $\ddot{u}(t)$ and the mass motion is $z(t)$. In the frequency domain, $Z(\omega)$ may be obtained by the mechanical acceleration response function $T_a(\omega)$ (2. 29)

$$T_a(\omega) = \frac{Z(\omega)}{\ddot{U}(\omega)} = \frac{-1}{\omega_0^2 - \omega^2 + 2ih\omega\omega_0} \tag{2.59}$$

The output z of the mechanical system acts as input of a displacement transducer, characterized by its response $T_l(\omega)$, which by now is simply a constant factor, $T_l(\omega)= D$, the sensitivity of the transducer (measured, e.g. in volts/meter). The voltage output of this causes a current $I=V/R$ to flow in the feedback coil, and this exerts a proportional force $\sigma I=\sigma V/R$ on the mass, in opposite sense to its motion. The effect is the same as if a stiffer spring had been added, and the system resonance frequency is now higher. The mass now senses the ground acceleration and the feedback acceleration (feedback force divided by the mass m). The effect of a ground motion upwards, is to push the mass m downward and the feedback force will oppose that, so both accelerations are summed with the same sign. Hence

$$Z(\omega) = T_a(\omega)\cdot(\ddot{U}(\omega) + \frac{\sigma}{m}\frac{V(\omega)}{R}) = T_a(\omega)\cdot(\ddot{U}(\omega) + \frac{\sigma}{m}\frac{DZ(\omega)}{R}) \tag{2.60}$$

As we are interested in the response function T_{af} between ground acceleration and output voltage,

$$Z(\omega)\cdot(1 - \frac{\sigma D}{mR}T_a(\omega)) = T_a(\omega)\cdot\ddot{U}(\omega) \tag{2.61}$$

And, since $Z=V/D$,

$$T_{af}(\omega) \equiv \frac{V(\omega)}{\ddot{U}(\omega)} = \frac{D\cdot T_a(\omega)}{(1 - \frac{\sigma D}{mR}T_a(\omega))} \tag{2.62}$$

For frequencies for which $\frac{\sigma D}{mR}T_a >> 1$, the closed-loop response is approximately

$$T_{af}(\omega) \cong -\frac{mR}{\sigma} \tag{2.63}$$

This result is important because, for a high loop gain, a) the elements within the main (forward) path in the loop (the mechanical system and the displacement transducer) have little influence in the closed-loop response, e.g. the transducer sensitivity D does not appear, and b) the response is controlled by passive (in this case), stable elements in the feedback path (the resistor and the coil).

The exact expression for T_{af} is thus,

$$T_{af} = \frac{-D}{\omega_0^2 + \frac{\sigma D}{mR} - \omega^2 + 2ih\omega\omega_0} = \frac{-D}{\omega_f^2 - \omega^2 + 2ih_f\omega\omega_f} \tag{2.64}$$

where ω_f and h_f have been calculated as

$$\omega_f = \sqrt{\omega_0^2 + \frac{\sigma D}{mR}} \quad \text{and} \quad h_f = h\frac{\omega_0}{\omega_f} \tag{2.65}$$

This is equivalent to the acceleration response from a mechanical seismometer with a natural frequency ω_f and damping h_f. It is seen that the equivalent natural frequency has been increased, while the equivalent damping has decreased.

An example of this response for a given parameter set is shown below (Figure 2.26). There is a strong peak at the new resonance frequency due to the low damping.

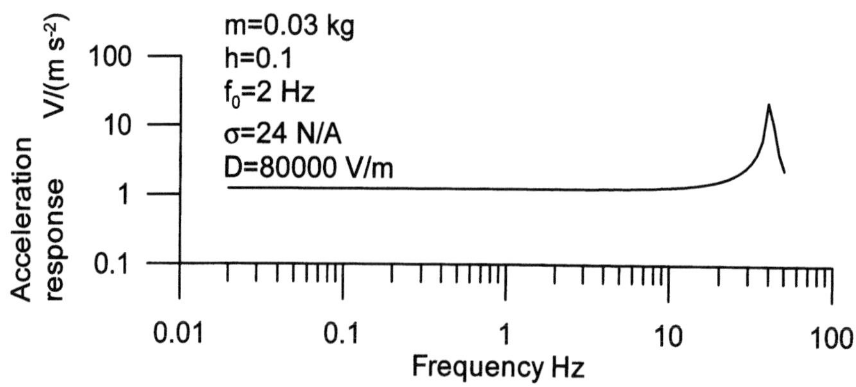

Figure 2.26 Example of the effect of force feedback on the sensor response.

The peak at the new resonance frequency in undesirable, because it also implies a ringing transient response. One possible way of damping these oscillations is to include an integrator with a short time constant, τ, as part of the displacement transducer response $T_1(\omega)$.

In this case, $T_1(\omega) = \dfrac{D}{i\omega\tau_1 + 1}$, which is a low-pass filter with corner frequency $\omega = 1/\tau$.

The closed loop response is now

$$T_{af}(\omega) = \frac{-T_1(\omega)}{\omega_0^2 + \dfrac{\sigma T_1(\omega)}{mR} - \omega^2 + 2ih\omega\omega_0} \tag{2.66}$$

This is represented in Figure 2.27 for a value of τ_1=0.02s (or a corner frequency of $f_1 = 1/2\pi\tau_1$=8Hz).

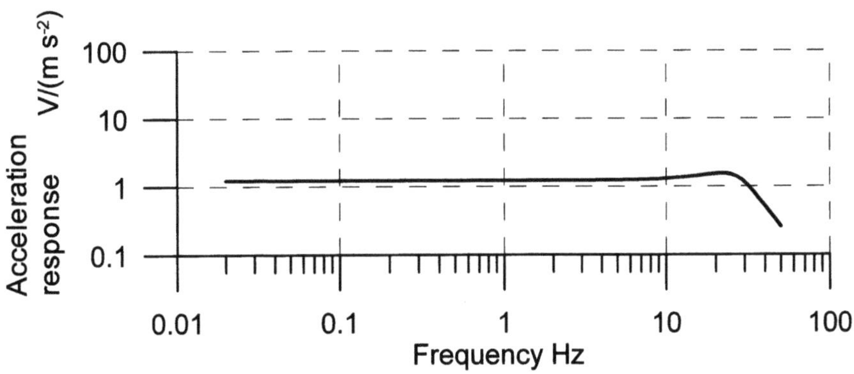

Figure 2.27 Frequency response of filtered FBA.

As it can be observed, the response is now almost flat, at the cost of a lower corner frequency. This is the compensation technique used in some accelerometers and BB seismometers based on the FBA principle. In practice, there exist several other compensation techniques owing to the servo control theory. Some of them use a proportional-plus-integral feedback and a phase-lead circuit in the feedback loop (e.g. Usher, Buckner and Burch, 1977; Usher, Burch and Guralp, 1979). Nevertheless, the above implementation illustrates the principle of force-balanced accelerometers (FBA).

2.14.3 BROADBAND FEEDBACK SEISMOMETERS

Once a response flat for ground acceleration is obtained, a simple integration outside the loop may yield another output flat for ground velocity to obtain a BB seismometer. This solution, though practical, is not near the optimal in order to achieve a high dynamic range. As a rule, it is better to shape the response within the loop using as few active components as possible. One of the implementations with best performance is the one of Wielandt and Streckeisen (1982). The start design is a basic FBA, but two additional feedback paths proportional to the time derivative of the mass displacement and to its integral are added to shape a closed-loop response that resembles a passive long-period electromagnetic seismometer. The block diagram is shown in Figure 2.28.

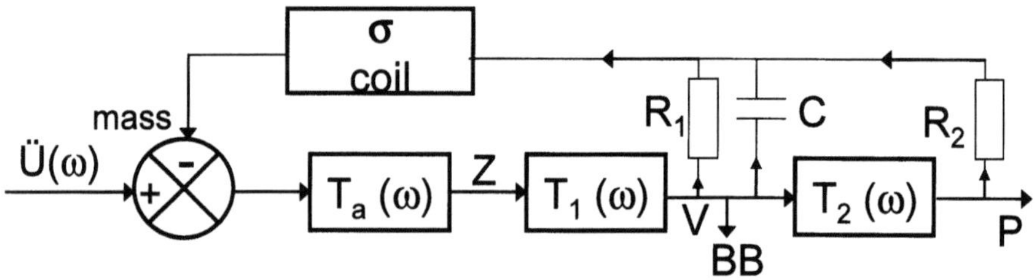

Figure 2.28 The FBA with two additional feedback paths. BB is the broadband output.

$T_1(\omega)$ represents the displacement transducer including an integrator, as in the example above, (first order low pass) with a time constant τ_1. $T_2(\omega)$ is a second integrator of the same type, however with a larger time constant τ_2.

$$T_2(\omega) = \frac{D}{i\omega\tau_2 + 1} \tag{2.67}$$

Both integrators are made with active circuits. BB is the velocity broadband output. The voltage $V(\omega)$ at the transducer output is given by $V(\omega)=Z(\omega)\cdot T_1(\omega)$. The current in the capacitor is (neglecting coil resistance) $I_C=CdV(t)/dt=V(\omega)\cdot Ci\omega$. Thus, the current $I(\omega)$ flowing through the feedback coil will be

$$I(\omega) = T_1(\omega)(Ci\omega + \frac{1}{R_1} + \frac{T_2(\omega)}{R_2})Z(\omega) \tag{2.68}$$

and the equivalent ground acceleration (the feedback force per unit mass) is $\sigma I/m$. Following the same scheme as above, the closed-loop response at the BB output is obtained as

$$T_{af}(\omega) = \frac{-T_1(\omega)}{(\omega_0^2 - \omega^2 + 2ih\omega\omega_0) + \frac{\sigma}{m}(\frac{R_1Ci\omega + 1}{R_1} + \frac{T_2(\omega)}{R_2})T_1(\omega)} \tag{2.69}$$

This is the response function for ground acceleration, *which is not flat* due to the presence of the capacitor. The amplitude and phase response for ground velocity T_{vf} is simply $T_{vf}=T_{af}i\omega$. This is represented below (Figure 2.29) for a very broadband sensor (solid line) (Steim and Wielandt, 1985), together with the ideal response (dots) of a passive electromagnetic sensor with a free period of 360s. The parameters used are D=80000 V/m, m=0.6 kg, σ=24 N/A, f_0=0.05 Hz, R_1=3096 kΩ, R_2=320 kΩ, C=10 μF, τ_1=0.02 s, τ_2=1036 s. From all of these, the relevant parameters for shaping the response are the parameters in the feedback loop, i.e. σ, R_1, R_2, C and τ_2. However, other parameters such as the transducer sensitivity are relevant for the overall system performance (e.g. self-noise level) (Rodgers, 1992 b).

As it is apparent, the response of the feedback seismometer matches the one of the passive sensor, except at high frequencies, where the feedback instrument response decreases as ω^{-2}. This is due to the low gain of the loop at high frequencies that makes the damping become ineffective. In this band the instrument is no longer force-balanced. By changing just R_1 to be 220 kΩ and τ_2= 3.2 s, the response is equivalent to a 20 s passive sensor, as it was in the original design of Wielandt and Streckeisen (1982). The passive components in the feedback have a nearly ideal behavior and, unlike amplifiers, do not saturate. This is a major key feature of the design and is mostly responsible for the high performance in terms of reliability and dynamic range of this instrument, together with the astatic leaf-spring suspension.

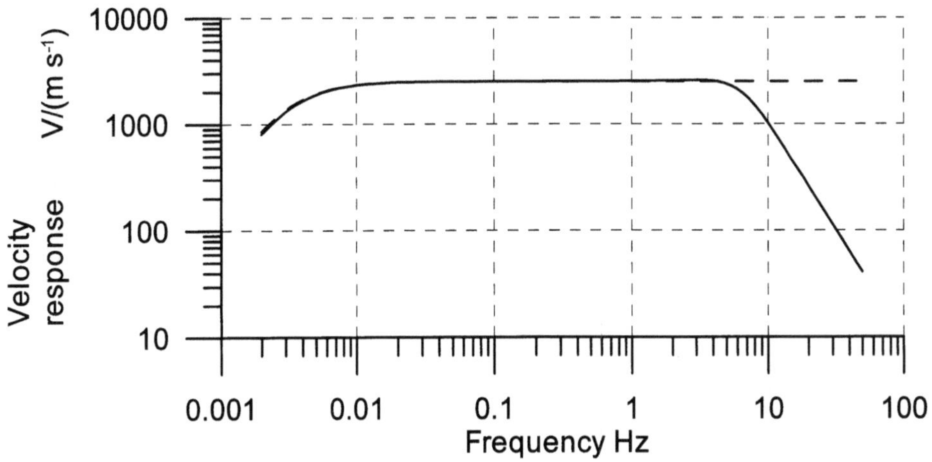

Figure 2.29 Frequency response of a BB system

The output at P on Figure 2.28 is the integral of the BB output, hence proportional to ground displacement in the band for which BB is flat to velocity. A voltage offset at P means that a DC voltage is being applied to keep the mass centered: a mechanical adjustment (a "mass centering") is required.

2.14.4 OTHER FEEDBACK TECHNIQUES

The use of AC-driven displacement transducers overcomes the low-frequency noise of amplifiers and, since actually inertial sensors respond to ground acceleration, it helps to improve low-frequency velocity response. Nevertheless, the response of sensors with conventional electromagnetic (velocity) transducers may be extended towards longer periods using feedback, although its inherently noisier transducer does not permit to reach the low frequency sensitivity of BB instruments based on displacement transducers.

One possible way is to differentiate the output and apply a feedback current proportional to the mass acceleration using a separate coil. This is equivalent to using a larger mass and thus increases the free period. Another technique uses a *negative resistance* as the sensor load, which actually is equivalent to applying a feedback current (and force) proportional to the mass velocity. This may be done on the same signal coil. Let us consider the circuit in Figure 2.30 in which S represents the sensor.

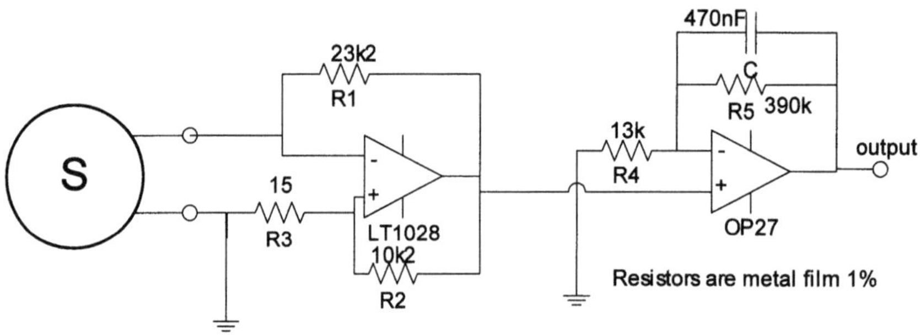

Figure 2.30 Typical feedback system for extending the response of a velocity sensor. The component values shown were calculated for a 4.5 Hz Mark L15 of 380 ohm.

The analysis of this circuit gives for the current flowing through the sensor coil with resistance R_g

$$I = V_s \frac{R_2}{R_2 R_g - R_1 R_3} \tag{2.70}$$

where V_s is the voltage generated in the sensor. But $V_S = G \cdot \dot{z}$ and the magnetic force opposite to the mass motion is F=GI, thus the total damping will be (see 2.4.1)

$$h = h_0 + \frac{G^2 R_2}{2(R_2 R_g - R_1 R_3) m \omega_0} \tag{2.71}$$

The effect of this over-critical damping is to split the corner frequency f_0 into two frequencies (actually a complex double pole is transformed to two single poles):

$$f_1 = f_0(h - \sqrt{h^2 - 1} \quad \text{and} \quad f_2 = f_0(h + \sqrt{h^2 - 1} \tag{2.72}$$

In the band between f_1 and f_2 , the response is proportional to acceleration. A damping h=5, e.g. on a 1 Hz sensor, will give f_1=0.1 Hz and f_2=9.9 Hz. A further equalization filter (the second stage in the circuit shown) may integrate just in this band and yields a wide flat velocity response. For the values in this circuit, a geophone Mark L15-B of f_0=4.5 Hz and R_g=380 Ω, the system is damped to h=3 and its response is extended to be flat above f_l=0.73 Hz .

Figure 2.31 shows the ground velocity response of the over-damped and amplified sensor (dots), the equalizing filter (dash) and the combined total response (solid). Vertical axis is V/(m·s^{-1}) and horizontal is frequency in Hz.

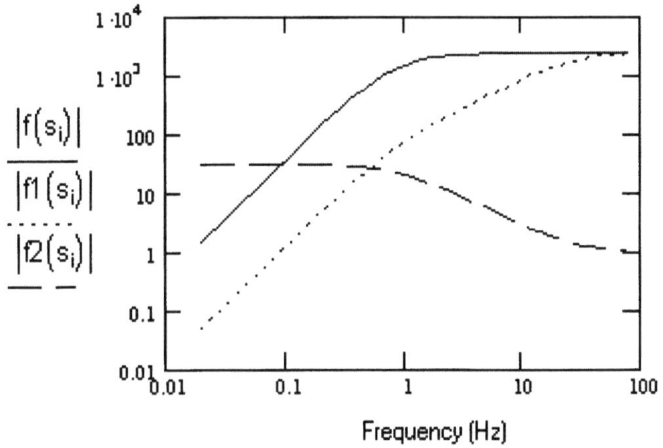

Figure 2.31 Velocity response of an overdamped and amplified sensor (dots), the equalizing filter (dash) and the combined response (solid). Note that the new corner frequency is 1 Hz. Vertical axis unit is V/(m·s^{-1}).

Sensors with extended response using this technique are available commercially and present almost the same performances of their equivalent higher-cost passive ones.

2.15 Sensor self noise

Theoretically, a standard seismometer or geophone will output a signal in the whole frequency range of interest in seismology. So why not just amplify and filter the seismometer output to get any desired frequency response instead of going to the trouble of making complex BB sensors? That brings us into the topic of instrument self noise. All electronic components as well as the sensor itself generate noise. If that noise is larger than the signal generated from the ground motion, we obviously have reached a limit. A very simple test of the self-noise in the recording system is to clamp the seismometer (or replace it with a resistor of the same value as the signal coil) and to measure the output of the recording system. This output would give a good idea of the noise generated within the recording system. Hopefully, that will be much less than the unclamped signal, otherwise we are just measuring recording system noise and not ground noise.

Active sensors of the force balanced types cannot be tested for instrumental noise with the mass locked. The self-noise must therefore be determined in the presence of seismic noise, for more details, see Chapter 10.

Instrument self-noise can be divided into two parts. (1) Brownian thermal motion of the seismometer mass-spring system, which is inversely proportional to the mass and directly proportional to the damping. Unless the mass is very small, this noise is so low that it has no practical importance. Even for a geophone with a mass of 23 g, this noise is below the level of the New Low Noise Model (NLNM, see Chapter 3) (Figure 2.32). (2) Electronic noise. This is roughly caused by current flowing trough different elements[2] (also the sensor coil and damping resistor) and semiconductor noise, for more details see Riedesel et al., 1990. The electronic noise is thus the real limiting factor in seismograph sensitivity. This is particularly a problem at low frequencies because sensor output is low and semiconductors tend to have a higher noise level at low frequencies. So even though the sensor might give out a signal measuring ground motion, this signal is so small that it drowns in the electronic noise. We can illustrate this by calculating the output of a 4.5 Hz sensor corresponding to some of the points of the Peterson noise curves (Peterson, 1993), which represents the ground motion limits we normally are interested in. Table 2.1 gives the displacement and velocity for some of the points on the NLNM and NHNM curves (see Chapter 3).

Table 2.1 Ground motion as derived from the Peterson noise curves and corresponding output from a geophone. When converting from dB to ground motion, a half-octave filter was used. The geophone has a generator constant of 30 V/ms^{-1} and a natural frequency of 4.5 Hz. The abbreviations are: F: Frequency, Level: Level in dB below 1 $(ms^{-2})^2$/Hz, Disp: Displacement, Vel: Velocity, Out: Output of sensor

F, Hz	Low noise model				High noise model			
	Level (dB)	Disp (nm)	Vel (nm/s)	Out (nV)	Level (dB)	Disp (nm)	Vel (nm/s)	Out (nV)
0.01	-185	11	0.7	3 10^{-6}	-132	4490	295	0.04
0.1	-163	4.2	2.6	0.001	-115	1050	660	10
1.0	-166	0.1	0.6	1	-117	26	167	246
10	-168	0.002	0.1	3	-91	16	1046	31000

It is very difficult to get electronic noise lower than 0.1 μV so from Table 2.1 it is seen that a 4.5 Hz geophone would not be able to resolve the NLNM at any frequency and barely the NHNM at frequencies below 0.1 Hz. This is also what is predicted theoretically (Figure 2.32).

Therefore, the noise level of the recorder or sensor in volts does not in itself illustrate the sensitivity limitations. What we are really interested in is how this limits the sensitivity to ground motion, as illustrated in Table 2.1. Since different sensors have different sensitivity, the total ground noise resolution of the seismograph will depend on the sensitivity of the sensor as well as the recorder. So, for a particular application, the recorder might be O.K. with one sensor, but not with another, although both sensors can

[2] Actually, the physical nature of mechanical noise is related to the coupling between mechanical and thermal energy. This exists only in dissipative elements such as viscous damping. Noise in passive electronic elements has a similar nature, like current heating of a resistor (Johnson noise). Non-dissipative elements as ideal capacitors or coils are free of intrinsic noise, as are undamped mass-spring systems.

record the ground motion. Since the Peterson curves give the limits of ground motion we normally are interested in, the simplest way of representing the electronic self-noise is to convert it to the equivalent ground motion for the sensor being tested and comparing to the Peterson curves. Any ground motion below this theoretical level will produce an output signal smaller than the electronic noise.

Figure 2.32 shows an example. For the standard 4.5 Hz geophone, it is seen that the predicted self noise is between the NLNM and NHNM in the frequency range 0.1 to 10.0 Hz and actually close to NLNM in the range 0.3 to 10 Hz. Only at 0.1 Hz will the noise be equivalent to the level of the NHNM, as also seen in Table 2.1. The actual measured noise was within a factor of two of the predicted noise (Barzilai et al, 1998). The amplifier used in this example is the LM27 from National Semiconductors. A more elaborate amplifier might lower the noise level further.

In the example shown, the equivalent ground noise curve appears like a mirror image of the sensor acceleration response, since it turns out that the electronic noise is almost constant with frequency. In other words, the sensor frequency response will shape the curve, except for a steeper slope at low frequency than expected, due to a slight increase of the electronic noise PSD in this band (see 2.16 for further details).

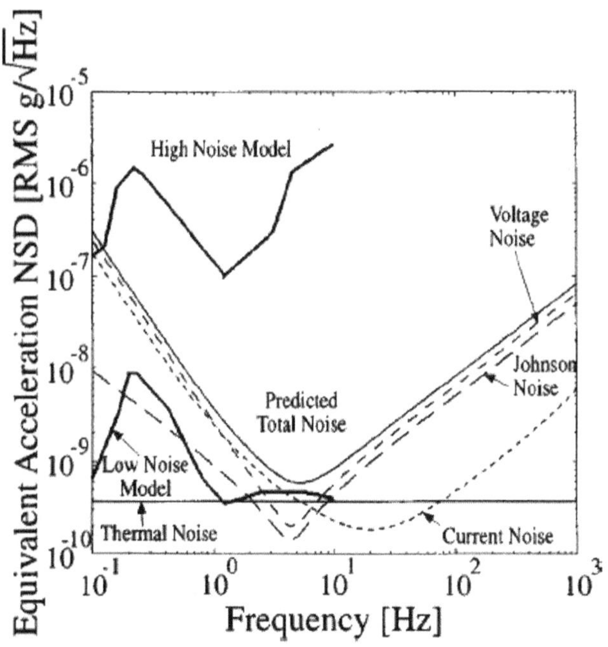

Figure 2.32 Predicted total noise equivalent acceleration of a standard electronic circuit with a 4.5 Hz geophone. The contribution from each element is shown and the Peterson noise curves are shown for reference. Thermal noise is the noise due to Brownian motion of the mass, Johnson noise is caused by thermal fluctuations of the electrons within any dissipative element in the electronic circuit, voltage and current noises are generated within the amplifier. Figure from Barzilai et al (1998).

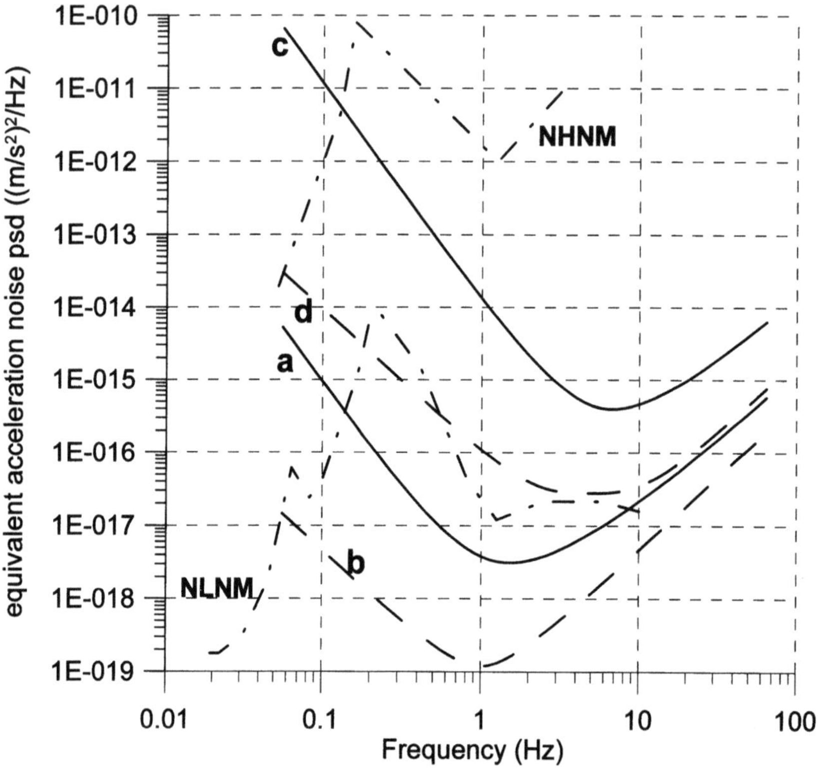

Figure 2.33. Predicted PSD of equivalent noise acceleration generated by 1 Hz and 4.5 Hz seismometers. a) Self noise of a standard 1 Hz seismometer with an amplifier. b) Noise generated by this seismometer alone. c) Self noise of a 4.5 Hz seismometer with amplifier included. d) Noise due to the same seismometer alone. The standard 4.5 Hz geophone presents an acceptable noise level down to 0.3 Hz and would be able to replace, at a much lower cost, the standard SP seismometer for application where there is no need for data below 0.3 Hz or the ground noise level is not low. This is the case for recording micro earthquakes and in the case of many aftershocks studies. If the flat level of the signal spectrum should be resolved, the corner frequency must be above 1.0 Hz, which corresponds to roughly a magnitude 4.5-5.0 event. This does not mean that larger events cannot be recorded, only that the bandwidth of the 4.5 Hz sensor is too narrow for spectral analysis or other analysis that require the complete signal to be recorded.

Figure 2.33 also shows the predicted noise equivalent acceleration expressed as its power spectral density (PSD) for a standard 1 Hz sensor and a 4.5 Hz sensor, both with amplifiers. The figure plots the contributions to noise of the sensors alone and the total noise. For the sensor of 1 Hz, an OP27 amplifier has been used, suitable for medium-impedance sources. For the sensor of 4.5 Hz, with lower coil resistance, the amplifier is a LT1028, with lower voltage noise, but higher current noise, has been used. As it can be observed, the major contribution to noise is due to the amplifiers. Should these same sensors be equipped with other kinds of transducers (e.g. displacement transducers) more sensitive and less noisy, the instruments would resolve ground motions of lower amplitude. Of course, this is the case of feedback BB seismometers.

So, why not just increase the generator constant of a 4.5 Hz geophone? This hypothesis is rather unrealistic: If G were to be increased by a factor of 10, the coil resistance would be 100 times larger (if using the same physical space) and this would raise the noise. Even if G is 10 times larger, and keeping the value of the coil resistance (by using thicker wire), the damping resistance would have to be much higher (see 2.31) and this again would increase noise. To avoid this, also mass has to be higher, and we would end up with a 1 Hz sensor! Most 4.5 Hz geophones, however, are available with different coil resistances and the corresponding generator constants, as a customer option: The choice should be made in each case according to the characteristics of the amplifier to be used and the expected main noise contribution. The self noise of the 1 Hz seismometer is well below the NLNM until 0.1 Hz. This means that, with adequate amplification, an SP seismometer should be able to record low amplitude 20 s surface waves, and this is in fact frequently observed with high gain digital SP stations.

Modern passive sensors have high generator constants. Remember that this constant is also the motor constant if a current flow through the coil. The current noise generated in the front-end of the preamplifier thus generates a counter-electromotive force in the sensor that contributes to the total noise. Therefore, a realistic estimation of the system noise has to be made with all the elements connected and not separately.

2.16 Noise in passive sensors coupled to amplifiers, theoretical aspects

It has been stated above that the direct contribution of the sensor to the total noise is of relative importance only for a very small suspended mass and with a high damping. Nevertheless, as most passive sensors are directly coupled to a preamplifier, which contributes with the most of the total system noise, part of the generated noise voltage and current flow back to the sensor and may move the mass, causing a non-negligible additional contribution.

Table 2.2 The formal similarity between the equations describing a second order mechanical and electrical system allows us to use an analogy between them. The table shows the equivalences.

Mechanical system	Electric system
$F = M \cdot \dot{v}_U = M\ddot{u}$ (inertia force)	$i = C \cdot \dot{v}$
$F = k\int \dot{z}dt = kz$ (restoring force)	$i = \frac{1}{L}\int v dt$
$F = d\dot{z}$ (damping force)	$i = \frac{1}{R}v$
$Z(\omega_0^2 - \omega^2 + 2hi\omega_0\omega) = -\omega^2 U$	$V_O(\omega_0^2 - \omega^2 + 2hi\omega_0\omega) = -\omega^2 V_S$
$\omega_0^2 = k/M \quad 2h\omega_0 = d/M$	$\omega_0^2 = 1/LC \quad 2h\omega_0 = 1/RC$

A useful approach to the study of such an electro-magneto-mechanical system, which simplifies calculus making use of electric network analysis, is to substitute the mechanical part by an equivalent circuit. That is, a circuit for which equations are formally the same as for the mechanical system. Several electric analogies have been proposed in the literature, but the one used by Byrne (1961) is especially meaningful, because the element values have a clear physical interpretation. In this analogy (Figure 2.34), as shown in Table 2.2, the current is equivalent of force, voltage source is equivalent of ground velocity, and mass motion is equivalent of output voltage.

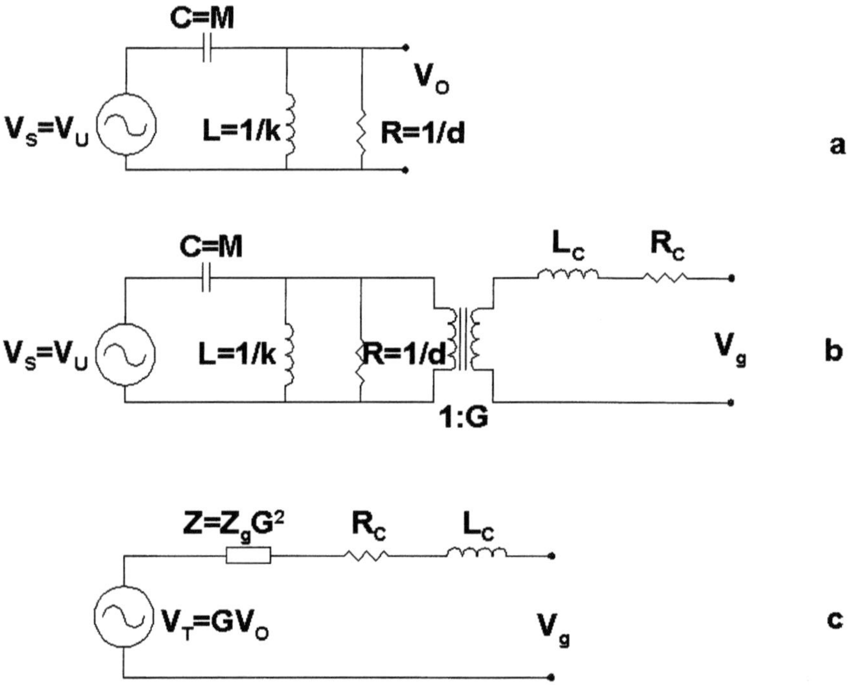

Figure 2.34 Equivalent circuit of a sensor. a) Mechanical sensor. b) The same with a velocity transducer. A transformer is used for coupling the equivalent circuit to the real charge, see text. c) The circuit of b where the virtual part (the transformer and everything on its left) has been replaced by its Thévenin equivalent.

The left column lists the equations for the mechanical system and the right the corresponding for the electrical circuit shown in Figure 2.34a. In this circuit, a voltage source V_S represents the ground velocity $V_U \equiv \dot{u}$, a capacitor C makes the function of mass, a coil L yields the equivalent of "restoring force" and a resistor R performs damping. As it can be seen in the Table 2.2, the final equations describing the relationships between ground motion and mass motion for the mechanical system and between voltage source and voltage output for the electric one are identical.

The damping devices in both systems are responsible of thermal dissipation, thus generating thermal noise. In the equivalent circuit, this noise is expressed as the voltage power spectral density (PSD) in the resistor (Johnson noise), $e^2=4k_BTR$ (see, e.g. Chapter 10 of Aki and Richards, 1980), where k_B is the Boltzmann constant and T the Kelvin temperature. This will produce a current PSD $i^2=e^2/R^2$. As current is equivalent to force, dividing by mass (squared) and by substitution of R by its mechanical equivalent, we found the PSD of noise acceleration of the mass $(\ddot{z})^2$ as

$$(\ddot{z})^2 = \frac{4k_B T}{M} \cdot 2h\omega_0 \tag{2.73}$$

Now, let us include to the mechanical system an electromagnetic (velocity) transducer with generator constant G, coil resistance R_g, and, to be general, also a coil self-inductance L_C. This is easily accomplished in the circuit of Figure 2.34b, by an ideal transformer with relation 1:G between primary and secondary number of windings. This transformer also links the equivalent circuit to the real circuit of the signal coil. From all points of view, as seen from the output terminals of this circuit, it is undistinguishable whether it is the output of an electromagnetic seismometer or the circuit of the figure.

The circuit of Figure 2.34b may be simplified by applying Thévenin theorem. All the elements to the left of the transformer output terminals are substituted by a source V_T (the voltage between these terminals in open circuit) and a series impedance Z (the impedance as seen from them). If we neglect the sensor coil inductance L_C, it can be shown by analysis of this circuit that this impedance is

$$Z = R_C \frac{\omega_0^2 - \omega^2 + 2h\omega_0 \cdot i\omega}{\omega_0^2 - \omega^2 + 2h_0\omega_0 \cdot i\omega} \tag{2.74}$$

where h_0 represents the seismometer open-circuit damping (i.e. air damping) and h is the short-circuit damping (see 2.4.1 on damping),

$$h = h_0 + \frac{G^2}{2RgM\omega_0} \tag{2.75}$$

Only the real part of this impedance is dissipative (resistive) and therefore contributes to the noise. So, the voltage (real, not equivalent) noise PSD at its output will be

$$v_G^2 = 4k_B T \cdot \mathrm{Re}(Z) \tag{2.76}$$

The equivalent noise ground acceleration PSD may be obtained if this is divided by the squared magnitude of the acceleration transfer function of the system.

We may consider now the noise model of an amplifier to which the sensor output will be connected. The most common input stage of a seismic preamplifier is an operational amplifier (following stages have lower contributions to noise, since the first stage input is the most amplified in the chain). A general noise model for this is shown in Figure 2.35 for a differential input configuration. The voltage sources e_{s1} and e_{s2} are the noise generators associated to each source impedance Z_{s1} and Z_{s2}. There is also a differential voltage noise generator e_n due to the amplifier. Two additional current sources i_{n1} and i_{n2} are included, one for each amplifier input. The values of these sources (frequency dependent) are specified by the manufacturer as typical values. The total noise contribution will depend on the particular configuration (inverting, non-inverting, differential) and on the components used. As an example, we use the amplifier circuit in Figure 2.35 (only essential components shown).

To obtain the total voltage PSD at the output of this amplifier connected to the seismometer, we include the noise sources of the noise model in the particular circuit. Since all the sources are independent, their contributions to the output PSD v_{no}^2 are simply added. For details of how to perform this analysis, see appendix A of Rodgers (1992a).

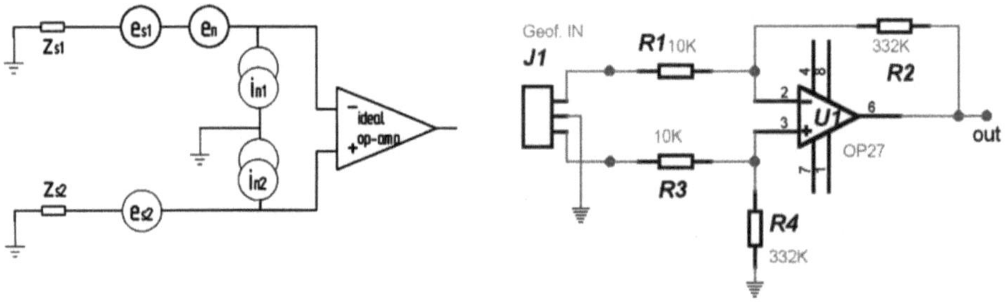

Figure 2.35 Left: A noise model of an instrumentation differential amplifier. e_n represents the internal noise voltage referred to the input, e_{s1} and e_{s2} are voltage noise sources associated to the signal sources, and i_{n1} and i_{n2} are current noise sources associated to each amplifier input. Right: A typical differential seismic amplifier, whose noise is estimated in the example below.

After this analysis, it may be shown that the result for the total output noise voltage PSD is

$$v_{no}^2 = \frac{4R_2^2}{\left|2R_1 + Z\right|^2} \cdot 4k_B T\left[2R_1 + \mathrm{Re}(Z)\right] + e_n^2 \left|\frac{2(R_1 + R_2) + Z}{2R_1 + Z}\right|^2 + 2i_n^2 R_2^2 + 8k_B T R_2 \tag{2.77}$$

(This equation has been simplified by using R_1=R_3 and R_2=R_4 in this case). The first term represents the voltage noise associated to the source impedance and the two input resistors R_1 and R_3; the second is the contribution of the voltage noise e_n, the third adds the effect of the current noise sources and the last is the Johnson noise generated in resistors R_4 and R_2.

The low frequency curve of e_n and i_n for a typical instrumentation amplifier decreases with frequency up to a certain corner frequency (within the seismic band) and then becomes nearly flat for the rest of the seismic frequencies. For the OP27, typical values are as follows (Analog Devices, 2003):

$$e_n^2 = e_{n0}^2(1+\frac{f_{ce}}{f}) \quad \text{with } e_{n0} = 3\cdot 10^{-9}\frac{V}{\sqrt{Hz}},\ f_{ce} = 2.7Hz \text{ , and} \tag{2.78}$$

$$i_n^2 = i_{n0}^2(1+\frac{f_{ci}}{f}) \quad \text{with } i_{n0} = 0.4\cdot 10^{-12}\frac{A}{\sqrt{Hz}},\ f_{ci} = 140Hz \tag{2.79}$$

With these values and the circuit components, using this amplifier for a sensor Mark L4-C (G=270 V/(m·s^{-1}), R_g=5500 Ω, f_0=1 Hz, h_0=0.28, M=1 Kg), the total voltage noise v_{no}^2 at the amplifier output is represented in Figure 2.36, as well as the contribution of sensor alone at the output of an ideal amplifier with the same gain.

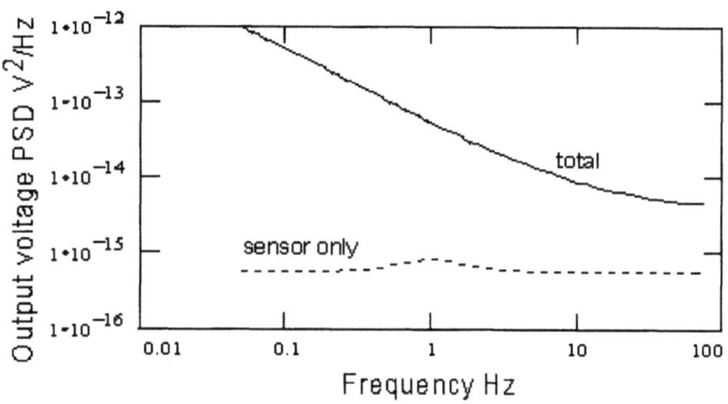

Figure 2.36 The output voltage noise PSD at the output of the amplifier due to the sensor noise and the total noise including the amplifier noise: It is clear that this latter is the most relevant contribution.

The amplifier gain is R_2/R_1 and its input resistance is $2R_1$, so the effective generator constant will be $G_e=G\cdot 2R_1/(2R_1+R_C)$. Now, using the transfer function for ground acceleration $T_a(\omega)$, including amplifier, we can find the PSD of the equivalent acceleration noise at input a_{ni}^2 as

$$a_{ni}^2 = \frac{v_{no}^2}{|T_a|^2}, \tag{2.80}$$

with

$$T_a(\omega) = \frac{i\omega}{\omega_0^2 - \omega^2 + 2h\omega_0\omega i}\cdot G_e \cdot \frac{R_2}{R_1}, \tag{2.81}$$

and where

$$h = h_0 + \frac{G^2}{2(R_C + 2R_1)M\omega_0} \tag{2.82}$$

This is represented in Figure 2.37, together with the equivalent input noise acceleration due only to the sensor.

A similar method can be followed for estimation of noise in feedback seismometers. For details and actual examples of this calculation, the reader is referred to Rodgers (1992b).

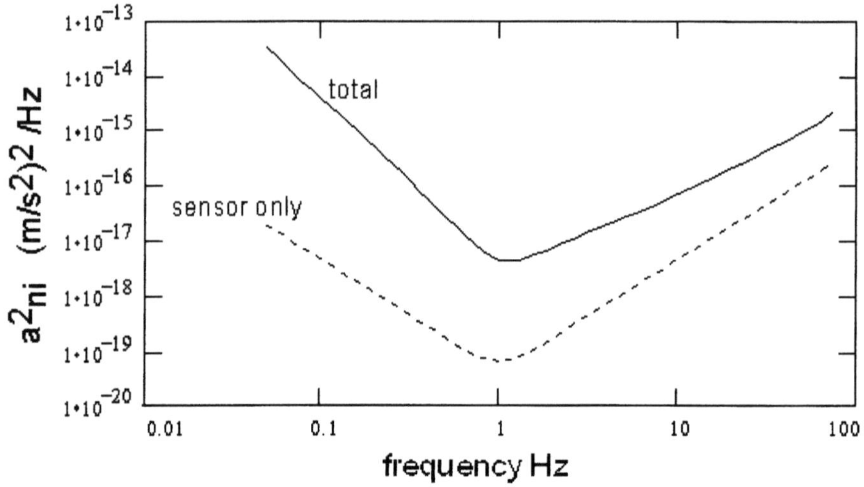

Figure 2.37 The same as in Figure 2.36, but in this case the noise is specified as an equivalent ground acceleration noise, that is, the ground acceleration that would produce the voltage output noise represented in Figure 2.36.

2.17 Some new trends in seismic sensors

2.17.1 SEISMOMETERS WITH ELECTROCHEMICAL TRANSDUCERS

Besides the traditional approaches, some newly developed concepts have led to prototypes and commercial models that use different techniques. One of them is the electrochemical seismometer (PMD Scientific, Inc., 2002, *www.eentec.com*; CME, 2003, *www.cmesci.rbcmail.ru/projects.html*). This design (Figure 2.38) uses a fluid as inertial mass and the motion of the fluid is detected by multilayer platinum electrodes with spacing of a few tenths of millimeter (R. Leugoud, 2003, personal communication). The fluid is an ion-rich electrolyte and is free to move. At both ends of the channel, an elastic diaphragm allows the fluid motion. When a DC voltage is applied to the electrodes, it produces an ion concentration gradient between them. Due to the electrolyte conductivity, the bias voltage and its associated current produce only an ion concentration gradient between the electrodes.

As the system is accelerated by a ground motion, the fluid flows relative to the electrodes and this yields a charge transfer and therefore a current change, proportional to the fluid velocity and to the ion concentration. The technique is called *Molecular*

Electronic Transfer (CME, 2003, *www.cmesci.rbcmail.ru/met.html*; R. Leugoud, 2003, pers. comm.). The transducer is essentially of velocity type. The symmetric arrangement of the electrode pairs improves the transducer linearity. The free period depends on the channel geometry and on the restoring force mechanism. The damping is influenced by the fluid viscosity and also by the channel geometry. Some instruments use three identical inclined sensors, forming an orthogonal frame.

The system response may be further linearized and shaped by a feedback loop, using a magnet-coil combination (PMD Scientific, Inc., 2002 *www.eentec.com*) to apply a force on the fluid mass. Eventually this feedback may be implemented by a magneto-hydrodynamic (MHD) (i.e. Lorentz) force on the conducting fluid (CME, 2003, *www.cmesci.rbcmail.ru/projects.html*).

There exist at present several commercial models of seismometers, including BB, using this principle, either open-loop or equipped with feedback. As they have no springs, hinges or moving mechanical parts, except the fluid, the manufacturers claim that they are rugged and suitable for field portable use. The BB models do not need any mass-centering, since the "center position" of the fluid is self-defined dynamically. They are relatively insensitive to leveling errors, a further advantage, for instance, in borehole or OBS installations. The sensing technique is also especially appropriate for pure rotational seismometers. Up to 2003, no paper has been yet published about tests and field performance of these sensors.

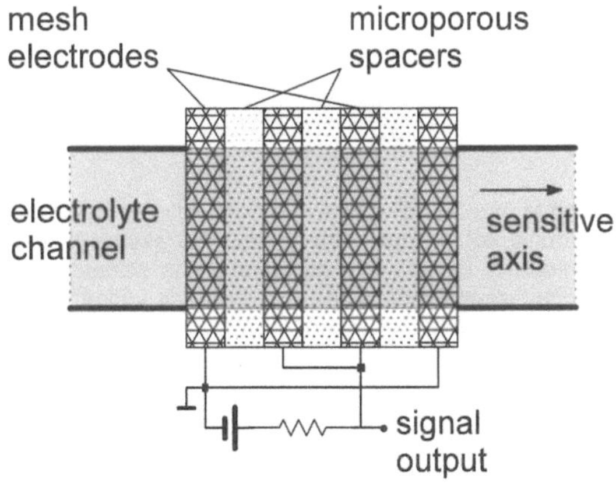

Figure 2.38 A simplified schematic of the electrochemical motion transducer used in MET seismometers (modified from R. Leugoud –PMD Scientific, Inc.-, 2003, personal communication). The electrolyte fluid is free to move in a channel. A set of platinum electrodes creates an ion concentration gradient by a small DC voltage. When the fluid moves due to acceleration, an additional current proportional to fluid velocity flows past the electrodes. The symmetric arrangement improves the linearity. The seismometer may operate with feedback to shape the response and increase linearity and dynamic range.

2.17.2 MICROMACHINED ACCELEROMETERS AND SEISMOMETERS

Another new trend is directed to the development of miniature or *micromachined* instruments or MEMS (for Micro Electro-Mechanical Systems). Originally, the technique was directed to the manufacturing of integrated accelerometers for industry and navigation (airborne) applications. The space exploration requirements, among other applications, have lead to the adaptation of some of these techniques to the production of very lightweight and low power sensitive accelerometers and even some seismometers (Center for Space Microelectronics Technology, 2002 *http://mishkin.jpl.nasa.gov/csmtpages /Technologies/Microseismometers*)

The idea is to build the mass-spring system within the multi-layer structure of an integrated circuit, using the same photo-lithographic techniques (Figure 2.39). As the suspended mass is then very small, the Q factor has to be very high (very low damping) for the Brownian noise to be acceptably low, see (2.71).

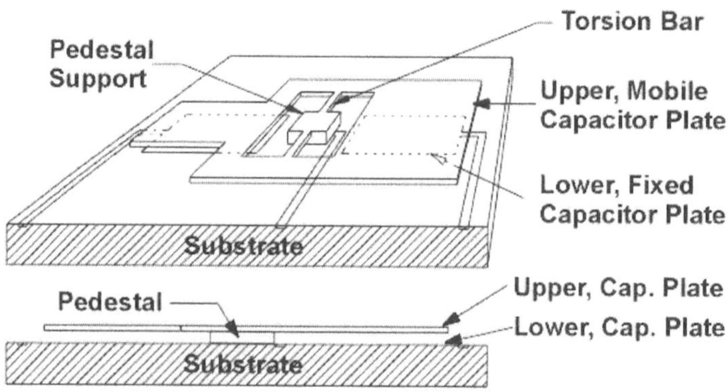

Figure 2.39 Principal elements of a MEMS (micro electro-mechanical system) accelerometer with capacitive transducer. The mass is the upper mobile capacitor plate which can rotate around the torsion bars. The displacement, proportional to acceleration, is sensed with the variance in the capacitance. For high sensitive applications, a feedback circuit is added which controls a restoring electrostatic force, thus we have a FBA. The size of the sensor above is about 2 mm. Figure from *www.silicondesigns.com/tech.html.*

Two kinds of displacement transducers have mainly been used for these devices: capacitive and tunnel effect (see below). Very small capacitive transducers are not practical in general, since the circuitry parasitic capacitances make the relative changes too small. Nevertheless, integrated transducers are possible with the associated circuit integrated within the same chip, thus reducing these parasitic capacitances to a minimum (see, for instance, Li, Lu and Wang, 2001). A capacitive (electrostatic) force feedback may also be applied to achieve the required response and linearity. Devices using this technique are currently manufactured by several companies (Silicon Designs, 2002; Maxwell, Tessman and Reichert, 2001) and have a noise level of 5-20 $\mu ms^{-2}/\sqrt{Hz}$. Although the performance does not yet reach the conventional sensor standards and their price levels are comparable, the field tests seem to show a promising future for these kind of lightweight, low power sensors (Holland, 2003).

A prototype of a microseismometer using capacitive transducer and electromagnetic feedback has been developed to be used in a Mars mission (Center for Space Microelectronics Technology, 2002 *http://mishkin.jpl.nasa.gov/csmtpages /Technologies/Microseismometers*). Its sensitivity is better than $0.1\mu ms^{-2}/\sqrt{Hz}$. This corresponds to a noise level of –140 dB relative to 1 $(ms^{-2})^2/Hz$ (see Figure 3.4) and corresponds to a low quality BB sensor, entirely appropriate for a bit noisy sites.

On the other side, the tunneling effect is a quantum-mechanic phenomenon used in modern atomic-level microscopy. A very sharp conductor pin (actually a needle with a single atom at its end) is at a distance of a few atoms from another conductor. The electronic clouds interact and a small current flows between them as a voltage is applied. This current varies strongly with the distance. As the range of such a displacement transducer is very narrow and its response is quite non-linear, an electrostatic feedback force is required to keep the distance almost constant (Liu and Kenny, 2001). The sensitivity of this kind of transducer is very high and currently has achieved a noise level as low as $0.2\mu ms^{-2}/\sqrt{Hz}$ (Stanford Micro Structures & Sensors Lab., 2002 *http://micromachine.stanford.edu/~chliu/MTA.html*).

All of these devices have two distinct parts within the same package: the mechanical sensor and the associated electronics, implemented by an ASIC (Application Specific Integrated Circuit). Some commercial models have a sigma-delta modulator included and so its output is a pulse-density digital signal, able to be directly interfaced to a microcontroller or computer. Including electronics and housing, the total weight of these sensors might be around 0.1-0.5 kg. If the development continues in the same track as now, it might be possible that most of the seismic sensors we know today might be replaced by highly reliable economical MEMS sensors in the future. For this to happen, production must be in large scale and at prices which can compete with the current geophones. MEMS devices are currently being tested by the oil exploration companies, which is a promising sign.

2.18 Sensor parameters

The most common seismic sensors have now been described. The market has a large range of sensors with a range of specifications and prices and it can sometimes be difficult to make the best choice, particularly since the sensor could be the single most expensive part of a seismic station. The main selection will of course depend on what the sensor is going to record, in most cases meaning frequency band. For further discussion of this topic, see the last section in this chapter. However, apart from the sensor frequency band, there are many other parameters describing a sensor and in this section we will look at bit on those parameters.

2.18.1 FREQUENCY RESPONSE

Most sensors are described to have a flat velocity or acceleration response within a given frequency band and for velocity sensors, even for active sensors, the frequency response at the low end can be described with the usual seismometer response function. The main thing to be aware of is that the dynamic range and/or sensitivity might be frequency dependent, so the sensor might have better specifications in a more narrow frequency band. E. g., a FBA23 accelerometer has a dynamic range of 135 dB from 0.01 to 50 Hz and 145 dB from 0.01 to 20 Hz. It thus has more noise above 20 Hz than below. This is an example with a small variation, but the difference might be much larger and it is not always specified.

2.18.2 SENSITIVITY

The term sensitivity can be understood in two ways. In some cases, sensitivity is given as the gain of the instrument, like 1000 V/ms^{-1}. However, this number does not tell us much about the smallest ground signal we can record, so here we will understand sensitivity as an expression for the smallest signal that can be resolved. This is today limited by the noise generated in the electronics (in sensor or recorder), while earlier the gain in the amplifiers was the limiting factor, so a seismometer with a high generator constant would also mean a more sensitive seismograph.

Ideally, a sensor should be able to resolve ground noise in its main frequency band; in other words, the noise should be below the New Low Noise Model (NLNM). This is only the case with the very best sensors; however many sensors, particularly passive seismometers, get below the NLNM in some part of their frequency band (see instrument noise section). The best way of specifying the sensitivity is to give the sensor acceleration power spectral density curve equivalent to the sensor noise. Figure 2.40 shows an example of sensor noise level.

In many cases the equivalent noise curve is not available and/or other measures are given. This could be the electronic noise output or the equivalent ground motion. The Lennartz LE-3D active 1 Hz seismometer has a specification < 3nm/s equivalent RMS noise at 1 Hz or about 1nm RMS displacement. Compared to Figure 3.3, this looks rather good. Assuming a 1/3 octave filter and a displacement of 1.3 nm, the spectral level is -144 dB, above the NLNM. So this sensor will not be able to resolve the NLNM at 1 Hz. On the other hand, a standard 1 Hz passive seismometer would easily be able to resolve ground noise below NLNM at 1 Hz (see section instrument self noise and Figure 2.40). This is not to say that the LE-3D is bad, since we rarely need to resolve ground noise at a very quiet site.

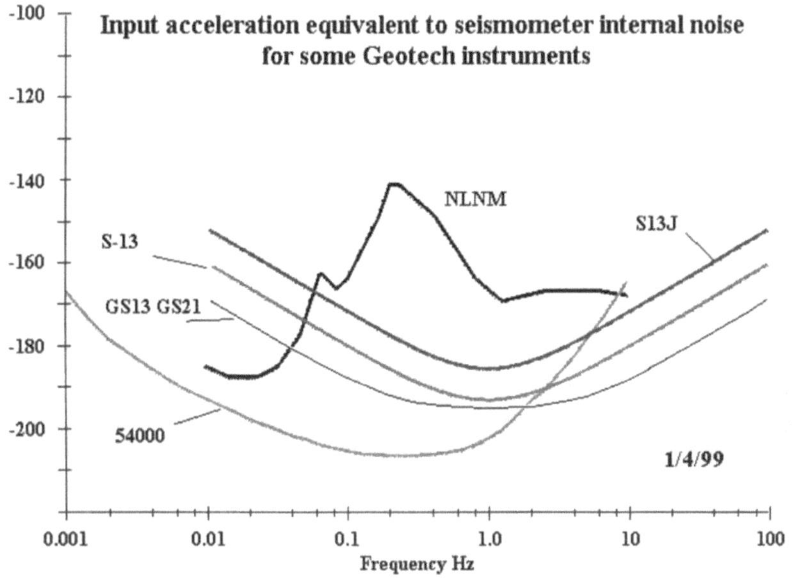

Figure 2.40 Instrument sensitivity of several Geotech seismometers (Table 2.3 and 2.4) ranging from the short period S-13 to very broad band 54000 seismometer. The curves show input acceleration equivalent to sensor internal noise in dB relative to $1(\mathrm{ms}^{-2})^2/\mathrm{Hz}$ (see also Chapter 3). NLNM is the Peterson New Low Noise Model (Peterson, 1993). Slightly modified from Geotech home page, www.geoinstr.com, with permission.

Some passive sensors may be supplied with different signal coils. Ideally, the larger the generator constant, the better the sensitivity. Nevertheless, for a given mechanical design, the space available for the coil is fixed, and a larger number of turns (and more length) has to be achieved using a smaller wire gauge. This increases resistance and thus self-noise (Johnson noise and the influence of current noise from the amplifier). Reidesel et al. (1990) have shown that for a given coil size and wire type, coil resistance is inversely proportional to the square of wire section, while the generator constant is inversely proportional to the section. Thus, the generator constant increases as the square root of the coil resistance. If a recorder or preamplifier with low current noise is used, a larger coil resistance may be favorable, but not in the case of an amplifier with low voltage noise and somehow larger current noise.

2.18.3 SENSOR DYNAMIC RANGE

The dynamic range is the ratio between the largest and smallest signal the sensor can record. Ideally this should be larger than 200 dB, but since no seismic recorder (read digitizer) can handle more than 155 dB and most are below 135 dB (see Chapter 4), there is obviously no point in even trying. What should we expect? If we use a standard 1 Hz sensor, it can easily resolve a displacement of 0.1 nm at 1 Hz. Assuming a generator constant of 300 $\mathrm{V/ms^{-1}}$, this would give an output of $0.1 \cdot 2 \cdot \pi \cdot 300$ nV =

188nV ≈ 0.2μV, which is about what good digitizers can resolve (see Chapter 4). Mechanically, the largest amplitude could be 10 mm, so mechanically the sensor has a dynamic range of 10^7nm/0.1 nm = 10^8 equivalent to 160 dB. This number would be frequency dependent since the sensor displacement response is frequency dependent (Figure 2.7). At 0.1 Hz, a 1000 times larger displacement is needed to generate the same voltage output (0.2μV), but the ground displacement needed for clipping is also higher 100 times, so the dynamic range would be 10 times smaller (with the same digitizer). At 10 Hz, only a 1/10 of the displacement is needed to generate the same voltage output so the dynamic range would be 10 times larger. The best active sensors have a dynamic range > 150 dB, while the majority of such sensors lie in the range 120-140 dB (Table 2.4), which is well adapted to the dynamic range of the average 24 bit digitizer.

As an example, Figure 2.41 shows an estimate of the frequency-dependent dynamic range of two commonly used sensors of natural frequencies 1 Hz and of 4.5 Hz and with velocity feedback in the latter. As it can be seen, the major limit is imposed by the amplifier, which is needed if the digitizer does not have enough sensitivity. This argument supports the use of high dynamic range digitizers even with simple 1 or 4.5 Hz sensors.

2.18.4 SENSOR LINEARITY

Ideally we would like the sensor to behave as a linear system, so that if we e.g. double the input, the output will also be doubled. This is where the passive sensors have a weakness, since it is not easy to make springs that behave linearly. The active sensors with feedback are much more linear than the passive sensors, since the mass hardly moves. A good sensor should have a linearity better than 1 %. Linearity is not always specified, so it might be difficult to compare sensors.

2.18.5 SENSOR CROSS AXIS SENSITIVITY

Cross axis sensitivity is defined as the sensitivity to motion at a right angle to the sensor measuring direction. So, e.g., a vertical sensor will have some sensitivity to horizontal motion. When three sensors are mounted together, this is sometimes described as cross talk, meaning that, e.g., a large amplitude in the Z-direction will also give a small signal on the horizontal components. The cross axis sensitivity typically is less than 2 %. Feed back systems have much less cross axis sensitivity than traditional sensors.

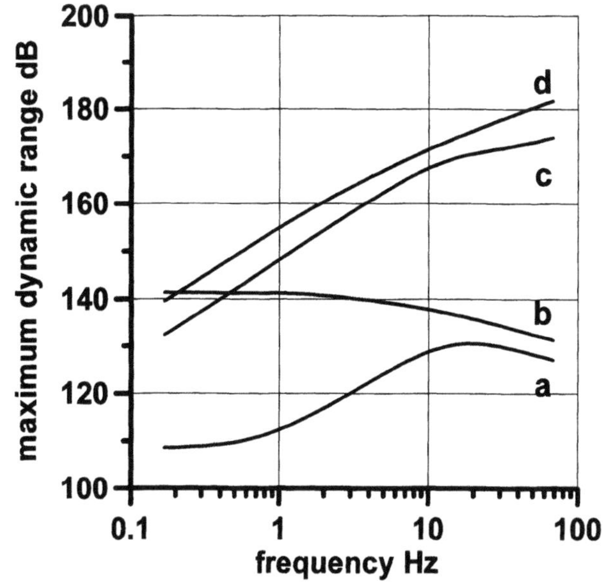

Figure 2.41 Frequency dependent dynamic range due to mechanical or electronic limitations. For the mechanical limitations, the dynamic range is estimated by the ratio of the output for the mechanically largest motion and the input noise of the amplifier. For the electronic limitation, the dynamic range is estimated as the ratio of the largest electrical output of the amplifier and the electrical noise output of the amplifier. a) Electronic limitation of a Mark L15 B (4.5 Hz) with velocity feedback for extending its response to an equivalent frequency of 1 Hz; b) Electronic limitation of a Mark L4-C (1 Hz); c) MarkL15 B considering mechanical clipping; d) Mark L4-C considering mechanical clipping.

2.18.6 SENSOR GAIN AND OUTPUT

The gain is given with the generator constant. For velocity sensors it is given in V/ms^{-1}, while for accelerometers, it is traditionally given in V/g. While earlier, a high gain was an advantage, this is no longer so important, since recorders mostly have sufficient sensitivity. What is more important is that the maximum output has a level fitting the recorder maximum input, so that the sensor dynamic range is optimally used. In some cases a 4.5 Hz sensor has too little output for standard recorders.

2.19 Examples of sensors

The following section will give examples of actual sensors in the market and the intention is to show sensors representative of different types.

2.19.1 EXPLORATION TYPE 4.5 HZ GEOPHONE

This geophone (Figure 2.42) with similar specifications is produced by several companies. It is small in size (about 3 x 5 cm) and only costs about $50, since it is produced in large numbers for seismic exploration. It typically has a mass of 20 g and the generator constant is nearly always around 30 V/ms^{-1}.

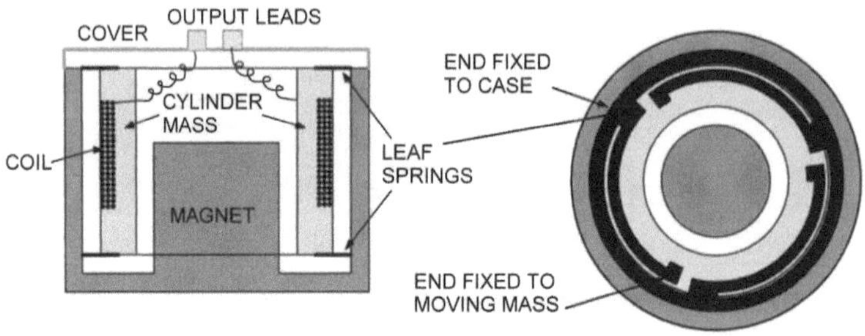

Figure 2.42 Typical construction of a geophone. Note the leaf spring suspension. The magnet is fixed to the case and the coil moves with the mass.

The sensor is very simple and robust to use. It is often used with circuits that will extend the frequency range, see above and the example of the Lennartz sensor below. Due to its low sensitivity, it requires more amplification than the standard sensors. Is has traditionally not been used for earthquakes recording as much as the 1 Hz sensor, however with more sensitive and higher dynamic range recorders, it is now possible to use it directly without any special filtering and obtain good recordings down to 0.3 Hz by post processing. This type of geophone has also been used to construct BB sensors with a period down to 30 s by mounting it with a displacement sensor (Barzilai, 2000).

2.19.2 SHORT PERIOD SENSOR, THE L4-C

This 1.0 Hz sensor from Sercel (earlier Mark Products) is probably one of the most used due to its compact design, reasonable price and good performance. The construction is similar to the geophone, however with the same drawback that the mass cannot be clamped. So, to transport it safely, the sensor has to have the generator coil shorted and be carried upside down (vertical sensor) or vertical (horizontal sensor).

2.19.3 ACCELEROMETER, THE KINEMETRICS EPISENSOR

This accelerometer is a standard FBA with a full range maximum acceleration between 0.25 and 4 g and a frequency response between DC and 200 Hz. The dynamic range is 155 dB, which is one of the better ones available. In the accelerometer section it was calculated that for the 0.25 g sensor, the noise signal resolutions at 1 Hz is 1 nm displacement and at 0.1 Hz it is 100 nm, which is quite comparable to many seismometers. No wonder then that it is called EpiSensor since it can equally well be used as seismometer and accelerometer. And it costs less than half of a 3-component standard SP seismometer from the same company. This sensor is a good indication of the trend in multi-use sensors.

Figure 2.43 The Kinematrics EpiSensor. The case contains a 3-component sensor. The diameter is 13 cm.

2.19.4 BROADBAND SENSORS, THE STRECKEISEN STS-1 AND STS-2 AND GÜRALP CMG-3T

The STS-1 (360 s period) and later the STS-2 (120 s period) were some of the first widely available broadband sensors of the FBA type and they are the prime sensors in the Global Seismic Network together with the Quanterra digitizers. Now other similar seismometers are available from other sensor manufacturers, but they are still the standard by which others are measured. The STS-1 is a single component sensor and each one is installed in vacuum, so it is quite a tricky sensor to install. It is no longer produced. The STS-2 is a 3-component sensor much easier to handle and it gives nearly as good a performance if properly installed (see Chapter 7). The normal arrangement for a 3-component sensor is to have 3 sensors oriented in the Z, N, and E directions. Since horizontal and vertical seismometers differ in their construction and it is therefore difficult to make them equal, the STS-2 is constructed in an alternative way, by using three identical sensors whose axes U, V and W are inclined against the vertical like the edge of a cube standing on its corner (Figure 2.44). Each sensor is made with an *astatic* leaf spring suspension (Figure 2.17). The angle of inclination is $\tan^{-1}\sqrt{2} = 54.7$ degrees, which makes it possible to electrically recombine the oblique components to X, Y and Z simply as

$$\begin{pmatrix} X \\ Y \\ Z \end{pmatrix} = \frac{1}{\sqrt{6}} \cdot \begin{pmatrix} -2 & 1 & 1 \\ 0 & \sqrt{3} & -\sqrt{3} \\ \sqrt{2} & \sqrt{2} & \sqrt{2} \end{pmatrix} \begin{pmatrix} U \\ V \\ W \end{pmatrix} \tag{2.83}$$

The STS-2 can be considered a pure velocity sensor with a period of 120 s (within 1.5% up to 10 Hz and within 15 % up to 50 Hz).

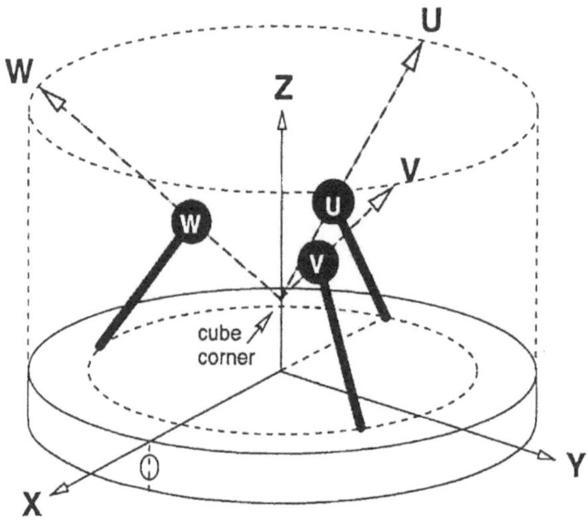

Figure 2.44 The triaxial geometry of the STS-2 seismometer. The oblique components are W, V and U. From Figure 5.13 bye E. Wielandt, in NMSOP, Vol. 1, Bormann (Ed.), 2002; copyright granted by IASPEI.

Figure 2.45 The internals of the Güralp CMG-3T BB sensor. Photo kindly supplied by Nathan Pearce, Güralp.

The CMG-3T is the Güralp three component sensor competing with the STS2 and has also been around for some time and has specifications comparable to the STS2. Figure 2.45 shows the inside of the unit, where it is clearly seen that modern BB sensors are delicate instruments.

2.19.5 NEGATIVE FEEDBACK SENSORS, LENNARTZ LE-3D

The negative feedback sensor is relatively new to the sensor market and one of the first models sold was made by Lennartz. The sensor is based on a standard 4.5 Hz geophone and is now produced in both a 1 and 3-component version (Figure 2.46). The sensor is very compact and has proven to be reliable over many years with little need for re-calibration. The advantage of the feedback sensor compared to a traditional passive 1 Hz sensor is that it can be made smaller, is very rugged (no need for clamping) and the output from the 3 components can be made nearly identical (hard to do with passive sensors). The dynamic range is only 120 dB, a bit lower than a good passive sensor. If the need is a traditional 3-component sensor, this type of seismometer probably offers the best compromise between price and quality. The sensitive accelerometer is another alternative. Lennartz now also make 5 s and 20 s sensors with the same principle, and other manufactures also offer this type of sensors (Table 2.4).

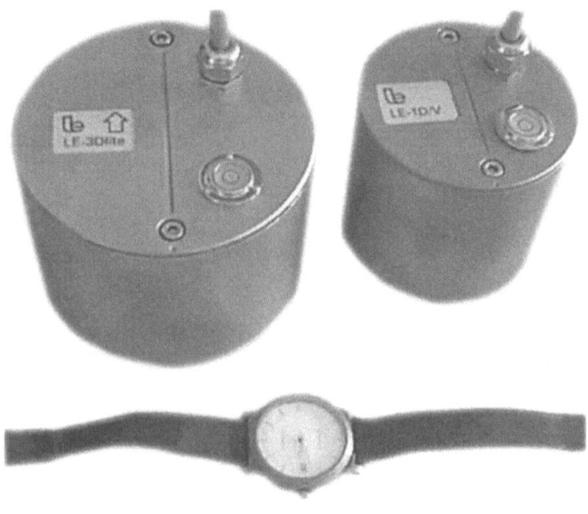

Figure 2.46 The Lennartz negative feedback 3 (left) and 1-component sensors based on 4.5 Hz geophones. From www.lennartz-electronic.de.

2.19.6 BOREHOLE SENSORS

It is well known that the ambient noise is reduced if measurements are taken underground (see Chapter 7). Most manufactures of sensors make special models that can be installed in boreholes. The sensors are essentially similar to the active sensors described so far, but they must be installed in special container in order to fit in a borehole. In addition, there must be special devices for clamping the sensor to the walls of the borehole. So borehole sensors are in general quite expensive to buy and install (Chapter 7). Figure 2.47 shows an example.

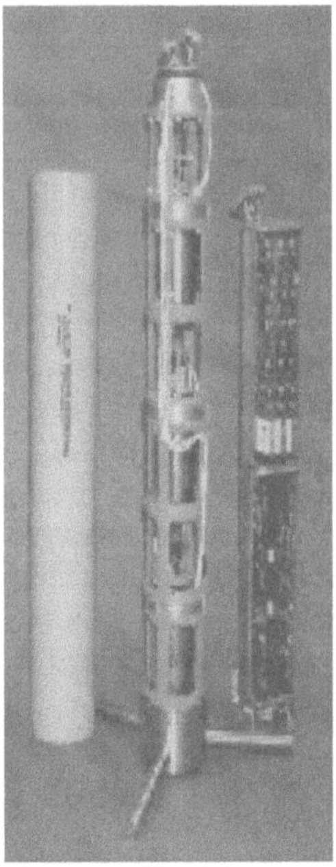

Figure 2.47 A BB borehole sensor model KS-54000 from Geotech Instruments. The height is 2.5 m. From www.geoinstr.com.

2.20 Summary of sensor specifications

Table 2.3 and 2.4 give an overview of some sensors. As it can be seen, there are far more active sensors on the market than passive sensors, which is limited to SP sensors. Borehole sensors and sensors used for OBS's are not listed since these sensors mostly are the same sensors as used for surface installation, except for special mechanical assembly and for OBS use, also particular low power options.

Most of the velocity-broad-band sensors will have the noise floor below the NLNM for some part of the frequency band, while only very few have it for most of the frequency band. Similarly the dynamic range is frequency dependent and the number given is mostly measured at the most favorable frequency.

The resolution calculations are approximate since not all information is available, but the number should give some indication of the sensor internal noise. Many sensors have several options for sensitivity and output range so only 'mid range' values are given.

Table 2.3 Overview of some passive sensors. Abbreviations are: C: Number of components, f_0: natural frequency (Hz), Dam: Open circuit damping, Mas: Moving mass (kg), R_g: Generator coil resistance (ohm), CDR: Critical damping resistance (ohm), R_c: Calibration coil resistance (ohm) (yes means coil is there with unknown resistance), G: Generator constant (V/ms^{-1}), K: Calibration coil motor constant (N/A), Dyn: Dynamic range (dB) at 1 Hz, Mov: Free motion of the mass. Some models have several options for specifications and only one model with typical specification is shown.

Sensor name	C	f_0	Dam	Mas	R_g	CDR	R_c	G	K	Dyn	Mov
Geodevice JC-V103	1	1.0	0.01		1000					100	5.0mm
Geodevice JC-V104	3	1.0	0.7		350		70				
Geodevice JC-45-3	3	4.5	0.02		300		yes				2.0mm
Geo Space GS1	1	1.0	0.54	0.70	4550			280			
Geo Space GS-11D (1)	1	4.5	0.34	0.023	380			32			1.8mm
Geo Space HS1	1	4.5	0.28	0.028	1295			45			
Geotech S-13	1	1.0	~0.02	5.0	3600	6300	23	629	0.198	164	3.0mm
Geotech S-13J	1	1.0		0.9	6400		20	344		140	1.5mm
Input/Output SM-6	1	4.5	0.26	0.016	375		---	28	---		4.0mm
Kinemetrics Ranger SS1	1	1.0	0.07	1.45	5000	6530	100	345	0.40		
Mark products L4C (2)	1	1.0	0.28	1.0	5500	8905	yes	276			6.2mm
Mark products L4A (2)	1	2.0	0.28	0.5	5500	8905		276			
Mark products L22 (2)	1	2.0						88			
Mark products L28B (2)	1	4.5	0.48	0.02	395			35			2.0mm
Sprengnether S6000	3	2.0		0.5	280			45	0.44		

(1) Optionally with 4 kohm coil and 101 V/m/s generator constant
(2) Now sold by Sercel

Table 2.4 Overview of some active sensors available on the market. Abbreviations are: C: Number of components, f-range: Frequency range in which the response is flat (Hz), Out: max voltage out (0-p,V), In: Supply voltage (V), W: Power used (W), G: Generator constant (V/ms^{-1} or V/g for accelerometers), Wt: Weight of sensor (kg), Dyn: Dynamic range (dB). Resolution in last column can be given in nm/s or acceleration. Most sensors are weather resistant, have some kind of calibration input and have a damping close to 0.7. If sensors are sold in both one and three-component versions, only the three- component version is listed.

Name	C	f-range	Out	In	W	G	Wt	Dyn.	Resolution
NEGATIVE FEED BACK									
Geodevice FSS-3B	3	1.0-40	8	12	0.6	800	12	120	
GeoSIG VE53	3	1.0-50	10	12	0.5	1000	2.5	120	
Geotech KS-10	1	0.05-20				500	3	140	
Kinemetrics WR –1	1	0.05-20	2.5	±12	0.3	160	5	125	
Lennartz LE-3Dlite	3	1.0-80	5.0	12	0.1	400	2	120	3nm/s, 1Hz
Lennartz LE-3D/5s	3	0.2-40		12	0.1	400	7	120	1nm/s, 1Hz
Lennartz LE-3D/20s	3	0.05-40	10	12	0.6	1000	7	120	2nm/s, 1Hz
ACCELEROMETERS									
Akashi V450	1	DC-6 Hz	6.6	±12		50.2	8.6		0.005 μg
Eentec EA-140	3	DC-100	5.0	12	0.4	5.0		140	2 μg

Geodevice BBAS-2	3	DC-	5.0	12	0.4	2.5	2	135	
GeoSIG AC63	3	DC-100	10	12	0.8	5.0	3	120	
GeoSIG AC23	3	0.1-50	10	±12	0.5	10.0	2.5	102	
Geotech PA-22	3	DC-50	4.5	±15	0.5	2.25	5	114	10 μg
Güralp CMG-5	3	DC-100	5.0	12	1.2	5.0	5	155	
Input/Output SF3000	3	DC-100	3.6	±12	0.4	1.2	0.5	120	0.3 μg
Kinemetrics FBA-23	3	DC-50	2.5	±12	0.2	2.5	2	135	
Kinemetrics EpiSensor	3	DC-200	10	12	0.4	2.5		155	
Sprengnether FBX23	3	DC-50	10	±12	0.4	10	1	90	11 μg
Sprengnether FBX26	3	DC-50	10	±12	0.2	10	1	135	0.4 μg
VELOCITY BB									
Eentec P-123	3	0.1-50	10	12	0.2	2000	5	130	
Eentec R1 rotational (1)	3	0.05-20	5	12	0.2	50	1	106	
Eentec EP-105	3	0.03-50	10	12	0.2	2000	5	135	
Eentec EP-300	3	0.017-50	7.5	12	0.4	2000	9.5	150	
Eentec PMD223	3	0.017-32	7.5	12	0.3	2000	11	146	
Eentec PMD103	3	0.033-50	10	12	0.5	2000	5.3	132	
Geodevice FBS-3	3	0.05-40	10	12	0.6	1000	12	120	
Geodevice MBS-1	3	0.017-50	10	12	2.0	1000	12	140	
Geodevice BBS-1	3	0.008-50	10	12	2.0	1000	14	140	
Geotech KS-2000	3	0.01-50	10	12		2000	7	160	
Geotech KS-54000IRIS	3	0.003-5	20	24	1.2	2400	66		(4)
Güralp CMG-1T	3	0.003-50	10	12	0.7	1500	14		
Güralp CMG-3T (2)	3	0.003-100	10	12	2.9	1500	12	180	
Güralp CMG3-ESP	3	0.1-50	10	12	0.9	2000	9	170	
Güralp CMG-6T	3	0.03-100	10	12	0.8	1500	3		
Güralp CMG-40T	3	0.03-50	10	12	0.6	3200	7	145	
Nanometrics Trillium	3	0.033-30	8	12	0.4	1500	11		
Sprengnether S-3000Q,	3	1.0-250		±12	0.1	278	2		
Sprengnether WB 2023	3	0.03-20	36	12	0.2	1000	5		
Sprengnether WB2123	3	0.016-50	36	23		1000	5		
Streckeisen STS-2	3	0.033-	20	12	1.8	1500		145	
Streckeisen STS-1 (3)	1	0.003-10							

(1) Rotational sensor, sensitivity in V/rad/sec
(2) If a D is added to sensor name (like CMG-3TD), then it is combined with a digitizer
(3) No longer produced
(4) Borehole sensor

2.21 Which sensor to choose

The number of sensors on the market is very large and range in price from \$50 for the one component geophone to \$ 35000 for the best one-component BB sensor, while the 'average' 3 component sensor cost around \$ 2000 - \$ 4000. Earlier most sensors were passive, but now only passive SP sensors are sold. The tendency is clearly to produce more active sensors and there is now a large range of active sensors available. This does not mean that there is something wrong with passive sensors; they are just too expensive to produce (and cannot be made for very low frequencies). For the price of three passive 1 Hz 'average' sensors, a decent BB sensor going down to 0.03 Hz can be bought! So for a 3 component SP installation, passive sensors have almost been squeezed out of the market. Some points to consider:

The frequency range and type of data to record is the main issue and the one that has the most implication on price. If intended study is small earthquakes or refraction experiments, there is no need for a broad-band sensor and in many cases (down to 0.3 Hz) a 4.5 Hz geophone is good enough. Most SP seismometers are passive sensors with a flat response to velocity above the natural frequency. They are easy to install and operate, and require no power.. They are relatively stable in a broad range of temperatures, which allows less demanding (and therefore inexpensive) vault designs. The electronic drift and mass position instability usually associated with active sensors is not a problem. They are, in short, a very practical solution for all applications where seismic signals of interest are not expected to contain significant components below 0.3 Hz and where demands to linearity and cross axis sensitivity is not very important.

The active SP sensors (electronically extended 4.5 Hz geophones or accelerometers) are often cheaper and smaller than a passive 3 component 1 Hz sensor. Their drawback is that they require power and are more complicated to repair.

Today, the broad-band sensors are a very popular choice. They provide complete seismic information from about 0.01 Hz to 50 Hz and therefore allow a broader range of studies than the SP records.

However, the BB seismometers are more expensive and demanding for installation and operation than SP seismometers. The BB seismometers require a higher level of expertise with respect to instrumentation and analysis methods. They are active feedback sensors and require a stable single or double polarity power supply. They also require very careful site selection in a seismo-geological sense, better-controlled environment in seismic vaults, and they are sometimes a bit tricky to install, although newer types are becoming more easy to use.

Since they do not attenuate the 0.2 - 0.3 Hz natural seismic noise peak, their raw output signal contains much more seismic noise than signals from a SP seismometer. Consequently, useful seismic signals are often buried in seismic noise and can be resolved and analyzed only after filtering to remove the background noise. So, for all but the largest earthquakes, filtering is required even for making simple phase picking. BB sensors are often perceived as the 'best choice', however there are several examples of networks being installed with BB sensors where SP or strong motion sensors could have served equally well, and thereby avoiding costs in installation, maintenance and processing.

The very broad-band sensors are utilized in global seismology studies. They are able to resolve the lowest frequencies, as they occur due to Earth tides and free oscillations of the Earth. Their primary purpose is the research of the deep interior of the Earth. Their only, however important, advantage, compared to the BB seismometers, is their ability to record frequencies around and below 0.001 Hz. They are expensive, require very elaborate and expensive seismic shelters, and are, as a rule, tricky to install. They are ineffective for seismic risk mitigation purposes and some also lack frequency response high enough for local/regional seismology. However, data from a very broadband station is very useful to the international scientific seismological community. It is also excellent for educational purposes. Site selection and preparation for a very broadband station requires extensive studies and often expensive civil engineering works. The cost of a single, good very broad band site preparation can exceed $100,000.

The long period passive sensors are not a suitable choice for new installations and are not sold anymore. These sensors have a corner frequency or 0.05 to 0.03 Hz and are in that respect inferior to most (but not all) BB sensors. Their dynamic range is in the order 120 dB. A LP sensor with a 24-bit digitizer still makes an acceptable low cost BB station provided the sensors and the vault are already available. Nevertheless, in the scope of new installations long period seismometers are of historical value only.

Several sensors now have the digitizer integrated and for some application this might be cost effective. It also makes for a compact design where the sensor and digitizer are optimally connected.

The remaining parameters are of less importance. Dynamic range and sensitivity is adequate for most sensors, but it is important that the sensor output matches the recorder input (see Chapter 5).

In conclusion: Carefully evaluate the lowest frequencies needed for analysis and then chose the appropriate sensor. If existing sensors are available, they might often provide a good continued service with a new high dynamic range recorder.

CHAPTER 3

Seismic noise

Recorded seismic signals always contain noise and it is important to be aware of both the source of the noise and how to measure it. Noise can have two origins: Noise generated in the instrumentation and 'real' seismic noise from earth vibrations. Normally, the instrument noise is well below the seismic noise although most sensors will have some frequency band where the instrumental noise is dominating (e.g. an accelerometer at low frequencies). The instrumental noise is dealt with in more detail in Chapter 2 on seismic sensors. So from now on in this section, it is assumed that noise is ground noise.

3.1 Observation of noise

All seismograms show some kind of noise when the gain is turned up and at most places in the world, harmonic-like noise (called microseismic noise) in the 0.1 - 1.0 Hz band is observed in the raw seismogram, unless obscured by a high local noise level (Figure 3.1). From Figure 3.1, it is also seen that, although the microseismic noise dominates (see noise sources later), there is also significant seismic noise in other frequency bands. So, obviously, the noise level must be specified at different frequencies.

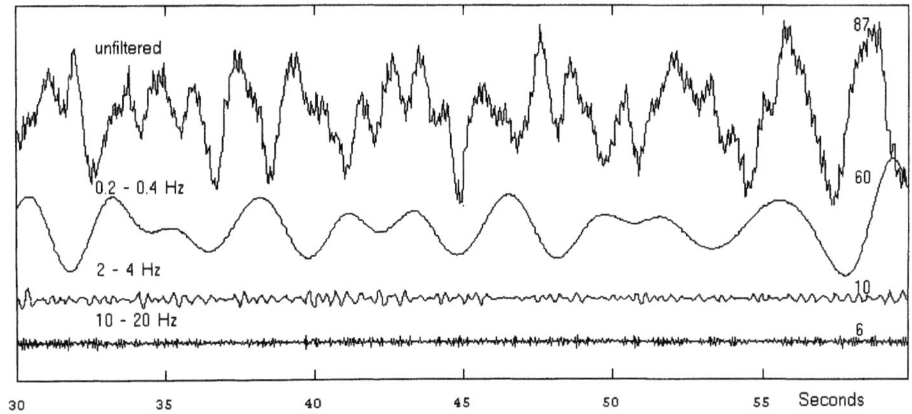

Figure 3.1 Seismic noise in different filter bands at station MOL in the Norwegian National Seismic Network The short period station (1 Hz) is situated about 40 km from the North Sea and the unfiltered trace clearly shows the high level of low frequency noise (~0.3 Hz) generated by the sea. All traces are plotted with the same scale and the numbers to the right above the traces are the maximum amplitude in counts.

Intuitively, the simplest way should be to measure the earth displacement in different frequency bands and plot the amplitude as a function of frequency or period. This was in fact the way it was done before the use of digital recording and an example of measurements from the old manual of Seismological Observatory Practice (Willmore, 1979) is shown in Figure 3.3.

The use of filtering and measurements in the time domain presents two problems:

(1) Bandwidth used is an arbitrary choice, (2) Getting average values over long time intervals. Both of these problems are solved by presenting the noise in the spectral domain, (see Chapter 6 for more on spectral analysis). Figure 3.2 illustrates the problem of the filter bandwidth. The figure shows the same signal filtered with an increasingly narrow filter. The amplitude of the signal decreases as the filter becomes narrower, since less and less energy gets into the filter band. In the example in Figure 3.2, the maximum amplitude at around 1 Hz varies from 23 nm to 6 nm depending on filter width. This decrease in amplitude is mainly due to making the filter narrower. However, for the widest frequency band (0.6 –1.7 Hz), relatively more low frequency energy is also present in the original signal. Comparing the amplitudes to Figure 3.3, the noise level can be considered worse or better than average depending on which filter band is used. So in order to do measurements in time domain, the noise values can only be compared if the same bandwidth of the filter is used.

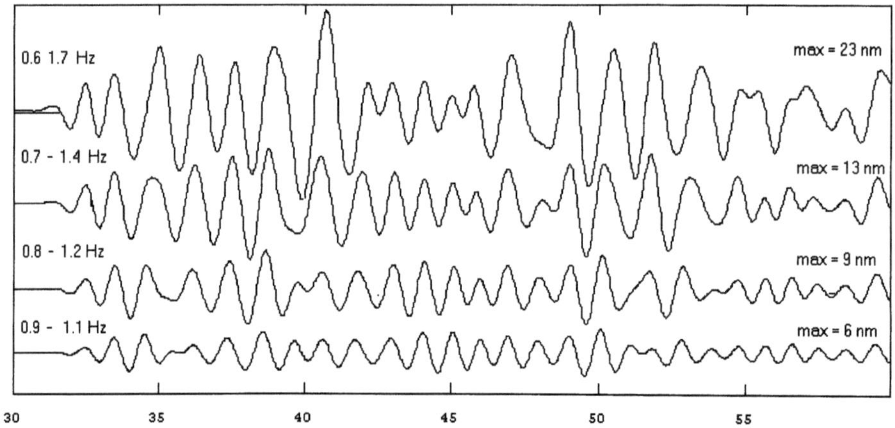

Figure 3.2 The signal from Figure 3.1 band pass filtered with different filter widths. The signals have been corrected for instrument response to show displacement. The maximum amplitude in nm is shown to the right on top of the traces.

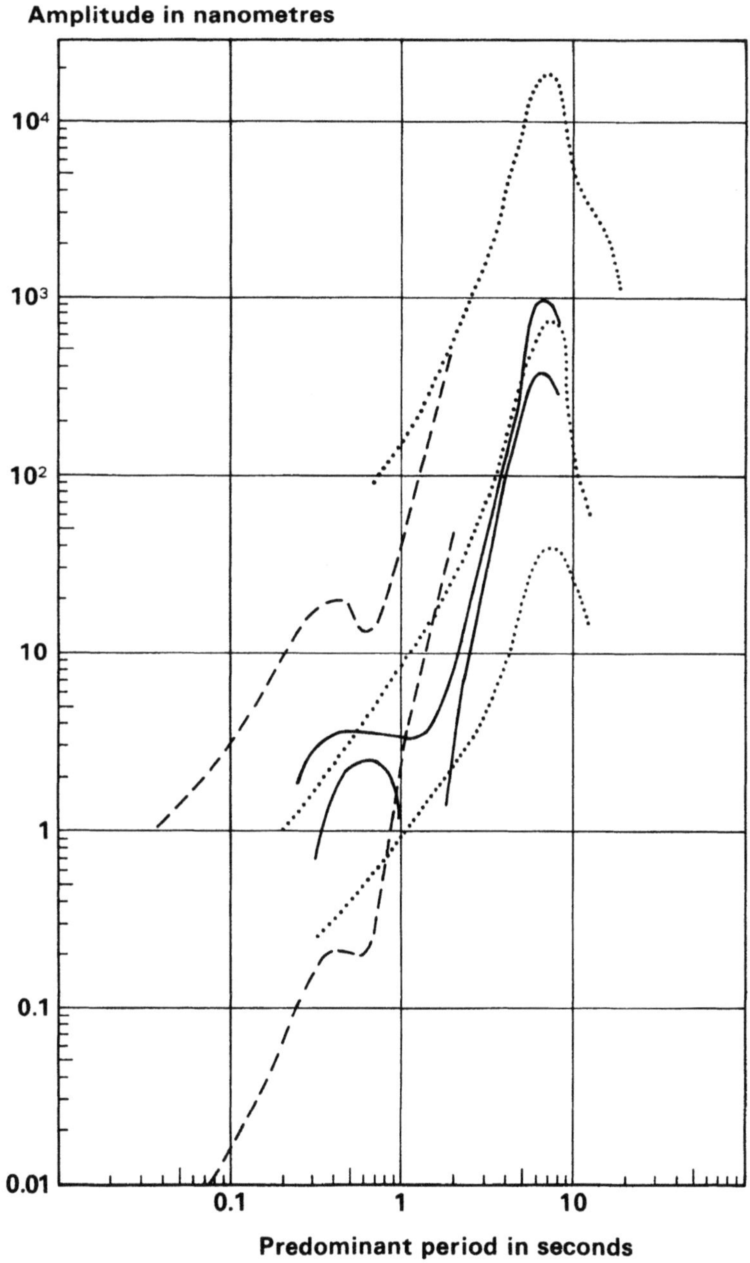

Figure 3.3 Noise curves in a rural environment. The 3 dotted lines correspond to the maximum, mean and minimum levels published by Brune and Oliver (1959), the dashed lines give two extreme examples observed in the US and the full line curves give the limits of fluctuation of seismic noise at a European station on bedrock in a populated area 15 km away from heavy traffic (figure from Wilmore, 1979).

3.2 Noise spectra

With digital data, it is possible to make spectral analysis, and thereby easily get the noise level at all frequencies in one simple operation. It has become the convention to represent the noise spectra as the noise power density acceleration spectrum $P_a(\omega)$, commonly in units of dB referred to 1 $(m/s^2)^2/Hz$. Noise Level is thus calculated as

$$\text{Noise Level} = 10 \log [P_a(\omega)/(m/s^2)^2/Hz] \tag{3.1}$$

Figure 3.4 shows the new global high (NHNM) and low noise models NLNM) (Peterson, 1993) and an example of noise spectra at a seismic station. The curves represent upper and lower bounds of a cumulative compilation of representative ground acceleration power spectral densities determined for noisy and quiet periods at 75 worldwide distributed digital stations. These so-called Peterson curves have become the standard, by which the noise level at seismic stations is evaluated. *A power density spectrum can be defined in different ways and it is important that the definition used is identical to the one used originally by Peterson (Figure 3.4) in order to compare the spectra. The exact definition is given in Chapter 6.*

3.3 Relating power spectra to amplitude measurements

The Peterson noise curves and the way of representing them have standardized the way of representing seismic noise. However, looking at such a curve, it is difficult to relate them to something physical as seen in Figures 3.1, 3.2 and 3.3 and a standard question is often: So what is the physical meaning of the Peterson curve? The old noise curve in Figure 3.3 can be directly related to the seismogram in Figure 3.2, while the Peterson curve cannot since one is a frequency domain measure and the other a time domain measure. However, as will be shown below, under certain conditions it is actually possible to go from one to the other (the following text largely follows Bormann in Chapter 4 of the NMSOP (Bormann (Ed.), 2002).

The problem is how to relate a spectral amplitude at a given frequency to a time domain amplitude in a given frequency band. The root mean squared amplitude, a_{RMS} of a signal in the time interval *0-T*, is defined as

$$a_{RMS}^2 = \frac{1}{T}\int_0^T a(t)^2\, dt \tag{3.2}$$

The average power of the signal in the time interval is then equal to a_{RMS}^2. The average power can also be calculated (Parseval´s Theorem) from the power density spectrum as

$$a_{RMS}^2 = \int_{f_1}^{f_2} P(\omega)df \approx P\cdot(f_2 - f_1) \tag{3.3}$$

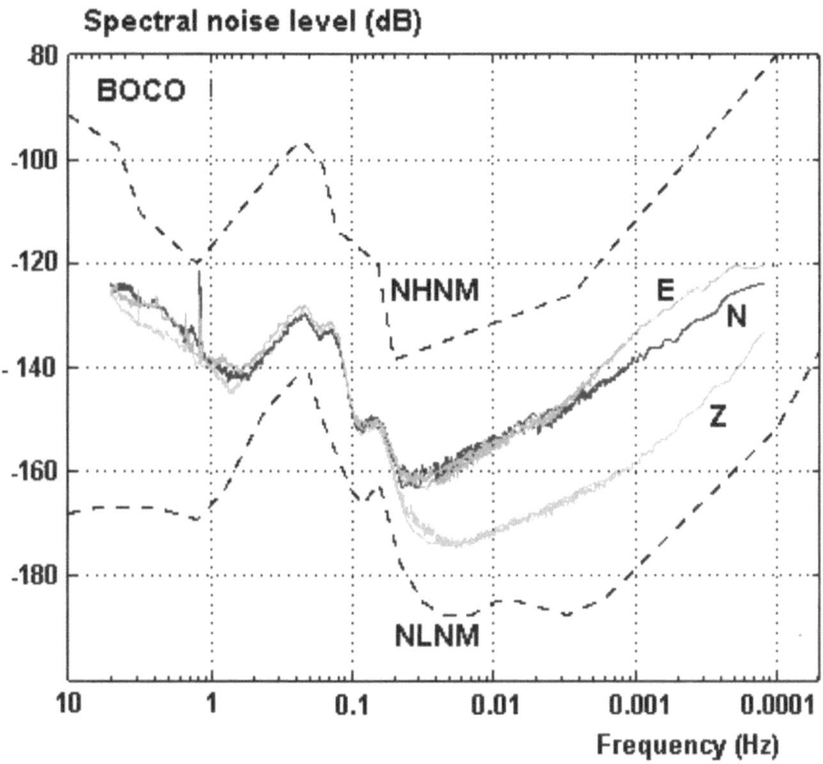

Figure 3.4 The Peterson noise curves and noise spectral level for the IRIS station BOCO. The noise level is in dB relative to 1 $(ms^{-2})^2$/Hz. The Peterson high and low noise models are shown with dashed lines. The noise spectra are shown for all 3 components. Note the lower noise level for the vertical (Z) component (note that x - axis represents decreasing frequency). Figure modified from FDSN station book, http://www.fdsn.org/FDSNpubs.htm.

under the assumption that the power spectrum is nearly a constant P in the frequency range f_1 to f_2, which is not unreasonable if the filter is narrow. In the general case, P would represent the average value of $P(\omega)$ in this frequency band. It is important that $P(\omega)$ is the normalized power spectral density as defined in Chapter 6 and that the power represents the total contribution at ω from both positive and negative frequencies. If, as is the usual practice, the power spectrum is calculated using only the positive frequencies as defined in the standard complex spectral analysis, a^2_{RMS} will have to be calculated as $2P(f_2\text{-}f_1)$. *The power values given by the New Global Noise Model by Peterson (1993) do already contain this factor of two, or said in other words, represent the total power.*

Under these assumptions, we then have a relationship between the power spectral density and the RMS amplitude within a narrow frequency band:

$$a_{RMS} = \sqrt{P \cdot (f_2 - f_1)} \tag{3.4}$$

There is thus a simple way of relating the power spectral density to amplitudes, as seen on a seismogram; however, note that relation (3.4) gives the RMS amplitude. There is statistically a 95% probability that the instantaneous peak amplitude of a random wavelet with Gaussian amplitude distribution lie within a range of $2a_{RMS}$. Peterson (1993) showed that both broadband and long period noise amplitudes closely follow a Gaussian probability distribution. In the case of narrowband-filtered envelopes, the average peak amplitudes are 1.25 a_{RMS} (Bormann, Chapter 4 in NMSOP, Bormann (Ed.), 2002). From measurement of noise using narrowband filtered broadband data, values of 1.19 to 1.28 were found (Peterson, 1993). Thus, in order to get the true average peak amplitude on the seismogram, a factor of about 1.25 can be used. (NB: For a pure sine wave, $a = a_{RMS}\sqrt{2}$, not so very different). We can now set up the relation between the power spectral values and the average peak amplitudes

$$a = a_{RMS} \cdot 1.25 = 1.25\sqrt{P \cdot (f_2 - f_1)} \tag{3.5}$$

The frequency band depends on instrument (mainly if analog), while for digital data the user can select the filter. A common way of specifying filter bands is to use the term octave filter. An *n*-octave filter has filter limits f_1 and f_2 such that

$$\frac{f_2}{f_1} = 2^n \tag{3.6}$$

For example, a half-octave filter has the limits f_1 and f_2=$f_1^{1/2}$ (e.g. 1-1.42 Hz). Many of the classical analog seismographs have bandwidths of 1-3 octaves and digital seismographs might have a bandwidth of 6-12 octaves. However, the signal bandwidth of many dominating components of seismic background noise might be less then 1 octave. Thus, the frequency of measurement is really a frequency range. However, for practical reasons, the average frequency will be used to represent the measurement. For the average frequency, the geometric center frequency f_0 must be used

$$f_0 = \sqrt{f_1 f_2} = \sqrt{f_1 f_1 \cdot 2^n} = f_1 \cdot 2^{\frac{n}{2}} \tag{3.7}$$

and

$$f_1 = f_0 2^{-\frac{n}{2}} \quad \text{and} \quad f_2 = f_0 2^{\frac{n}{2}} \tag{3.8}$$

For narrow filters, the geometric center frequency is almost the same as the average frequency. The filters used in Figure 3.2 have average frequencies of 1.2, 1.1, 1.0 and 1.0 Hz while the geometric center frequencies are all 1.0. For the filter 10-20 Hz, the geometric and average frequencies are 14 and 15 Hz respectively.

In comparing time and frequency domain signals, there was no mention of units since this has no importance on the relations. However, the Peterson curves are in acceleration, so unless the time domain signal is in acceleration too, the power spectrum must first be transformed to acceleration. The most common unit for the original seismogram is velocity, sometimes acceleration and rarely displacement. If the power spectra of acceleration, velocity and displacement are called P_a, P_v, and P_d respectively, the relations are:

$$P_v(\omega) = P_d(\omega) \cdot \omega^2 \tag{3.9}$$

$$P_a(\omega) = P_v(\omega) \cdot \omega^2 = P_d(\omega) \cdot \omega^4 \tag{3.10}$$

Now, some examples. Figure 3.2 has a peak amplitude of 13 nm for the frequency band 0.7-1.4 Hz. This can be converted to P_d

$$P_d = (13 \cdot 10^{-9}/1.25)^2/(1.4 - 0.7) = 1.55 \cdot 10^{-16}\ \mathrm{m^2/Hz} \tag{3.11}$$

The center frequency is $\sqrt{0.7 \times 1.4} = 1.0\,\mathrm{Hz}$ and the acceleration power density is

$$P_a = 1.55 \cdot 10^{-16} \cdot 16 \cdot \pi^4 \cdot 1^4 = 2.4 \cdot 10^{-13}\ (\mathrm{m/s^2})^2/\mathrm{Hz} \tag{3.12}$$

or –126 dB relative to 1 $(\mathrm{m/s^2})^2/\mathrm{Hz}$. Figure 3.5 shows the complete noise power density spectrum for the time window used in Figures 3.1 and 3.2. We can see that the level (–126 dB) is high compared to the spectral level which is around –136 dB. How can that be explained? We have to remember that (3.5) is based on the assumption that the amplitude is the *average peak amplitude,* while what has been used is the largest amplitude in the window. If the average peak amplitude is 3 times smaller than the maximum amplitude, the level would be corrected by –10 log (3^2) = -9.5 dB and the time domain and frequency domain measures would be similar.

This discussion should demonstrate that, by far, the most objective way to report the noise level at a given site is to use the power density spectrum, although it is nice to be able to relate it, at least approximately, to some amplitude measure.

The spectrum in Figure 3.5 shows a relatively high noise level at lower frequencies relative to high frequencies. This is not surprising, considering the general high microseismic noise level along the Norwegian West coast. The noise level above 3 Hz is quite low since the station is on granite in a rural area (20 m from nearest house).

The noise spectral level can be calculated for all 4 cases in Figure 3.2, see Table 3.1. It is seen, that although the amplitudes were quite different in the 4 filter bands, power spectral levels are nearly equal except for the filter band 0.6 – 1.7 Hz where the amplitudes in the relatively wide filter is influenced by the stronger background noise at lower frequencies.

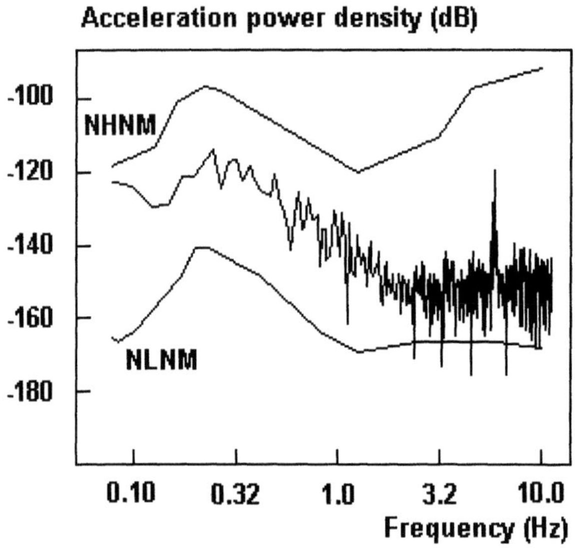

Figure 3.5 Noise power density spectrum of the raw signal seen in Figure 3.1, top trace. The acceleration power is in dB relative to 1 $(m/s^2)^2/Hz$. The limits of the Peterson noise model are indicated (NHNM and NLNM). The spectrum has not been smoothed.

Table 3.1 Maximum amplitudes of signals (Figure 3.2) in the different filter bands and corresponding noise levels.

Filter bands (Hz)	Amplitude (nm)	Noise power level (db)
0.6 –1.7	23	- 123
0.7 – 1.4	13	- 126
0.8 – 1.2	9	- 127
0.9 – 1.1	6	- 127

From Figures 3.4 or 3.5, we see that the NLNM has a level of –166 dB at 1 Hz. Assuming a 2-octave filter and using (3.5) and (3.8), the corresponding average peak displacement is 0.3 nm. From Figure 3.3 it is seen that the lowest displacement is 1 nm at 1 Hz so there is a reasonable agreement considering the uncertainty of whether RMS values or average peak values have been used. At 10 Hz, the NLNM gives 0.01 nm, which also agrees well with the values on Figure 3.3. So a rule of thumb (and easy to remember) is: *A peak displacement of 1 nm at 1 Hz means a good site in terms of ambient noise.*

One point to be underlined is that noise is of random nature (any "predictable" disturbance is not noise as defined here!). The power spectrum estimated by the simple Fourier transform of a sample window of raw data is also random. The standard error of such an estimate is very high (see, e.g. Blackman and Tukey, 1966; Press et al., 1995). Actually, when we express the noise level as the power spectrum density or RMS within a certain frequency band, we are implicitly assuming that noise is a stationary process, which means that its statistical characteristics are not time-dependent or at least vary slowly enough to be considered constant within a certain time interval. Therefore, if a reliable estimation of the noise level, at a given site, is needed, an average of power density estimation on several overlapping sample time-windows or some spectral smoothing will decrease the estimate variance (e.g. Blackman and Tukey, 1966).

3.4 Origin of seismic noise

Man made noise. Often referred to as "cultural" noise, it originates from traffic and machinery, has high frequencies (>2-4 Hz) and die out rather quickly (m to km), when moving away from the noise source. It propagates mainly as high-frequency surface waves, which attenuate fast with distance and decrease strongly in amplitude with depth, so it may become almost negligible in boreholes, deep caves or tunnel sites. This kind of noise usually has a large difference between day and night and can have characteristic frequencies depending on the source of the disturbance. The noise level can be very high.

Wind noise. Wind will make any object move so it will always generate ground noise. This noise is usually high frequency like man made noise, however large swinging objects like masts and towers can generate lower frequency signals. Trees also transmit wind vibrations to ground and therefore seismic stations should be installed away from them. In general, wind turbulences around topography irregularities such as scarps or rocks generate local noise and their proximity must be avoided.

Ocean generated noise. This is the most widespread noise (called microseisms or microseismic noise), and it is seen globally, although the interior of continents has less noise than coastal regions. Long period ocean microseisms are generated only in shallow waters in coastal regions (NMSOP, Bormann (Ed.), 2002), where the wave energy is converted directly into seismic energy either through vertical pressure variations, or smashing surf on the shores. They therefore have the same period as the water waves ($T \approx 10$ to 16 s). Shorter period microseisms can be explained as being generated by the superposition of ocean waves of equal period traveling in opposite directions, thus generating standing gravity waves of half the period. These standing waves cause perturbations which propagate without attenuation to the ocean bottom. The higher frequency microseisms have larger amplitude than the lower frequency microseisms (Figures 3.3-3.5). During large storms, the amplitudes can reach 20000 nm at stations near the coast and make analog seismograms useless.

Other sources. Running water, surf and volcanic tremor (an almost harmonic noise associated to fluids motion, often lasting hours or days) or background activity are other local sources of seismic noise. Man made noise and wind noise are usually the main source at high frequencies and since the lower limit is about 0.01 nm at 10 Hz, very small disturbances will quickly get the noise level above this value. For more details on seismic noise generation, see NMSOP, Chapter 4, Bormann (Ed.) (2002).

CHAPTER 4

Analog to digital converter

Since the computers are the heart of all data processing, we must convert our analog signals to numbers. The process of converting a continuous analog signal to a series of numbers representing the signal at discrete intervals is called analog to digital conversion and is performed with analog to digital converters (ADC). Figure 4.1 shows a signal, where the amplitude is measured at regular intervals Δt. In its simplest form, we could simply envisage measuring the amplitude of a slowly changing signal with an analog voltmeter and write down the numbers. Alternatively, if we have an analog plotted signal like in Figure 4.1, we just measure the amplitudes with a ruler and enter the numbers into a computer. In principle, this way of getting digital data is used when older paper seismograms are digitized manually, except that the process is automated by using a digitizing table, which automatically reads the position of a cursor on a table.

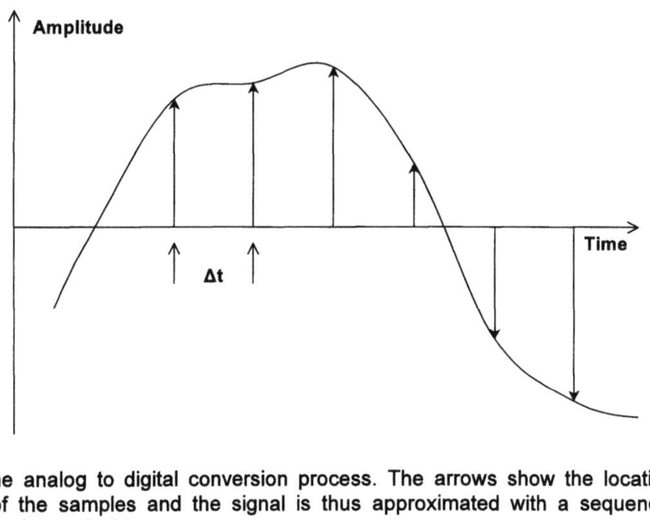

Figure 4.1 The analog to digital conversion process. The arrows show the location and values (amplitudes) of the samples and the signal is thus approximated with a sequence of numbers available at time intervals Δt.

The process of analog-digital conversion involves two steps: First the signal is sampled at discrete time intervals, then each sample is evaluated in terms of a number (usually integer, but in any case with finite resolution) and output in form of a code.

This process will introduce errors into the data and limits the information in them, simply because we replace a continuously varying signal, which can have any value between two limits, with a discrete set of points, information between the points is lost, and with a limited number of possible values (quantization). This creates errors in both the amplitude and frequency content and much of the efforts in improving the ADC process are related to minimizing these errors. Before getting into that discussion, it is illustrative to describe how the most common ADC's work in principle.

Let us first look at how the analog signals are represented by numbers. An ADC will give an output binary number when we put in a specific voltage. In a computer, we usually represent numbers by a 2-byte word or 4-byte word (one byte = 8 bits), hence 2^{16} or 2^{32} possible values. Since numbers are positive and negative, the corresponding ranges are $\pm 2^{15}$ and $\pm 2^{31}$, respectively (more exactly, $+2^{15}$ to $-2^{15}+1$ or $+2^{31}$ to $-2^{31}+1$, because the 0 value has to be included in the 2^n possible codes and the so-called two's complement code is used).

Just to get an idea of the numbers involved. The best ADC's sold for seismology can resolve a step of 0.1μV and have a range of $\pm 2^{24}$ although the most common maximum is $\pm 2^{23}$ ($\approx \pm 8 \cdot 10^6$). We should have this in mind when describing the ADC's.

4.1 Example of a simple analog to digital converter, the Flash ADC

One of the simplest ADC's to understand is the flash ADC (Figure 4.2).

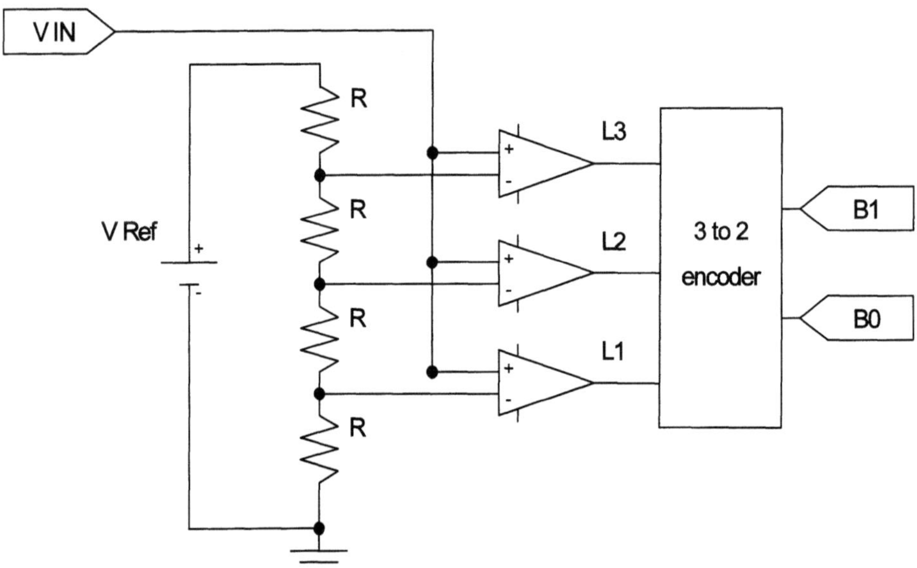

Figure 4.2 Flash ADC. *Vin* is the signal to be digitized, *R* are resistors of equal size, *Vref* is a reference voltage. The circuits at right are comparators. L1 to L3 are the direct outputs of the comparators, which are coded to a two bit output, B0 and B2, since only four values are possible. This example is actually a two-bit digitizer (see later).

The reference voltage creates three reference voltage levels, which are input to the 3 comparators (minus pin). If *Vin* is zero, all comparators are off. As the voltage increases to Vref/4, the first comparator goes on. As *Vin* keeps increasing in steps of $V_{ref}/4$, successive comparators switch on. The possible states of (L1, L2, L3) are: (0,0,0),

(1,0,0), (1,1,0), (1,1,1). If the reference voltage is 1.0 V, we can then detect 3 transition levels of input voltage, 1/4, 2/4, 3/4 V and four possible intervals (0-0.25, 0.25-0.5, 0.5-0.75 and >0.75) corresponding to the numbers 0-3 (Table 4.1). In the two's complement binary code, this corresponds to the numbers 00, 01, 10, 11. Since the data values are contained in a 2 bit word, using the whole range available, we call this a 2 bit converter. The number out of an ADC is commonly called counts.

In the above example, only positive voltages were digitized, but our signals also contain negative voltages. We can get the negative signals by adding another ADC with a negative reference voltage so one ADC would give out counts for the negative signals and the other for the positive signals. The other alternative is to add a voltage of half the reference voltage, to the input signal so that the voltages reaching the ADC never become negative. The ADC now should have the output range +2 to -2 counts instead of the 0-3 counts (when the offset of 2 has been subtracted). But this would be five states, not four! With 2 bit representing positive and negative values, we only have the values +2 to −1 to use (as defined in the two's complement code). Actually, in bipolar converters, since the binary full scale is not exactly symmetric, the offset subtracted is not exactly half of the scale, but half a count less, to avoid this bias. Table 4.1 gives the input and output levels for this case.

For high resolution (many bit) converters, one can simply subtract half of the full scale without significant error.

V input	V center	V input to ADC	ADC out	Output code
-3/8 to −1/8	-0.25	0.00 to 0.25	0	11 (-1)
-1/8 to +1/8	0	0.25 to 0.5	1	00 (0)
+1/8 to +3/8	0.25	0.5 to 0.75	2	01 (1)
+3/8 to +5/8	0.5	0.75 to 1	3	10 (2)

Table 4.1 Input and output to flash ADC. First column gives the voltage in; second column, the center voltage of each interval; third, the voltage input into the converter itself after adding (0.5 V - 0.125 V) (half the full scale minus half a count); third column, the numbers out and the last column the two's complement code and its decimal value in parenthesis. Numbers in column 3 and 4 correspond to the example with input of only positive numbers.

Flash ADC's are extremely fast and very expensive if a high resolution is needed since one comparator is needed for each level, so if the signal is going to be resolved with 10000 levels, 9999 comparators and precision references are needed. Flash ADC's are not much used in seismology but illustrate some of the characteristics of the ADC.

4.2 Basic ADC properties

We will now define some of the basic properties of ADC's to have in mind when continuing the description of common ADC's :

Resolution. The smallest step that can be detected. In the above example, the smallest step was 0.25 V, which is then the resolution corresponding to one change of the *least significant bit* (LSB, the rightmost bit). For a high dynamic range digitizer, this could be 0.1 to 1 μV. ADC resolution is also labeled ADC sensitivity. The higher the resolution, the smaller a number is given. As it was described in the section on sensors, the output from a passive sensor can be in the nV range in which case many digitizers will need a preamplifier. The number of bit is also often referred to as resolution. Most ADC's have an internal noise higher than one count: In this case, the *number of noise-free bits, rather than the total bit number, limits the effective resolution.* For instance, one count corresponds to 0.3μV in a 24 bit ADC with full-scale of ±2.5V, but it may have a noise of 2μV peak-to-peak, and signals under this level may not be resolved in practice.

Gain. The sensitivity expressed in counts/V. It can be derived from resolution. If e.g. the resolution is 10 μV, the gain would be $1\text{count}/(10^{-5}\text{V}) = 10^5$ counts/V.

Sample rate. Number of samples acquired per second. For seismology, the usual rates are in the range 1 to 200 Hz (or, more specifically, samples per second – sps) while for exploration seismology, sample rates can be more than 1000 Hz. In general, the performance of the ADC degrades with increasing sample rate.

Maximum input or full-scale (FS). The maximum input for the ADC. Using any higher input will result in the same output. In the example above, the maximum input is 1.0 V or in the bipolar mode ±0.5V. Typical values are ±1 to ±30 V.

Dynamic range. Defined as the ratio between the largest and smallest value the ADC can give. In the above example, the dynamic range = 4/1 = 4, or in dB, 20·log(4) = 12 dB. This can be a bit misleading since both negative and positive numbers are input and the ADC has to work in bipolar mode. So the real dynamic range is only half, in this case 6 dB. However, dynamic ranges given in dB for ADC's are sometimes given for the full range. For some digitizers, the lowest bits only contain noise, so the dynamic range is defined as the ratio between the largest input voltage and the noise level of the digitizer. This number can be substantially smaller than the theoretical largest dynamic range of a digitizer and may depend on the data rate or sampling frequency. So, to give one number for the dynamic range, a frequency bandwidth should ideally also be given.

Dynamic range in terms of bit. The dynamic range can also be given as number of bits available in the output data sample. An n-bit converter then gives the numbers $0\text{-}2^n$ or in bipolar mode $\pm 2^{n-1}$. In the example in Table 4.1, we have a 2-bit converter. This way of giving the ADC dynamic range is the most common and there is no confusion about what the meaning is. In seismology we mostly use 12, 16 and 24 bit converters. As seen later, the 24 bit converters give out 3 bytes, but in many cases only the 17-22 more significant bits are noise free. So, like for dynamic range, the usable dynamic range for a 24-bit converter could be given as e.g. 18 bit.

Accuracy. The absolute accuracy is a measure of all error sources. It is defined as the difference between the input voltage and the voltage representing the output. Ideally this error should be ±LSB/2 and this is achieved by several low and medium-resolution commercial ADC's (it is more difficult for higher resolution ones). The error only due to the digitization steps (±LSB/2) is also called the quantization error.

Noise level. Number of counts out if the input is zero (subtracting DC offset). Ideally, an ADC should give out 0 counts if the input is zero. This is usually the case for low dynamic range digitizers 12-14 bit, but rarely the case for high dynamic range digitizers (see section on 24 bit digitizers). The noise level is most often given as an average in terms of RMS noise measured over many samples. A good 24-bit digitizer typically has an RMS noise level of 2 counts.

Conversion time. The minimum time required for a complete conversion. Often it is expressed by the maximum data rate or sampling frequency. Due to the finite time required to complete a conversion, many converters use as input stage a *sample and hold* circuit, whose function is to sample the analog signal before the start of a conversion and hold the converter input constant until it is complete to avoid conversion errors. This is not required with sigma-delta converters (see later), as they track the signal continuously and their output represents an average of the input signal value during the conversion interval.

Cross talk. If several channels are available in the same digitizer, a signal recorded with one channel might be seen in another channel. Ideally this should not happen, but it is always present in practice. The specification is given in dB meaning how much lower the level is in the neighboring channel. A 24 bit digitizer might have cross talk damping of 80 dB or a factor of 10 000 damping. If the input in channel 1 is at the maximum giving $\pm 2^{23}$ counts out and channel 2 has no signal, the output of channel 2 caused by cross talk would still be $\pm 2^{23}/10000=839$ counts. This is well above the noise level for most 24-bit digitizers, so cross talk creates a clear artificial signal in this case. In practice, the signal shape and level is often similar in the different channels (e.g. for a 3 component station), so the problem might not be as bad as it sounds, but it certainly should not be ignored. A good 24-bit digitizer has 120 dB of damping or better (see Table 4.6). Cheaper multichannel digitizers use a single ADC and an analog multiplexer, which connects different inputs sequentially to the ADC input. This limits the cross-talk separation because analog multiplexers have limited performances. For high resolution digitizers, one digitizer per channel is preferred.

Non-linearity. If the analog input is a linear ramp, this parameter is the relative deviation of the converter output from the ideal value. It is expressed with relation to full scale (FS), e.g. 0.01% of FS. For high dynamic range converters, it is important because a poor linearity may cause two different signals at the input to be intermodulated (the amplitude of one depends on the other) at the output. Usually it is not a problem with modern sigma-delta converters.

Input impedance. The input impedance (ohm). Ideally it should be as high as possible in order to have little influence on the sensor or other connected equipment. A typical value is 1 Mohm.

Offset. If the input is zero, the offset is the DC level (the average) of the output. This could also be called the DC shift of the ADC. There is nearly always some offset, either caused by the ADC itself or caused by the components connected to the ADC. The ADC might have a possibility of adjusting the offset by changing some reference voltage. A small offset is of no importance, but any offset will limit the dynamic range since the ADC will reach its maximum value (positive or negative) for smaller input values than its nominal full-scale. Figure 4.3 shows an example where the effect of the offset is to reduce the dynamic range by a factor of 2. The offset will be temperature dependent, but usually this is a small problem with a drift of typically less than 1 μV/°C.

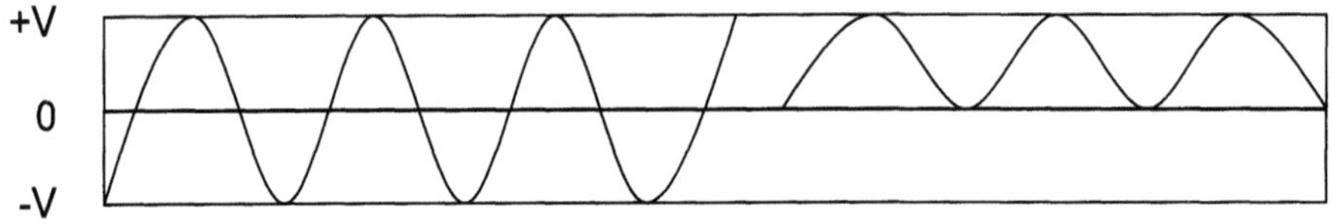

Figure 4.3 Effect on dynamic range when a large offset is present. The range of the ADC is ±*V*. The large amplitude signal (left) has no offset and amplitude is ±*V*. The smaller amplitude signal (right) has an offset of *V*/2 and the input signal that can be recorded by the ADC is now limited to +*V*/2 to −3*V*/2.

Now that the properties of ADC's have been given, we can continue to describe common ADC's.

4.3 A typical ADC, the Ramp ADC

One of the simplest approaches of implementing an ADC is the ramp ADC. Figure 4.4 shows a simplified diagram. The control logic sends a signal to the ramp generator to start a conversion. The ramp generator then generates a ramp signal starting from level 0 (seen on left). The ramp signal enters the comparator and once the ramp signal is larger than or equal to the input signal, the output from the comparator switches from zero to 1. At the same time as the ramp generator starts, the counter will start to count the number of levels on the ramp. When the comparator switches to level 1, the control logic will stop the counter and the number reached by the counter is then a measure of the input voltage. After some time, the counter is reset and a new sample can be taken. A cheaper variant of this type of digitizer generates a ramp by integrating a reference voltage while a clock signal drives a counter. When the ramp voltage reaches the input voltage, the digitizer just reads the time elapsed as the counter output. The counter and integrator are then reset for a new conversion.

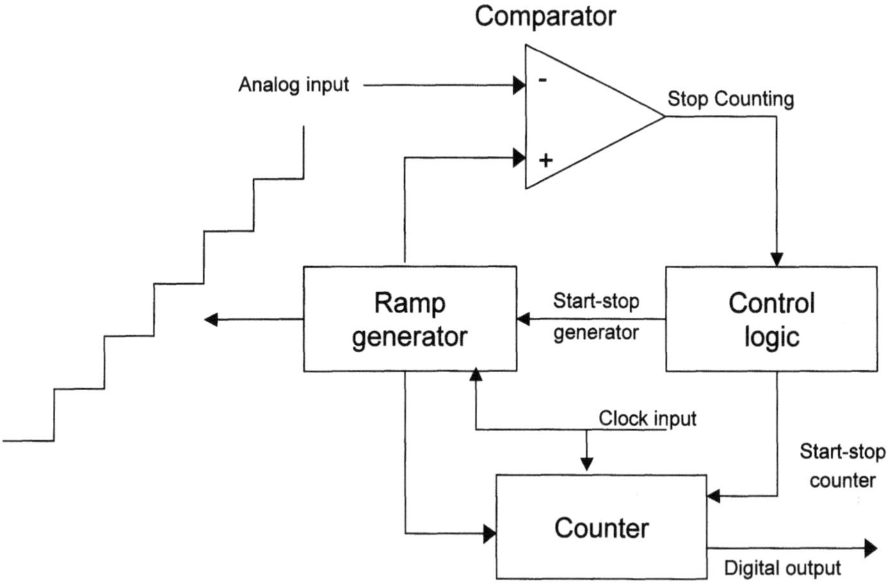

Figure 4.4 Ramp ADC, see text.

The Ramp ADC is relatively slow and even slower if a high resolution is required. It also becomes slower as the number increase in size since the counter has to count longer for large amplitudes. An improvement of the ramp ADC is the successive approximation ADC, which is almost identical to the ramp ADC except that it has a more sophisticated control circuit. The converter does not test all levels, but first tests if the input level is below or above half the full scale, thus the possible range has been halved. It then tests if the input level is above or below the middle of this new range etc. The conversion time is much smaller than for the ramp ADC and constant. This design is the most popular of the classical type of ADC's. A typical 16 bit digitizer of this type may have a conversion time of 20μs, which is fast enough for multichannel seismic data acquisition. Nevertheless, it requires a sample and hold circuit.

We have all used a digital multimeter. It contains a digitizer sampling a few times a second which can be seen by how often the numbers change on the display. Most multimeters have a range of ±2000 (V, mV etc). Why 2000 and not ±10 000 which would be a more convenient range? Simply because a cheap 12 bit converter is used, which has a range of ±2048 counts.

4.4 Multi channel ADC

We usually have more than one channel to digitize. For three component stations there are 3, while for telemetric networks or small arrays, there might be up to 100 channels. The simplest approach is to have one ADC for each channel. However, this might be overkill depending on the application and in addition quite expensive. There are several ADC cards for PC's (and other computers) on the market that have up to 64 channels with 16 bit resolution and sampling rates in the kHz range. These cards only have one ADC. How is this possible? The ADC has, in the front, a so called multiplexer which connects the ADC to the next analog channel as soon as a conversion is finished. The input signals are therefore not sampled at the same time and there is a time shift, called skew, between the channels. If the ADC is fast, the skew might be very small, but in the worst case, the ADC just has time to take all the samples and the skew is the sample interval divided by the number of channels. For many applications, like digitizing the signal from a network, skew has no importance, but in other application where a correlation between the traces will be made like for arrays or three component stations, the samples should be taken at the same time. The standard in seismic recorders now is to use one ADC per channel, while multi channel ADC cards only are used in some analog networks.

The skew should be known and, for some types of data analysis, has to be corrected for. For example, using multiplets (very similar earthquakes with clustered hypocenters), relative precise locations (e.g. Stich et al., 2001), need channel relative timing with tenths of milliseconds accuracy.

4.5 Digitizers for a higher dynamic range

The digitizers described have a practical limit of 16 bit dynamic range. This is not enough for most applications in seismology. Imagine a network recording local earthquakes. A magnitude 2 earthquake is recorded at 100 km distance with a maximum count value of 200 which is a lower limit if the signal should be recorded with a reasonable signal to noise ratio. What would be the largest earthquake at the same distance we could record with a 16-bit converter before clipping? A 16-bit converter has a maximum output of 32768 counts or 164 times larger. Assuming that magnitude increases with the logarithm of the amplitude, the maximum magnitude would be 2.0 + log (164) = 4.2. With a 12-bit converter, the maximum magnitude would be 3.0. So a higher dynamic range is needed. In the following, some of the methods to get a higher dynamic range will be described.

In earlier designs, gain ranging was frequently used. The principle is that in front of the ADC there is a programmable gain amplifier. When the signal level reaches e.g. 30 % of the ADC clipping level, the gain is reduced. This can happen in several steps and the gain used for every sample is recorded with the sample. When the input level decreases, the gain increases again. In this way, a dynamic range of more than 140 dB can be obtained. The drawback with gain ranging is that when a low gain is used, the resolution goes down, so it is not possible to recover a small signal in the presence of large signals. In addition, many designs had problems with glitches occurring when the gain changed. One of the well known models is the Nanometrics RD3 (not sold anymore, but many still in operation). 24 bit digitizers have now completely taken over the market from gain ranging digitizers.

4.6 Oversampling for improvement of the dynamic range

The method of oversampling to improve the dynamic range of a digital signal consists of sampling the signal at a higher rate than desired, low pass filter the signal and resample at a lower rate. Qualitatively what happens is then that the quantization errors of the individual samples in the oversampled trace are averaged over neighboring samples by the low pass filter and the averaged samples therefore have more accuracy and consequently a higher dynamic range.

Lets look at some examples to better understand the principle. Figure 4.5 shows an ADC where the first level is at 0.0 V and the second level at 1 V. The signal is sampled at times *t1, t2, etc.* The input signal is a constant DC signal at 0.3 V. Thus the output from the converter is always 0 and no amount of averaging will change that. A saw tooth signal is now added to the DC signal to simulate noise. For the samples at *t1* and *t2*, the output is still 0 but for *t3*, it brings the sum of the DC signal and the noise above 1.0 V and the output is 1.The average over 3 samples is now 0.33 counts and we now have a better approximation to the real signal. Instead of an error of ±0.5 V we now expect an error of $\pm 0.5/\sqrt{3} = \pm 0.29$ V and we have increased the resolution and dynamic range by a factor of $\sqrt{3}$. The square root comes from the assumption that the error in an average is reduced by the square root of the number of values averaged.

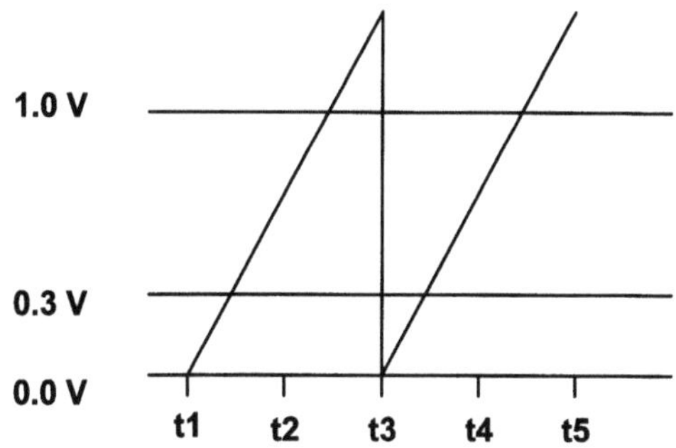

Figure 4.5 Improving dynamic range by oversampling in the presence of noise, see text.

In addition to getting a higher resolution, we are able to get an estimate of the signal level even when it is below the level of the LSB. The value 0.33 will be the best we can get in this case, even if we continue to average over many more samples. If we sample 10 times in the time interval *t1* to *t3,* the average will be 0.30 and the error $\pm 0.5/\sqrt{10} = \pm 0.16$ counts. In the first case we have an oversampling of a factor 3 and in the last case 10.

In this example we have a rather special noise. In real ADC's, it is the normal noise that is doing the job of the saw tooth 'noise' and it is assumed that the noise is uniformly distributed such that in the above case, 70% of the digital outputs would be 0 or smaller and 30% would be 1 or larger such that the average output would be 0.3.

Real signals are not constant. So, in addition to the effect of the noise, there will also be an averaging effect of the varying signal. If we have a signal which increases linearly with time, and we make a running average of 4 samples, we get the result shown in Figure 4.6.

The samples are taken at times shown with a black dot and the running averages over 4 samples (associated with the average time) are associated to *t1...t4...* Again remember that the real process doing this is low pass filtering and resampling, so only every 4th sample would be used and high frequency information would be lost. This example corresponds to 4 times oversampling and as it can be seen, the quantization steps in the averaged signal is now 0.5 counts instead of 1.0 counts in the original signal. This is what we would expect since we have the √4 effect on the quantization error.

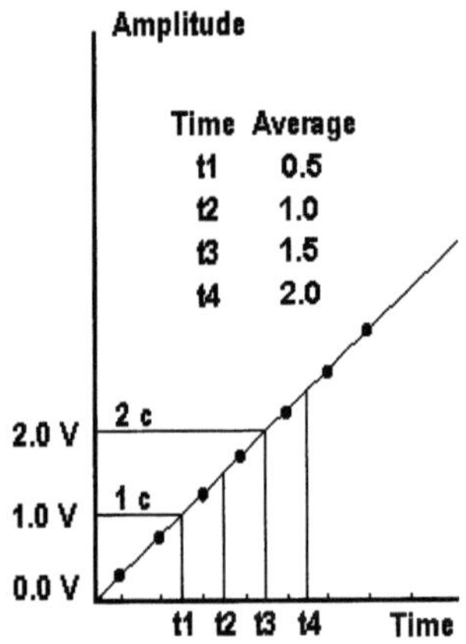

Figure 4.6 Digitization of a linearly increasing signal. The ADC has levels 0, 1 and 2 V corresponding to counts(c) 0,1 and 2. The running average of 4 samples represented by times *t1..t4* is given in the table under 'Average'.

The two ways of getting a higher dynamic range, by using a varying signal and noise superimposed on a signal, are very similar. However, with a completely constant, noise free signal, oversampling would not be able to increase dynamic range as we saw in the first example. Normally it is no problem to have noise in the signals, rather the contrary. However, in some designs, a known noise is added to the signal and then subtracted after digitization.

From the discussion it seems that the improved sensitivity is proportional to √n, where n is the decimation factor, which is in fact what is predicted theoretically, given certain assumptions like uniform distribution of the quantization errors (for details, see Scherbaum 2001)). Thus, for every time the sampling rate is doubled relative to the desired sample rate, the dynamic range is improved by a factor of √2 or 3 dB. So we should think any dynamic range could be obtained by just doing enough oversampling. Unfortunately, it is not that simple. No electronic circuits are ideal, and so limitations in the accuracy of the actual components will present limits of oversampling. For instance, the input analog amplifier will limit the sensitivity, so oversampling that brings the theoretical LSB below the noise of the amplifier will no longer produce any improvement in the dynamic range.

As an example, we will look at digital recording of seismic background noise, see Figure 4.7.

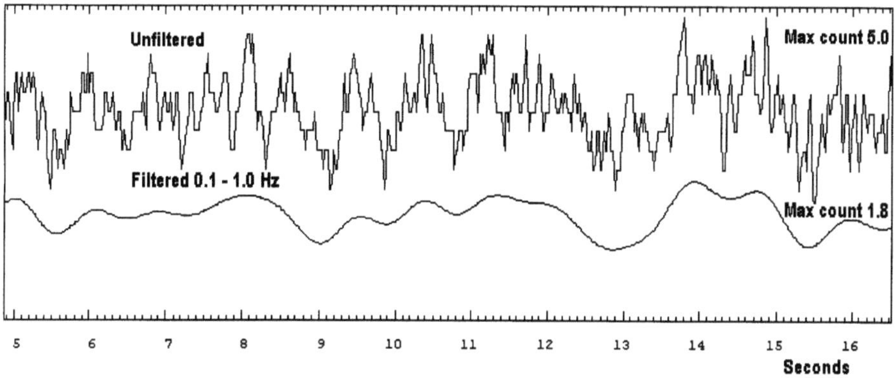

Figure 4.7 Unfiltered and filtered record of seismic background noise in a residential area in Western Norway on a hard rock site. The recording is made with a 4.5 Hz geophone and a 16-bit ADC at a sample rate of 50 Hz (GeoSIG GBV116). The filter is an 8-pole Butterworth filter with zero phase shift.

The top trace shows the unfiltered record which has a maximum amplitude of 5 counts. It is possible to see that there is a low frequency signal superimposed on the cultural noise signal. After filtering, a smooth record of the microseismic noise (3.4) with a typical period of 3 s is clearly seen. Although the maximum amplitude is only 1.8 counts, it is clear that the resolution is much better than one count. Since this example corresponds to a decimation of a factor 50, the theoretical resolution should be $1/\sqrt{50}$ =0.14 counts for the filtered signal below 1 Hz which does not look unreasonable from the figure. This is a simple example of the effect of decimation and shows that a 16-bit recorder at low frequencies will have a larger dynamic range than predicted from the 16-bit ADC.

We can theoretically expect that for each factor of two the oversampling is increased, the dynamic range improves by 3 dB. A doubling in the sensitivity is thus 6 dB, which also corresponds to 1 bit change in ADC specification. We can now compare that to what real ADC's can deliver (Table 4.2).

Table 4.2 Effective resolution of 2 different ADC's as a function of sample rate. F is the output data rate (samples per second), AD7710 is a chip from Analog Devices and Crystal is the Crystal chip set (see text).

	AD7710		Crystal	
F(sps)	ADC number of Bits	Dynamic range (dB)	ADC number of bits	Dynamic range (dB)
25	20.0	120		
31			22.1	133
50	19.5	117		
62			21.9	132
100	18.5	111		
125			21.6	130
250	15.5	93	21.1	127
500	13.0	78	20.8	125
1000	10.5	63	20.1	121

The AD7710 is a low priced chip that has been on the market for several years and used in lower resolution ADC's. The Crystal (CS5323 and CS5322) chipset is the main standard used in well known so called 24 bit recorders from e.g. Nanometrics, Reftek, GeoSIG and Güralp. Both ADC's are of the Sigma Delta type (see below). As we can see from the table, none of these ADC's delivers 24 useful bits, even at the lowest sample rate. The Crystal ADC improves its dynamic range as a function of sample rate almost as predicted until a rate of 62 Hz. Going from 62 to 31 Hz, there is only an improvement of 1 dB. The initial sample rate is in the kHz range. The AD7710 seems to get more improvement in the dynamic range than predicted by resampling theory, or alternatively, it becomes much worse than theory predicts when the sample rate increases. This is because the change in dynamic range in the AD7710 is not only a question of resampling but also of the degradation of the performance of the electronic circuits due to high sample rate. So not all sigma delta ADC's will improve performance as predicted just by the theory of oversampling.

This description of oversampling is very simplified. For more details, see e.g. Scherbaum (2001), Oppenheim and Schafer, (1989) and Proakis and Manolakis (1992).

4.7 Sigma Delta ADC, SDADC

All ADC's will digitize the signal in steps, so, even with the highest resolution, there will be a quantization error. The idea behind the SDADC is to digitize with a low resolution and get an estimate of the signal level, add the quantization error to the original signal, get a new estimate etc. This process will continue forever and the actual value of the input signal is obtained by averaging a large number of estimates. In this way a higher resolution can be obtained than is possible with the original ADC in much the same way as described with oversampling. Most SDADC's are based on a one-bit ADC; however, in order to better understand the principle of sigma delta ($\Sigma\Delta$) converters, we shall first look at a SDADC with a normal ADC (Figure 4.8).

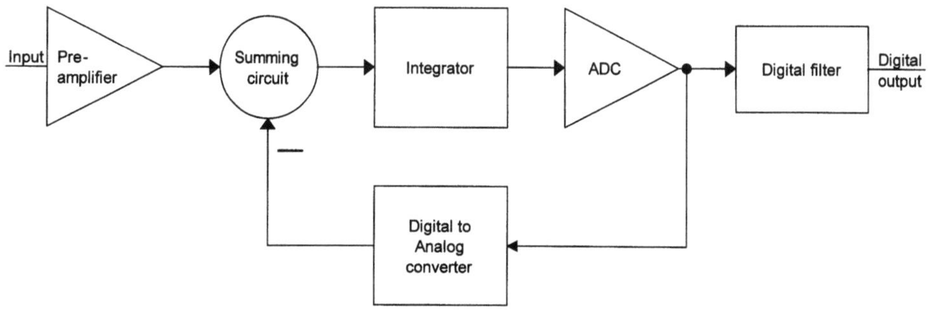

Figure 4.8 Simplified overview of a sigma-delta ADC. The signal from the ADC is converted back to analog with the digital to analog converter (DAC). The summing circuit subtracts the ADC signal from the input signal. The digital converter ADC, the integrator and the digital to analog converter are all synchronized and controlled by a logic circuit (not shown).

The signal first enters a preamplifier, followed by the so-called analog modulator loop, which consists of a summing circuit (just a differential amplifier) and an integrator. After the integrator, the signal is digitized. The digitized signal is going into the digital filter, which calculates a running average of the ADC-values and does resampling or decimation to a lower data rate. The integrator may be in practice the same as a first order low pass filter (actual ΣΔ converters use a higher order filter, this is the order of the modulator). The ADC values also go to a DAC (digital to analog converter). The output from the DAC is the analog signal level corresponding to the digitized signal (after integration or addition) and is subtracted from the original signal. In the first iteration, this difference is the quantization error. So, by adding this to the previously digitized value, the next value to digitize will have the quantization error added.

In order to better understand the principle, let us look at some examples. In the first example we will use an ADC with ±10 levels corresponding to the range ±1 V. This ADC is set up a little different from the previous examples since 0-0.09 V, corresponds to 1 count, (previously 0 count), -0.1 to –0.01 to –1 count etc. This is because, in the end, we want to use a 1-bit converter, which only has the levels –1 and 1 for negative and positive signals respectively.

The input signal is put at a constant 0.52 V. We can now follow the output of the ADC as well as the averaged output to see how the SDADC approaches the true signal level, see Table 4.3. In the first clock cycle (0), it is assumed that the DAC is turned off, so the integrated amplitude is also 0.52 and the ADC will give out the value 6 and the DAC 0.6 V, which is to be used for the first full clock cycle (1). This value of 0.6 V is now subtracted from the input voltage and the resulting quantization error is added to the previous amplitude estimate. Since the quantization error is negative, the new corrected input is smaller at 0.44 V. Digitizing this gives 5 corresponding to 0.5 V and the average is 0.5 which is our first estimate of the input signal. The new quantization error is now 0.02. The quantization error is the same for the next 2 conversions until the integrated signal again is ≥0.5 V. The average is the exact input value after 5 cycles and this pattern repeats itself for every 5 cycles and the overall average approaches the true value. The 5 cycle sequence is called the duty cycle or limit cycle.

This process is actually very similar to the process of oversampling. By continuously integrating the quantization error, we get an effect as illustrated in Figure 4.5 and thereby count how many ADC output values are above and below a particular ADC value. The average ADC values will therefore reduce the quantization error like for oversampling. In the above example, the true value has been reached within 0.001 after 520 samples. A resolution of 0.0001 would require 5000 samples. In practice, there is no need to use 5000 samples since the digital data that represent the analog input is contained in the duty cycle of the output of the ADC which will be recovered by the digital processing.

Table 4.3 Input and output of SDADC with a normal 20 level ADC. Abbreviations are: I: Cycle number, Diff inp: Difference between input and output of ADC, Sum: Sum of Diff inp and previously digitized signal, DAC out: Output from DAC and Average: Running average of output from ADC.

I	Diff inp.	Sum	DAC out	Average	
0	0.5200	0.5200	0.6000		
1	-0.0800	0.4400	0.5000	0.5000	
2	0.0200	0.4600	0.5000	0.5000	
3	0.0200	0.4800	0.5000	0.5000	
4	0.0200	0.5000	0.6000	0.5250	
5	-0.0800	0.4200	0.5000	0.5200	exact value
6	0.0200	0.4400	0.5000	0.5167	
7	0.0200	0.4600	0.5000	0.5143	
8	0.0200	0.4800	0.5000	0.5125	
9	0.0200	0.5000	0.6000	0.5222	
10	-0.0800	0.4200	0.5000	0.5200	exact value
11	0.0200	0.4400	0.5000	0.5182	
12	0.0200	0.4600	0.5000	0.5167	
13	0.0200	0.4800	0.5000	0.5154	
14	0.0200	0.5000	0.6000	0.5214	
15	-0.0800	0.4200	0.5000	0.5200	exact value
16	0.0200	0.4400	0.5000	0.5187	
...	...	...	...	...	
519	0.0200	0.5005	0.6000	0.5200	
520	-0.0800	0.4205	0.5000	0.5200	
521	0.0200	0.4405	0.5000	0.5200	
522	0.0200	0.4605	0.5000	0.5199	

The beauty of this process is that we can use ADC's with a low resolution as long as we use enough oversampling. Obviously we need to take more samples if the ADC resolution is decreased. Using an ADC with ±2 levels would reach the accuracy of 0.001 and 0.0001 within 1600 and 15000 samples respectively and the duty cycle is 25. This is not so different from the ±10 levels ADC. So the logical solution is to use the very simplest ADC, the 1 bit converter that in reality is just a comparator that can determine if the level is negative or positive. This ADC can be made very fast and accurate and is essentially linear, since two points determine a unique straight line! The number of samples needed for the above accuracies are now 10000 and 95000, respectively, and the duty cycle is 25. Again we do not need that many samples since only the number of samples in the duty cycle is needed. Table 4.4 shows how the output looks for the 1 bit converter with the same input of 0.52 V as in the example above.

Table 4.4 Input and output of SDADC with a one-bit ADC. Abbreviations are: I: Cycle number, Diff inp: Difference between input and output of ADC, Sum: Sum of Diff input and previously digitized signal, DAC out: Output from DAC and Average: Running average of output from ADC.

I	Diff inp.	Sum	DAC out	Average
0	0.5200	0.5200	1.0000	
1	-0.4800	0.0400	1.0000	1.0000
2	-0.4800	-0.4400	-1.0000	0.0000
3	1.5200	1.0800	1.0000	0.3333
4	-0.4800	0.6000	1.0000	0.5000
5	-0.4800	0.1200	1.0000	0.6000
6	-0.4800	-0.3600	-1.0000	0.3333
7	1.5200	1.1600	1.0000	0.4286
8	-0.4800	0.6800	1.0000	0.5000
9	-0.4800	0.2000	1.0000	0.5556
10	-0.4800	-0.2800	-1.0000	0.4000
11	1.5200	1.2400	1.0000	0.4545
12	-0.4800	0.7600	1.0000	0.5000
13	-0.4800	0.2800	1.0000	0.5385
14	-0.4800	-0.2000	-1.0000	0.4286
15	1.5200	1.3200	1.0000	0.4667
16	-0.4800	0.8400	1.0000	0.5000
17	-0.4800	0.3600	1.0000	0.5294
18	-0.4800	-0.1200	-1.0000	0.4444
19	1.5200	1.4000	1.0000	0.4737
20	-0.4800	0.9200	1.0000	0.5000
21	-0.4800	0.4400	1.0000	0.5238
22	-0.4800	-0.0400	-1.0000	0.4545
23	1.5200	1.4800	1.0000	0.4783
24	-0.4800	1.0000	1.0000	0.5000
25	-0.4800	0.5200	1.0000	0.5200
26	-0.4800	0.0400	1.0000	0.5385
27	-0.4800	-0.4400	-1.0000	0.4815

The duty cycle would increase to 167, if the input had been 0.521. For an input of 0.0 V, the duty cycle would be 2 and the exact input is the average of every two output samples (Table 4.5).

Table 4.5 Input and output of SDADC with a one-bit ADC when input is 0.0 V. Abbreviations are as in Table 4.4.

I	Diff inp.	Sum	DAC out	Average
0	0.0000	0.0000	1.0000	
1	-1.0000	-1.0000	-1.0000	-1.0000
2	1.0000	0.0000	1.0000	0.0000
3	-1.0000	-1.0000	-1.0000	-0.3333
4	1.0000	0.0000	1.0000	0.0000
5	-1.0000	-1.0000	-1.0000	-0.2000
6	1.0000	0.0000	1.0000	0.0000

Real SDADC's can be very complicated and there are many variations of the design compared to the description here. In practice, long limit cycles cause a problem: an output *idle tone* (a periodic signal) appears for a low amplitude or constant input. This undesirable effect is avoided, e.g., by using several stages of integrators (modulators of three or higher order) (e.g., Aziz et al., 1996). One of the main parameters characterizing the quality of a SDADC is the order of the integrator, which often can be 2, 3 or 4. In general, for an N-order modulator every doubling of the oversampling ratio provides an additional ($6N + 3$) dB of SNR. A typical SDADC uses several stages of digital filtering and decimation, so there are many ways on how to get the final signal. Most SDADC's use a one-bit ADC but few reach full 24-bit resolution as shown in previous section. The well known Quanterra digitizer has for many years been the de facto standard in broadband recording with a full 24-bit resolution. It uses discrete components and a 16-bit converter instead of the one bit converter. However the Quanterra digitizer uses a lot of power compared to single or two-chip 24 bit converters (with 120 dB dynamic range) which typically consumes 50 mW. There are now other digitizers with similar or better specifications than the Quanterra on the market like the Earth Data digitizer.

4.7.1 HOW SIGMA-DELTA IMPROVES DIGITIZATION NOISE: THEORY

The high effective resolution of sigma-delta ADC converters is due to the reduction of digitization noise in two steps: In the sigma-delta modulator and in the low pass digital filter, that operates on the oversampled 1 bit stream prior to the decimation.

Let us analyze the sigma-delta modulator schematics of Figure 4.9.

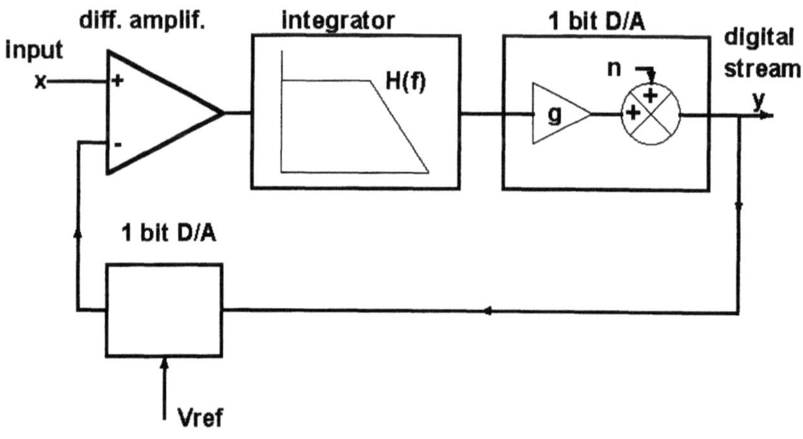

Figure 4.9 Schematics of a sigma-delta modulator.

The analog signal x is the non-inverting input of a differential amplifier. Its output amplifies the difference Δ between the input and the feedback. This difference is applied to a low pass filter, which normally is composed of several integrator stages DC may cause trouble in integrators, so in fact the signal may be integrated by a low-pass filter only above a corner frequency much lower than the slowest usable sampling frequency. The number of poles of this filter is the modulator order. Its response, including the gain of the amplifier, is $H(f)$. The output is then oversampled by a digital

1-bitADC, which is inherently linear, since its transfer characteristic is fixed by two points. Its gain is g (in count/V). The quantization noise n is actually included within this component, due to its low resolution. At its output, we get the oversampled digital 1 bit sequence y. This is now converted back to analog and fed to the inverting input of the input amplifier. With this scheme, we may write for the Fourier transforms

$$Y = (X - Y \cdot Vref) \cdot H(f) \cdot g + N \tag{4.1}$$

where we have assumed that the D/A output is $\pm Vref$. Then

$$Y = \frac{X \cdot H(f) \cdot g}{1 + Vref \cdot H(f) \cdot g} + \frac{N}{1 + Vref \cdot H(f) \cdot g} \tag{4.2}$$

If the loop gain $Vref \cdot H(f) \cdot g$ is high, this equation may be approximated as

$$Y \cong \frac{X}{Vref} + \frac{N}{Vref \cdot H(f) \cdot g} \tag{4.3}$$

The quantization noise has been reduced mainly in the low frequency band, by the effect of this noise-shaping filter. The digital filter that always follows this modulator achieves a further reduction of high frequency noise. Thus, in summary, the SDADC's use oversampling, noise shaping filter and digital filter together to yield low-frequency, high-resolution digital output using a low-resolution high-speed sampler.

4.8 Aliasing

We have seen that there is a quantization error in amplitude due to the discrete resolution. Similarly, errors can also be introduced due to the discrete steps taken in time. Figure 4.10 shows an example of a 5 Hz signal digitized with a rate of 2 Hz or a sample interval of 0.5 s. The 2 Hz digitization rate is missing out on several oscillations that are simply not seen. If the samples happen to be taken on the top of the 2 Hz cycles, the digitized output signal will be interpreted as a 1.0 Hz sine wave. If the samples were taken a bit later, the output would be a constant level. From this example it is clear that, in order to "see" a sine wave with frequency f, the sampling rate must be at least $2f$. In the above example, the sample interval must be at least $0.1\ s$ or a sample rate of 10 Hz in order to "see" the 5 Hz signal. Thus the general rule is that we must sample at a rate twice the frequency of the signal of interest. Or, in a given time series we can only recover signals at frequencies half the sampling rate. This frequency is called the Nyquist frequency.

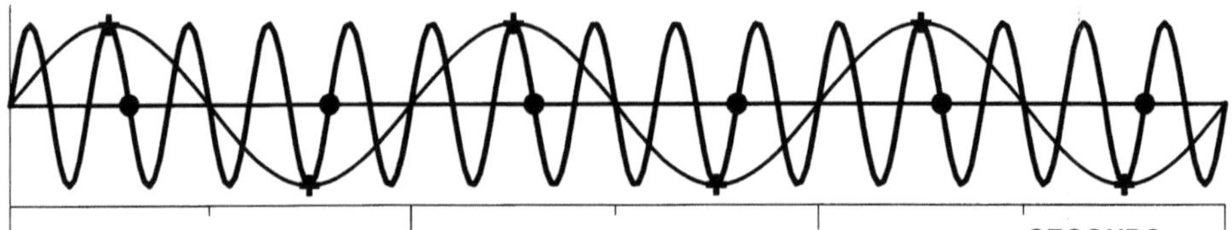

Figure 4.10 A 5 Hz signal is digitized at a rate of 2 Hz. The digitization points are indicated with black dots. Depending on where the samples are taken in time, the output signal can either be interpreted as a straight line (points in middle) or a 1 Hz sine wave.

The effect of sampling with a too low rate is not just that information of higher frequency signals is lost, but more seriously that the high frequency signals injects energy into the signals at frequencies lower than the Nyquist frequency. In the above example, the pure 5 Hz signal creates a non existing 2 Hz signal of the same amplitude as the 5 Hz signal so the digitized output would be completely distorted and have a wrong frequency content. This effect is called aliasing. The only way to avoid this problem is to make sure that the input signal does not contain energy above the Nyquist frequency. So an ADC must have a low pass filter (an anti alias filter) to remove unwanted high frequency energy above the Nyquist frequency prior to the sampling.

4.9 Anti alias filters

For traditional ADC's, the anti alias filtering is done using an analog filter before digitization, while for ADC's with built in oversampling and processing, most of the anti alias filtering can be done digitally, which, as we shall see, has the effect of increasing the useful passband of the digitizer. Ideally, the anti alias filter should remove all energy above the level of the LSB at the Nyquist frequency. For analog filters, this has the effect of limiting the passband severely. Let us look at an example. A 16 bit ADC samples at 100 Hz, so the Nyquist frequency is 50 Hz. The dynamic range (positive or negative signal only) is 90 dB, so at 50 Hz, the anti alias filter should attenuate the signal by 90 dB. Analog anti alias filters are mostly Butterworth filters (they have the flattest response on their passband) and an order of 8 is commonly used, since it is a compromise between the practicality of construction and steepness of the filter cutoff. It can be calculated that the corner frequency must then be 13.7 Hz. At 25 Hz, well within our frequency band of interest, the signal is attenuated 42 dB or a factor of 100. So 'real' signals at 25 Hz with amplitudes before filtering of less than 100·LSB, would be lost. Using a 4'th order filter, the same corner frequency would have to be 13.7 Hz/4 = 3.4 Hz and the attenuation at 25 Hz would be 66 dB, so a 4'th order Butterworth filter is not a good choice! In practice, analog anti aliasing filters are not set up to limit the signal to the Nyquist frequency to one LSB, since this clearly limits the frequency band too much. A common practice is to use an 8 pole Butterworth filter at half the Nyquist frequency. In the above example, this would give an attenuation of 48 dB at the Nyquist frequency and the attenuation would be down to 90 dB at about 90 Hz. Signals in the 50 to 90 Hz range could then cause aliasing problems, if present. In most cases this would not be a problem since seismic signal amplitudes decrease rapidly with increasing frequency. This is true for both the natural background noise (Figure 3.3) and signals from earthquakes where the signal level is proportional to f^{-2} for frequencies f, above the source corner frequency (Figure 8.2). So, only electrical noise in the 50 to 90 Hz band should cause problems. The net frequency of 50-60 Hz is a good candidate! If the sample rate is 200 Hz, an antialias filter is rarely needed for ′normal′ earthquake recording provided electrical noise is not present.

A large inconvenience with analog antialias filters is that they have a significant phase shift in a large part of the pass band, which has to be corrected for when analyzing the signals after recording.

For 24 bit digitizers, an analog anti alias filter would have been even more restrictive. For a 24-bit converter with a dynamic range of 138 dB (positive or negative signals only), an 8'th order Butterworth filter would have the corner frequency at 6.8 Hz in order to get a damping of 138 dB. Fortunately, all ADC's with more than 16 bit use some form of oversampling and decimation so that the main anti alias filtering can be done digitally. There is still a need for an analog anti alias filter for the ADC with the primary high sampling rate (e.g. 200 kHz). Due to the high frequency, this filter is simple to make and need not be so sharp since there are no seismological signals of interest at the high frequencies, so band limiting is not a problem here. A simple RC filter can be used. The 'real' anti alias filtering is then performed digitally before resampling. In the case of sigma-delta converters, the modulator averages the signal on a sample interval, which is equivalent to a low-pass filter, so an additional analog filter is not required. Furthermore, some of them (e.g. AD 7710) use an input stage with charge-coupled devices to implement programmable gain; this actually prevents the use of capacitive sources, or serious errors may be introduced otherwise.

When working with digital signals, we do not have the same limitations as given by discrete analog components and there are digital time domain filters available with a very steep transition band. In practice such a filter is specified by a series of filter constants. If the input signal is x_i, the output signal y_i, where i goes from minus infinity to infinity (we digitize for ever, hopefully!), the filter operation is simply

$$y_i = \sum_{k=0}^{N-1} x_{i-k} \cdot w_k \tag{4.4}$$

where w_k indicates the N filter constants. We can illustrate that with a simple example of a filter with 3 coefficients w and 5 input samples x:

$$\begin{array}{cccccc} x_1 & x_2 & x_3 & x_4 & x_5 & x_6 \\ & & w_2 & w_1 & w_0 & \rightarrow \\ y_1 & y_2 & y_3 & y_4 & y_5 & y_6 \end{array} \tag{4.5}$$

$$y_5 = x_5 w_0 + x_4 w_1 + x_3 w_2$$

The filter moves forward one step at a time in the direction of the arrow and the sum of the products of overlapping x and y is output. Sample 5 is explicitly calculated. This operation is called a convolution of x with w. The output signal for a given sample i can therefore be considered as a weighted average of the i'th sample and the N previous samples and the weighting function is the filter coefficients. A very simple filter could have 3 coefficients of value 0.33. This filter would simply make a running average over 3 samples and thereby filter out higher frequency variations and is a simple low pass filter. This also has the effect of moving some of the energy to a later time. If we e.g. have a signal

$$x = ...0, 0, 0, 0, 0, 1, 1, 1, 1, 1, 1......$$

which would be an onset of a square wave; we can convolve that with the filter

$$w = 0.33, 0.33, 0.33$$

and the result is

$$y = ...0, 0, 0, 0, 0, 0.33, 0.66, 1, 1, 1, 1...... \tag{4.6}$$

We see that the sharp onset has been smoothed and there is now a phase shift, however, the onset time has not changed.

The filter coefficient series w is also called the impulse response of the filter since convolving a time signal consisting of only one non zero sample (an impulse) ...0, 0, 0, 0, 1, 0, 0, 0, 0,... with w, would give back w. From signal processing theory (e.g. Scherbaum, 2001) it is known that the Fourier transform of the impulse response for a linear operator (here the filter) is the frequency response of the system and vice versa. From calibration of seismographs, we are familiar with the impulse response since the routine calibration consists of injecting a constant current into the seismometer calibration coil, which corresponds to a step in ground acceleration. What we get out of the sensor is then its step response (see Chapter 10), which is a non-symmetric pulse.

The filtering (4.4) can be done in two ways. If we strictly follow (4.4), the output of the filter is always determined by the N original samples and the filter is always stable since it only depends on the weighted average of N samples. This kind of filter is called a Finite Impulse Response filter (FIR). If the filter operates on the already filtered amplitude values as well as unfiltered values, the filter output will depend on all the previous samples, in theory an infinite number, it is recursive and is called an Infinite Impulse Response (IIR) filter. The filter has the potential of being unstable since we forever add a fraction of previous samples. An example of an IIR filter is the recursive Butterworth filter as implemented in many processing systems. An 8 pole filter can be implemented with just 8 coefficients while a similar FIR filter could require many more coefficients. Today, processors are so powerful that there is no reason to use IIR filters in order to save computation load.

The ideal antialias filter is very steep in order to get the widest useful frequency range of the ADC possible, and it distorts the signal as little as possible. This ideal is mainly limited by the number of coefficients that can be used.

Most digital antialias filters belong to the class of linear phase filters, which cause no phase distortion of the signal and only give a constant time shift. If the delay is zero or corrected for, the filter is called a zero phase filter. Normally, in order to make a linear phase filter one would run the convolution both forwards and backwards, in order to cancel the effect of the phase shift. This can obviously not be done on real time data, so instead a symmetric impulse response can be constructed, that achieve the same. In place of (4.4), we get

$$y_{i-\frac{N}{2}} = \sum_{k=0}^{N-1} x_{i-k} \cdot w_k \tag{4.7}$$

and the output is now dependent on samples both before and after the current sample. Using the same example as in (4.5), we get:

$$\begin{array}{cccccc} x_1 & x_2 & x_3 & x_4 & x_5 & x_6 \\ & & w_2 & w_1 & w_0 & \rightarrow \\ y_1 & y_2 & y_3 & y_4 & y_5 & y_6 \end{array} \tag{4.8}$$

$$y_4 = x_5 w_0 + x_4 w_1 + x_3 w_2$$

As it can be seen, output sample 4 will be affected by the following sample 5, so some energy in the signal will be moved forward in time. A disadvantage with the linear phase filter is, therefore, that an impulsive signal might be obscured by precursory oscillations, as it will be shown in Figure 6.10.

A so called minimum phase filter will have no precursors, however the signal would then have to be corrected for the phase shift for any other analysis than determining the first onset.
Some of the advantages an disadvantages of FIR and IIR filters are given below (following Scherbaum (2001):

FIR filter. They are always stable. If a very sharp response is needed, they require many coefficients, although special design procedures are available to overcome this problem. Linear and zero phase filters are easy to implement.

IIR filter. They are potentially unstable and, because they are recursive, the round-off errors may propagate in an unpredictable way. Steep filters can easily be implemented with few coefficients due to the recursive nature of the filter, so filtering is much faster than for FIR filters. Zero phase filters are difficult to implement exactly and, as a consequence, IIR filters always have phase distortion within the passband of the filter.

The most common filter used is the FIR filter and it is often implemented as a linear-phase or zero-phase filter. Some digitizers (e.g. from Earth Data) have optional minimum phase filters (also called causal filters) so that there will be no precursors to the first onset. The disadvantage of a minimum phase filter is that there is a phase shift, which must be corrected for, while the zero phase filter has no phase shift. Nearly all digitizers use zero phase FIR filters. The problem of how to correct for or avoid the precursory problem in zero phase filters is discussed in Scherbaum (2001). See also 6.6 and *www.geo.uin-potsdam.de/Software/FIR/.*

Figure 4.11 shows an example of an antialias FIR filter. The filter has been designed for a Nyquist frequency of 50 Hz where the rejection is 150 dB. The filter has 48 coefficients.

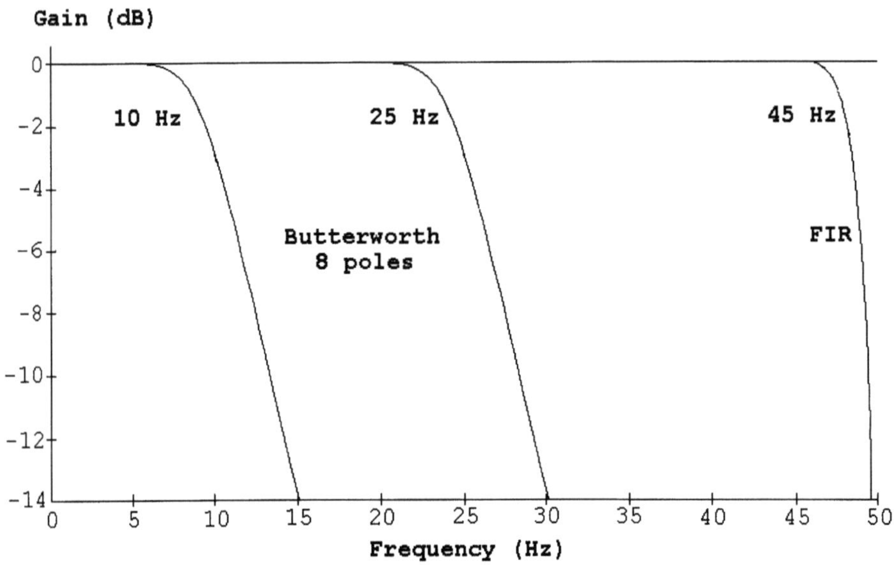

Figure 4.11 A FIR filter compared to Butterworth filters. The FIR filter has a gain of -150 dB at the 50 Hz Nyquist frequency, while the 10 and 25 Hz Butterworth filters have gains of –42 and -112 dB respectively.

As it can be seen, the FIR filter corner frequency is at 45 Hz, so the filter is extremely sharp. Figure 4.11 also shows two Butterworth filters for comparison. The 25 Hz filter has an attenuation of 42 dB at 50 Hz, while the 10 Hz filter has an attenuation of 112 dB. So, we obviously get a much larger bandwidth using FIR filters compared to using Butterworth filters and in addition there is no phase shift in the pass band.

4.10 Examples of digitizers

The quality of digitizers is constantly improving and the prices are going down, particularly for same quality products. So any specific comparison of products might quickly be outdated, but will give an idea of some of the digitizers available in 2004.

Many dedicated recorders can also be used as digitizers since they might have a real time output. The units are described in Chapter 5 and listed in Table 5.3. The distinction between what can be considered a digitizer and a recorder is getting more and more blurred, since once the digitizer, CPU and GPS are available, the digitizer suddenly becomes a recorder with a bit of software and some extra memory. So some of the digitizers mentioned below, can also be used as recorders. Some sensors also have digitizers built in, see Table 2.4.

Table 4.6 shows a comparison between some digitizers. The specifications for digitizers might not be completely comparable, particularly dynamic range, sensitivity and cross-talk. These measures might be given as RMS ratios or ratios of amplitudes. In the table, we have tried to uniformly represent the values as amplitudes. For ratios it should not make much difference, since the RMS values will be smaller by a factor of about 1.3 relative to the amplitude (see 3.3). The noise and dynamic range are also, apart from the sample rate, dependent on the frequency band used for measurement and usually improves with a narrower band. For a digitizer to have a dynamic range larger than 138 dB implies that the noise level is less than 1 count and obviously we then have to calculate the RMS of the noise in order to get the dynamic range. Most specifications given are 'typical' and since dynamic range changes with sample rate, most specifications of dynamic range are given at 100 Hz or in a few cases at 50 Hz.

Table 4.6 Comparison between some digitizers. Specifications are mostly given at 100 Hz unless otherwise noted. Abbreviations:
NCH: Number of channels.
Sensitivity: The lowest signal level that can be resolved considering the self-noise. The value is given at gain 1. This is higher than LSB in most cases, see also text. Mostly given at 100 Hz sample rate.
Dyn: Dynamic range, measured in dB, see text.
Input: The maximum zero to peak input voltage at gain 1.
G factor: The programmable (or jumper settable) gain levels.
Cross-talk: Cross-talk between channels (dB).
Alias: This is the attenuation at the Nyquist frequency (dB).
R: The input impedance.
Sample rate: Range of sample rates, usually only a few discrete values can be used.
Power: Power usage in W, the power is for the 3 channels unless otherwise indicated, it does not include the GPS. Power is calculated at 12 V, even when digitizer can operate at a lower voltage.
Out: Output port: 232: RS232, 422: RS422, T: Ethernet with TCP/IP protocol, lpt: parallel port.
Drift/GPS: The drift in ppm when digitizer is not synchronized with an external clock, GPS is written if a GPS can be connected, pps means synchronized with a pps signal.
Buf Tri: Buffer/memory, trigger: Y if memory is built in, Y if a trigger is built in.

Manufacturer and/or name	NCH	Sen-sitivity, μV	Dyn dB	Input V	G factor	Cross talk, dB	Alias DB	R M ohm	Sample rate, Hz	Power W	Out-put	Drift ppm GPS	Buf Tri
Earth Data PS6-24	3-6	1.0	140	8	1-10	120	120	1.0	1-1000	1.5	232 485	GPS	N N
Geodevice EDAS-24L	3-6	0.5	130	10		110	135		1-500	1.0	232	GPS	N N
Geotech DR24	1-6	4.8	129	20	1-256	80	130		10-1000	4.0	232	0.5 GPS	

Geotech 49.65 16 bit	1-3	140	90	4.5			(84)	10.0	50, 100	1.4	232		N N
Güralp DM24	3-6	3.3	129	10	1		140	1.0	1-200	1.5	232 422	1 GPS	Y Y
Güralp DM-16-R8, 16 bit	8	305	90	10	1		140	1.0	1-200	3.0	232 422	1 GPS	Y
Hakusan LS-7000Xt	6	2.0	135	10	1-10		~ 130		1-200	1.2	232 T	GPS	Y Y
Kinemetrics Q330	3-6	4.7	135	20	1-30		~ 130		1-200	< 1	232 T	GPS	Y N
Kinemetrics Q730BL	1-3	1.9	142	20	1	145	~ 130		1-250	10	232 T	GPS	Y N
Lennartz M24	3			10					1-500	< 2	232 422	pps	N N
Nanometrics HDR 24	3-6	7.2	130	14	1	80	140	1.0	20-500	4.4	232 422	GPS	Y Y
Nanometrics RD (1)	3-6	0.5	130		1				25-500		232 422	pps	N N
Nanometrics Trident	3	0.7	142	10	0.4-8	130	100		20-200	1.5	(2)	GPS	
RefTek 72A	3-6	2.2	130	100	1-32	80	140	2.0	1-1000			0.5 GPS	Y Y
SARA SADC10	4	305	90	10	1	120		0.06	0.01-200	0.4	232	1.6 GPS	N N
SARA SADC20	3	0.8	130	2.5	1	140	14 (3)	1.0	10-200	0.5	232	1.6 GPS	N N
Public domain SEISAD18 (4)	3	9.5	108	2.5	1-16	120	14 (3)	2.5	25-200	0.4	232 422	10 pps	N N
Symmetric Res PAR8CH	8	10	120	10	1-16	140		0.02	1e-5 5000	1.1	lpt	GPS	Y N
TAIDE enterp. TDE-324F	3	0.7	142	10	1-2	140	130	1.0	20-200	3.0	232	GPS	N N

(1) A 16 bit gain ranging digitizer, not sold anymore
(2) Special Nanometrics bus
(3) This converter has a simple filter with moving averages. The response has several notches and only provide some degree of antialias protection, so a high sample rate should be used.
(4) Public domain design and software available at www.geo.uib.no

Ideally, all digitizers should have been tested in an identical manner; however, that is probably not going to happen. One example of a very rigorous test is for the high performance Kinemetrics Q730Bl, for which a test report is available (Kromer et. al., 1999). Other digitizers might have passed similar tests without us being aware of it.

Several digitizers have programmable gain. A higher gain does not automatically mean a more sensitive digitizer, since the noise level also tends to increase. The dynamic range might actually decrease when the gain is increased. The lowest sensitivity is therefore only an estimate from parameters given in specifications; however, it is important to know if the digitizer is sensitive enough for a standard passive sensor, in which case it should be better than 1 μV for a 1 Hz sensor and better than 0.1 μV for a geophone. Very often, the noise level is given in terms of voltage or counts, which can be converted to the maximum sensitivity. If e.g. the maximum input voltage is 8 V and the dynamic range is 130 dB, then the noise level is 8 V/$10^{130/20}$ = 2.5 μV, which is too high for its use with a standard passive seismometer, unless at a noisy site. So either a higher gain or a higher dynamic range ADC (with the same max input voltage) must be used.

CHAPTER 5

Seismic Recorders

Seismometers give out electrical signals, which can be measured with lab instruments. In order to preserve these signals, we need a device for permanently recording them. Since the seismometer output signals generally are small, and maybe noisy, they have to be amplified and filtered before recording. This is the analog signal preparation. In its simplest way, a recorder is any device plotting the signal in a permanent way, like a chart or drum recorder, and we get a seismogram. However, the present most common way of recording is to digitize the signal with an analog to digital converter (ADC) and record it in some computer device.

A recorder and a sensor is called a seismograph and consists in most cases of separate units, while in earlier mechanical seismographs, like the Wiechert seismograph, they were integrated units. The reason for the separation is that the sensitive seismometer has to be placed at the most low noise site possible, with little disturbance from the recorder and people operating it. For field use, the ease of deployment makes it desirable to have one compact unit; in particular strong motion instruments -less sensitive- are sold as complete units with sensor and recorder integrated.

The recorder must not only record the signal, but also the time of the signal, and time keeping is a very important task in any recorder. While earlier it was difficult to get a good external time reference, this task has been much facilitated by the availability of GPS (Global Positioning System) signals. However, we should not think this is the end of all timing problems, since timing failure still is happening quite often, either due to technical failure or temporary unavailability of GPS signals.

This chapter will describe the elements of the recorder and give examples of available recorders. The intention is to cover both recorders used for permanent seismic stations and networks as well as recorders for field work, since they have very similar design. The most important difference is that field recorders must be designed for portability and low power consumption, while recorders at permanent seismic stations usually do not have to take this into consideration.

5.1 Analog amplifier

The output signal from sensors can be very small, as e.g. from geophones, or moderately high, as from a broadband sensor. If ,e.g., the ground velocity is a sinusoid of 5 Hz and amplitude 1 nm/s (typical background noise at a low noise station), a standard 4.5 Hz geophone with a generator constant of 30 V/ms^{-1} would give an output of 30 nV, while a broadband sensor with a generator constant of 1500 V/ms^{-1} would give 1500 nV. If the frequency is 1.0 Hz, the geophone output is down to 30 $nV \cdot (1/4.5^2) = 1.5$ nV (see seismometer response equation, Chapter 2). Such small voltages can usually not be recorded with any recording device directly and must first be amplified. This is not as easy as it sounds since making such a low noise amplifier, which furthermore covers the large frequency range used in seismology, is a fine art. Particularly making stable low noise amplifiers at low frequencies is difficult. In addition, the amplifier must not only have low noise but must also be free of amplitude and phase distortions. Since this is the front end of the recording system, usually this electronic part determines or limits the sensitivity of the seismograph more than the sensor itself (see also 2.16 on sensor self noise). The noise of an amplifier will here be defined as the noise referred to the input. This means that the noise level is the smallest input signal which can be resolved. A simple way of getting an idea of the noise is to short the amplifier signal input and measure the output[1]. If, e.g., the output noise is 1 mV for an amplifier with gain 1000, the equivalent input noise is 0.001 mV. The lower practical limit seems to be around 50 nV at 1 Hz (bandwidth of 1/3 octave) and increasing at lower frequencies (see e.g. Reidesel et al, 1990). Thus, it will normally not be possible to resolve the theoretical output of 1.5 nV from the geophone in the example above with the available recorders (Table 5.3).

The analog input amplifier also has the purpose of coupling the sensor to the recording system. This is not critical with broadband sensors that have high output level and low output impedance, so the critical input stage has in reality been moved from the recorder input to be part of the sensor electronics. For passive sensors, the amplifier input resistance must be large enough, so that the sensor is not damped too much, and this further makes it difficult to make a low noise amplifier, since a higher input impedance creates a higher noise. Ideally, the input stage of an amplifier should be optimized for the sensor it is to be connected to.

The amplifier itself is not the sole source of non seismic noise at the output. Another main source of noise may come from the coupling between the seismometer and the amplifier. If long cables are used, noise induced from an electrically noisy environment can also yield a high noise input. In order to reduce this noise, differential amplifiers are used nearly exclusively, not because they generate less intrinsic noise, but because they are better at reducing external noise picked up with equal amplitude in both input terminals (see Appendix on electronics). Cables must obviously be shielded, but even then, it can be difficult to avoid noise and, in practical installations, experiments must sometimes be made to avoid noise from cables. This often involves trying to ground the system at different points. Remember that grounding should only take place at one point to avoid ground loops.

5.1.1 DIFFERENTIAL INPUT-OUTPUT

The use of differential input and output can create confusion in terms of influence on signal levels and the way the signal should be connected. In seismology, we find

---

[1] A more accurate procedure is to connect a resistor equivalent to the sensor output impedance (e.g. the coil resistance in the case of a geophone) to the amplifier input, since noise current may contribute significantly to the total noise. Even so, the noise may be underestimated, since the back-electromotive force of the geophone, excited by the noise currents generated at the amplifier input, may contribute significantly (Rodgers, 1992a) to the total noise.

differential circuits in two places: Output from some active sensors and input to most digitizers. Since a differential amplifier has 3 terminals (+, - and ground), there are different possibilities for connection if single ended devices are used with differential devices. Depending on how the connection is done, this might change the signal level by a factor of 2. The different possibilities are seen in Table 5.1.

Table 5.1 The different possibilities for connecting differential and single ended devices together.

Output → Input ↓	Differential	Single ended
Differential	+ → + – → – ground → ground gain = 1.0	+ → + – → ground ground → ground gain = 1.0
Single ended	+ → + – → NC ground → ground gain = 0.5	+ → + ground → ground gain = 1.0

Note that in all cases except differential to single ended, there is no gain change. The most common problem is therefore when a differential sensor is connected to a single ended amplifier/digitizer. This has created confusion in the way manufacturers specify the sensor output. Some will specify e.g. 1000 V/ms^{-1} for a differential output device in which case the output is reduced to 500 V/ms^{-1} when connecting to a single ended digitizer. Others will specify 2 x 500 V/ms^{-1} for the same device and the generator constant is then 1000 V/ms^{-1} when used differentially.

5.2 Analog filters

Modern amplifiers are usually DC coupled, meaning that any DC shift in the input also will appear in the output. This means that any DC shift in either amplifier or sensor (passive sensors do not have a DC) will show up amplified in the output. This can usually be adjusted to zero for both sensor and amplifier. In some amplifiers, a high pass filter is installed, so the output has no DC. This was quite common in earlier designs to ensure stability and since low frequencies were not needed for many SP stations. This filter could mask serious problems[2] with either sensor or amplifier and gives an unfortunate phase shift at lower frequencies. Due to these problems, a general desire to acquire data at low frequencies and since modern amplifiers are more stable, high pass filters are rarely used anymore.

---

[2] E.g., some seismologist tried to find a "seismic interpretation" for a long transient superimposed on a seismogram of a local earthquake, which was explained simply by a preamplifier clipping, masked by a high-pass filter following it.

Low pass filters are part of all amplifiers. For amplifiers used with digital systems, there must always be also an anti alias filter to remove unwanted higher frequencies (see Chapter 4) and, in general, for stability reasons and noise bandwidth limiting, a low pass filter is built in.

In connection with analog recording, additional low and high pass filters might be available in order to record the most noise free signal.

Analog filters are usually Butterworth type, since these are easy to construct and have nice characteristics (see Chapter 6). The number of poles for the filter varies depending of the application. A DC blocking filter only needs to have one pole, while an anti aliasing filter might have 8 poles. Signal shaping filters usually have 4 poles. It should be kept in mind that a low-pass analog filter (e.g. a Butterworth type) causes a phase distortion at frequencies well bellow its corner frequency, even if its amplitude response is flat in this zone.

An example of a long surviving amplifier is the Sprengnether single ended AS110, which is used in analog recorders and analog telemetry systems:

Gain:	60 –120 dB in 6 dB steps
Fixed filter:	0.15 – 30 Hz
Variable filters:	Low pass: 5 and 10 Hz, 4 poles High pass: 5 and 10 Hz, 4 poles
Noise:	100 nV p-p with filters 0.15-30 Hz and input impedance 4 kΩ.
Maximum output:	20 V p-p

By modern standards, this amplifier has a reasonably low noise level, however that is achieved at the expense of having a band-pass filter. The amplifier can regulate the gain in 6 dB steps, or a factor of 2 between each. This is typical for most analog amplifiers with variable gain. We can also calculate the theoretical dynamic range. If gain is set at 60 dB, the noise output will be $1000 \cdot 100$ nV = 0.1 mV and the dynamic range is 20/0.0001 = 200 000 or 106 dB. At a gain of 120 dB, the noise output is 60 dB higher, but maximum output is the same, so the dynamic range is only 106-60 dB = 46 dB. As we shall see, none of these values are good enough for high dynamic range digital recorders, but they are ok for analog recorders.

Some common specifications for amplifiers are explained in 5.2.1.

Recent recorders (usually digital) have in general high resolution, so they do not require the signal to be amplified so much before it is input, if amplified at all. In these cases, a single-stage preamplifier is enough for passive sensors with a low-level output, such as the one shown in 5.2.1.

5.2.1 AMPLIFIER SPECIFICATIONS AND NOISE

The selection of a suitable preamplifier for a given use (e.g. a geophone) would require several technical specifications.

The noise of a preamplifier following a sensor may set a limit to the whole system sensitivity. It may be due to intrinsic or internal noise of the circuit components, essentially unavoidable, although it may be minimized by a careful design, and to induced noise from external sources, which in general may be avoided or minimized.

A usual way of characterizing intrinsic noise is the peak-to-peak or RMS input equivalent voltage noise. As noise is of random nature, these values both depend on bandwidth. Assuming that noise is Gaussian, a handy rule-of-thumb for relating these values is as follows: RMS multiplied by six gives a peak-to-peak value that will not be exceeded 99.7% of the time (see also Chapter 3). If only voltage noise is given, it must be referred to a given signal source impedance, since noise is generated either as voltage or current. Actually some amplifiers are optimized for low impedance sources (i.e. they have low input voltage noise, but perhaps not so low current noise) and others are suitable for medium-to-high source impedance (input current noise must be low in this case).

A rule of design is that amplifier bandwidth must be not much larger than signal bandwidth, as noise amplitude grows with bandwidth. For this reason and also because of the so-called pole compensation techniques (to assure stability against oscillations), a low pass filter is often included in the preamplifier with corner frequency slightly above the band of interest.

A more complete way of specifying internal noise in an amplifier is by its voltage noise spectral density, including the contribution due to noise current sources through the input and source impedance. This power density is almost flat, except for low frequencies at which it increases proportionally to 1/f. For a DC or low frequency amplifier it is thus important that this noise corner frequency is as low as possible. For an instrumentation low-noise amplifier, this should be less than 10 Hz. Some very low frequency amplifiers, chopper amplifiers, use the technique of spectral shifting the signal to higher frequencies by means of sampling it and then amplify before recovery. With the present state of the art of semiconductors, comparable levels of low frequency noise are achieved by direct amplification at less power consumption.

Single ended or *common mode* amplifiers are in general less noisy than differential ones, because one of their inputs contributes less to the noise (e.g. Rodgers, 1992a). In spite of that, most seismic preamplifiers have a differential input stage to facilitate the use of long cables and minimize pick-up noise from external sources. In such a case, the cable must be a balanced type, such as twisted pair with external shield. Magnetically induced disturbances from nearby transformers or mains power lines also cause pick-up noise. A proper shielding of the amplifier itself and keeping it away from these sources may avoid this kind of noise.

The output of amplifiers is also often differential. If this is the case, the best choice is to keep the preamplifier as close as possible to the sensor and use a cable as long as necessary from amplifier to recorder. An unshielded twisted pair type of cable, such as the ones used for phone lines or computer networks, may do the job for distances of up to 1km or more. For differential amplifiers, an important parameter is the *common mode rejection ratio* (CMRR), which indicates the fraction of a common mode signal referred to the ground and present in both input terminals that will be amplified up to the output. A value of -100 dB is acceptable.

Another source of possible external noise comes from the power supply. In particular DC-DC regulators usually have a non-negligible high-frequency ripple voltage at its output. The rejection of the amplifier to this noise is characterized by the PSRR (*power supply rejection ratio*).

Classically, an *instrumentation amplifier* (Figure 5.1) has differential input, high input impedance and low offset referred to its input. In the case of high dynamic range signals such as seismic signals, the input equivalent noise has to be very low. This last characteristic may be in conflict with the others and particularly the input impedance should be kept as low as possible, but high enough to avoid overdamping of the geophone. As stated above, the differential preamplifiers are slightly noisier than single-ended. And, due to the inherent character of inertial sensors, DC-component is never part of the true seismic signal and may be removed afterwards if some offset is present.

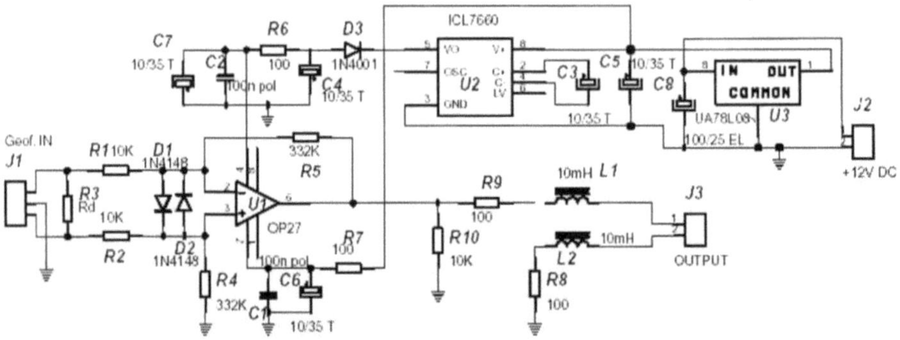

Figure 5.1 A single-stage, low noise, low offset preamplifier. It is composed of a suitable operational amplifier in differential mode, and a power supply regulator that generates also in-circuit the negative voltage from an unregulated supply of 12 V. The voltage gain is *Av=33.2 (R5/R1=R4/R2)*. The input resistance is *Ri=20 kohms*, which in parallel with Rd is the external damping resistance *Rp* (*Rp=Ri·Rd/(Ri+Rd)*). Therefore, for a geophone of generator constant G and coil resistance Rg, the total magnification factor, including the amplifier, will be G·Av·Rp/(Rp+Rg). The output of the amplifier is balanced, so that it may be connected to a long twisted-pair cable.

5.3 Analog recording

Recording in analog form on paper continues to some degree despite the advance of digital technology. No analog recording on tape is done anymore (see Lee and Stewart, 1981, for more details). Very few analog recorders are sold nowadays, but there are probably hundreds, if not thousands, still in operation. This might seem strange considering the advances in digital recording technology and digital processing methods. There are several reasons: (1) There are still a lot of us who like the look and feel of a 'real' seismogram, a wonderful way of getting, at a glance, the last 24 hours activity displayed. This can of course also be reproduced digitally, but it is rarely available in the same format and has a delay of up to 24 h, (2) Analog recorders do not get worn out or outdated like digital recorders, which seem to have to be replaced every few years, (3) An analog recorder provides a simple, very reliable real time back up recording. Thus, many observatories continue to have a few stations recording on analog recorders and will probably continue to do so until either the recorders or the 'older' seismologists die out, so analog recorders might be around for another 10-20 years! A brief description of analog recorder will therefore be included here.

All analog paper recorders are based on the principle of recording on a rotating drum with a pen which moves along an axis parallel to the rotating axis[3] to provide a continuous trace for the whole recording period, usually 24 hours. For this reason, these recorders are often called helical drum recorder or helicorders. A block diagram of a complete system is represented in Figure 5.2.

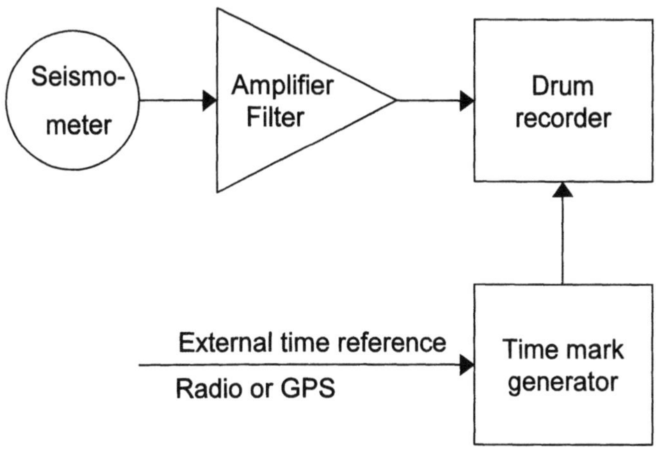

Figure 5.2 A complete analog recorder. The time mark generator is an accurate real time clock which might be synchronized with an external time reference.

The time marks (Figure 5.2) are usually generated at each minute, hour and 24 hr and have different length. The time mark generator can usually be synchronized with an external time reference. This type of system can be quite expensive due to the high cost of the drum recorder and the time mark generator. The motor operating the drum

[3] *Old photographic recorders usually moved the drum itself along its rotating axis.*

rotation can optionally be controlled by the clock to ensure that the time it takes to make one revolution is constant. This will make the time marks aligned in the vertical direction (Figure 5.3).

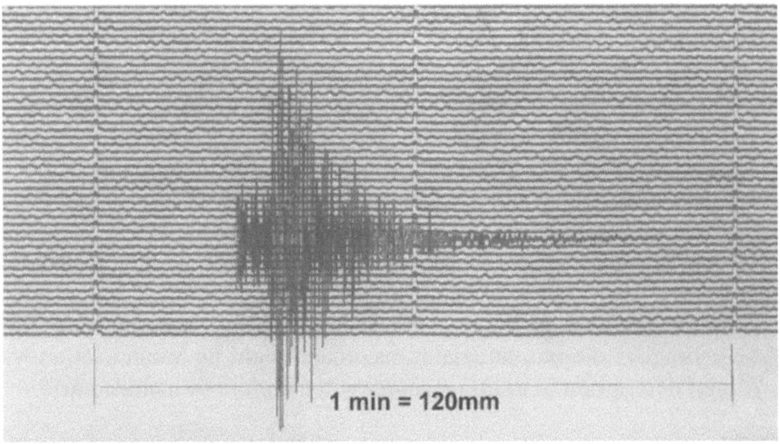

Figure 5.3 An example of an analog recording from a SP recorder. The figure shows part of a seismogram. The time marks are 60 s apart.

The earlier recorders were recording on photographic paper, like the famous *World Wide Standardized Seismographs System* (WWSSN) system (for details, see Willmore, 1979), while all recorders now record with either ink or a thermal pen on heat sensitive paper. Some WWSSN stations do in fact continue in operation today with analog recording on heat sensitive paper. Figure 5.3 shows an example.

Very early recorders also recorded on smoked paper (like the Wiechert seismograph) which later was fixed with varnish. The pen is a sharp needle, so the plot is a white trace on black background. Smoked paper has in recent times also been used for field recording. A few analog field recorders are still in use, both with smoked paper and ink. Heat sensitive paper has never been used for field recording due to the high power consumption. The smoked paper gives the clearest records of all and is very robust for field recording, but obviously not very practical by today's standards.

The standard recorder records 24 hours with a speed of 15 mm/minute for LP records and 60 mm/minute for SP records on a 'standard size' seismogram of 90 x 30 cm. This leaves 10 mm and 2.5 mm between the traces for LP and SP records respectively. Both translation and rotation speed can be changed. For some models, this requires changing mechanical gears and/or motors while the later models this is done by push button operation. Figure 5.4 shows an example of a modern version with push button operation. The recorder itself usually has a variable gain amplifier and the frequency response is usually flat from DC to around 50 Hz. In common recorders, as the input signal varies, the pen describes an arc centered on the pen motor axis, but some recorders have a rectilinear mechanism for the pen, instead of an arc-shaped, at the expense of a higher pen effective inertia. This limits the high frequency response, unless quite a powerful pen motor is used, which can be either a galvanometric type (open loop) or a servomotor (closed loop).

Apart from not recording digitally, the big drawback with the helicorder is its low dynamic range. The maximum amplitude is usually 100 mm and the minimum discernible amplitude is about 0.5 mm, so the dynamic range is 200 or 46 dB. This can be compared to digital recorders, which achieve up to 150 dB dynamic range. This corresponds to an amplitude of 15 km on the paper recorder!

Figure 5.4 An example of a helicorder. The unit consists of a rotating drum with paper and a pen mounted on a pen motor that translates from left to right as the drum rotates. The gain of the pen amplifier is variable to adjust to different noise conditions. This recorder is made by Geotech (Model Helicorder RV-301B). From Geotech home page, www.geoinstr.com.

One of the most well known analog field recorders still sold, is the Sprengnether MEQ800 which can record with either ink or smoked paper. The self contained unit with amplifier, helicorder and internal clock is shown in Figure 5.5. This recorder might seem very outdated today, but it gives instant display of recording in the field without the use of computers and it has a very low power consumption of 0.4 W. There is no digital field recorder on the market yet which has such a low power consumption for continuous recording and the MEQ800 has been on the market for more than 30 years! The lowest power consumption for the next generation of continuously recording digital system is going to be 1-2W.

Figure 5.5 MEQ800 analog field recorder made by Sprengnether. The amplifier and filter controls are on the back, and to the right the clock display is seen.

5.4 Introduction to digital recorders

Replace the analog recorder with a digitizer and a digital storage device and we have a digital recorder. The digitizer itself (ADC) is discussed in Chapter 4 and we can consider the ADC as a component providing digital data readable by a computer, in analogy with an amplifier providing an analog signal to be 'received' by some analog storage medium. Our problem is now how to deal with this large amount of data. One day of 24 bit (3 bytes) data of one channel, 100 Hz sampling rate, gives $3600 \cdot 24 \cdot 100 \cdot 3 \approx$ 26 Mb of data. For one month, a 3 channels recorder would then record 2.4 Gb. The first requirement is that we need a computer to read, check and perform some kind of storage of the data and possibly throw out data without earthquakes. Until very recently, this computer was in general different for portable field systems and recorders used for permanent stations or networks. The main reason was power consumption. General-purpose computers could not (and still cannot with a few exceptions) be used for portable equipment so that microprocessors have to be used. A similar problem arises with respect to storage, where general-purpose disk and tape drives were too bulky and used too much power for field use. Whereas making a data acquisition system for a general-purpose computer is a relatively straightforward task, it becomes more of a specialist task with microprocessors. Currently there is therefore two types of data acquisition systems based on general-purpose computers and microprocessors, respectively. Most portable recorders sold until 2000 were based on microprocessors.

The main elements of a digital seismic recorder are seen in Figure 5.6.

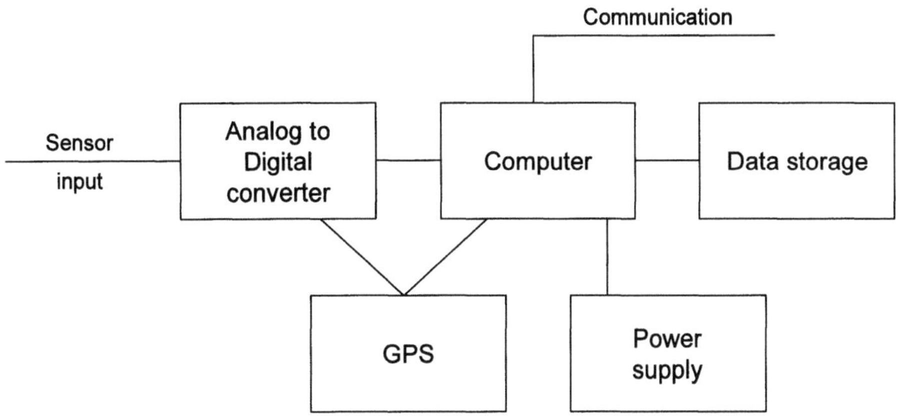

Figure 5.6 Main units of a seismic recorder. There are no flow arrows between the units since all can have two way communications. The GPS can be connected to the digitizer or the recorder. The power supply may be common for all elements or each may have its own regulator, but usually the power source is unique (e.g. a battery).

The main tasks of the seismic recorder can be summarized as follows:

- Read data from one or several digitizers, digitizer can be internal or external.

- If digitized data do not have a time stamp, the recording computer must perform the time stamping of the data. This means that a time reference must be connected to the computer, usually a GPS receiver.

- Store data in a ring buffer on a storage medium. Ring buffer means that after some time (hours to months depending on the size of the storage medium), the data is overwritten. For limited memory systems, there might not be a ring buffer, but just recording of declared events with some pre-event memory.

- Check incoming data for seismic events (perform triggering) and store only real events. This might not be needed if the system has a large ring buffer.

- Provide communication with the outside world for change of parameters and download of data. This can be performed through a communication port and/or by replacement of the storage medium such as tape, CD or hard disk.

Each of these tasks will now be discussed in some detail.

5.5 Digitizing

Portable recorders usually have 3 or 6 channels of data with the digitizer built in and the data is read directly by a microprocessor port. Often the microprocessor drives the digitizer directly. When a general-purpose computer is used, the digitizer can be a built in card controlled by the CPU (like in the W. Lee system (Lee, 1989)). These cards can have up to 64 channels and a resolution of up to 16 bit and are often used with array installations or seismic networks, see network section. Plug-in card digitizers have difficulties with achieving a higher resolution due to the electro-magnetically noisy environment inside a computer. Alternatively, the digitizers can be free standing, sending data out on a serial port (RS232 or RS485), sometimes on a parallel port. Examples of such commercial digitizers are Earth Data and Nanometrics 24 bit units with 3-6 channels (see Chapter 4). Several of these digitizers can then be connected to the general-purpose system to form a multi channel recorder. Remote digitizers can also be connected to the serial line via cable, telephone, radio or TCP/IP connections and digitizers can then be thousands of km away. In this case, it is important that the digitizer carries its own time stamp, since it might not be possible to predict the delay in the transmission system.

Free standing digitizers usually send out the data in some buffered form with a header and then some binary data following. A format used by many units is to send a data frame every second. Data compression is commonly used and the header gives a checksum of the data transmitted. The checksum is a unique number associated with the block of data transmitted, like in the simplest case, the sum of all data samples (creates overflow but still a unique number). At reception, the checksum is recalculated and if it does not match the received checksum, this is interpreted as an error in transmission. The data acquisition software can now ask for a retransmission of the data if the digitizer has this capability, or discard the data if not. It is however important to get rid of bad blocks since just one bit wrong can suddenly generate samples with very large values and thereby wreck havoc to the trigger system. Only the simplest digitizers do not have checksum and retransmit facilities. This might not be so important if the digitizer is located close by, but if the data is received by a long cable or over some remote transmission link, this facility is essential for stable operation.

5.6 Time stamping of data

Accurate timing of seismic data is one of the real headaches in all recording systems and has caused endless frustrations for seismologists. One should think that in these days of cheap and accurate GPS receivers giving out a pulse per second (1 pps) with μs accuracy, these problems were past, but we still have to be careful in order to get accurate timing. There are several issues involved in how the actual timing is done.

Absolute accuracy needed. As accurate as possible, however there is a tradeoff with price and power consumption. For regional and global networks, 100 ms is accurate enough considering the rest of the errors getting into the data and the accuracy of the travel time tables.

Relative accuracy needed. When a small network, an array or stations in a refraction experiment are used, the relative accuracy becomes very important and it should be better than 1 ms. An absolute accuracy of 100 ms might still be acceptable if no accurate channel correlation is needed, like for relative earthquake location. Within the same recorder with several channels, relative accuracy is not an issue since these channels will always be very accurately synchronized by using the same oscillator for digitizing. Nevertheless, in a multi-channel system with a single ADC, there exists a time delay between the precise times of sampling consecutive channels, as they are sampled and converted one by one. This is called time skew delay and should be accounted for if more relative accuracy that the interval between samples is required. Of course, if the digitizer has one ADC per channel, this skew is null.

Most recorders specify timing accuracy to better than 1 ms, at least when synchronized with a GPS, so what is the problem? In order to understand that, we have to look a bit more in the details of how the timing systems work. All recorders and/or digitizers will have a real time clock, which is doing the actual time stamping. So there are two main issues involved: Keeping the internal real time clock on time and the time stamping of the digitized data.

The real time clock can be based on an accurate crystal controlled oscillator, which is kept synchronized with the 1 pps from the GPS. Since the GPS signals disappear from time to time depending on the location of the antenna, the overall accuracy depends on the drift of the free running real time clock. Typical drift of quality oscillators are in the range 10^{-6} to 10^{-9} (or 1 ppm to 0.001 ppm, parts per million), meaning that they drift 1 s in 10^6 to 1 s in 10^9 s, respectively. Therefore, if the GPS was out 2 hours, these clocks would drift 0.007 and 7 ms, respectively. For one week, the same numbers would be 600 ms and 0.6 ms, respectively. So with the good oscillator, once every 10 days would be enough to keep an absolute accuracy of 1 ms while for the low accuracy oscillator, the station would have to be synchronized every 20 min. In Table 5.3, it is seen that the best accuracy is 0.1 ppm and the typical accuracy is 1 ppm for standard recorders. Accuracy of 0.01 to 0.001 ppm requires special high cost oscillators like an oven-controlled oscillator (has a very stable temperature obtained by thermostatically controlled heating), which might use as much power as the recorder and cost just as much. The question of which oscillator to use is a trade-off between cost, power consumption, accuracy required and availability of GPS signals. Some recent recorders use a frequency-locked oscillator which is frequency corrected by the GPS pps periodically (each time that GPS receiver is switched on). This technique saves power, while keeping a high time accuracy.

In some systems, there is no separate real time clock unit and the computer clock is used, so a real time operating system is required and the time must be updated frequently by the GPS. A real time operating system means that the response time to an instruction, in this case timing the arrival of data, is guarantied. Examples of real time operating systems: Microprocessor operating systems, QNX, DOS (not strictly, but with some limitations), but not Windows (Windows 2000 may also perform limited real-time tasks). Internet connected computers can also be kept synchronized to one of the international network time servers (SNTP – Simplified Network Time Protocol) and an accuracy of 1-50 ms can be obtained. PC's have notoriously bad real time clocks and it is often better to use the GPS built-in real time clock as a master clock. A typical accuracy, when not receiving, is 1-5 ppm (Garmin), depending on temperature stability.

Temporary field stations might have special requirements. The most special is the ocean bottom seismograph where no time synchronization can be done, so a high accuracy clock is needed. In this case, it is also important that the drift of the clock is linear (good long term stability), since it is then possible to correct for the time drift at the end of the experiment by knowing the time before and after the experiment. For short term deployment, small compact low power self contained recorders are used without GPS, except at initial installation and a typical drift rate is 0.1 ppm corresponding to a drift of 8 ms a day (temperature at the sea bottom is quite stable anyway).

Time stamping the digital data using the real time clock is the next issue. Again, there are various ways of doing it. The ADC itself will have an oscillator controlling the sampling, which can be synchronized with the real time clock.

Option 1. The oscillator is living its own life and the data must be time stamped by the computer at arrival. The accuracy of the sample rate is then determined entirely by the accuracy of the digitizer sample rate, which typically is no better than 10^{-6}, so if e.g. data is buffered for the number of samples corresponding to one second, the buffer times will change according to the drift of the digitizer oscillator. The timing accuracy might be better than 1 ms, especially if the digitizer and computer are built together. However, if the digitizer is external, there will be delays in transmitting the data, which have to be corrected for, and it might be difficult to obtain a 1 ms accuracy. Another disadvantage of doing timing this way is that two recorders participating in the same experiment might have 1 ms absolute accuracy, but the samples will be taken at different times and sample rates will be slightly different. Particularly for long time series, this is not desirable for processing. Example of a recorder working this way is Quanterra (Table 5.3).)

Option 2. The digitizer sampling is synchronized with the real time clock. This has many advantages. The sample rate is constant and nearby recorders will obtain the samples at the same time within the accuracy of the real time clock. It is now possible to start recording and extract data at the exact onset of the second, which facilitates processing. Particularly for long time series, like continuous recording, this way of time stamping makes life easier during the processing. Recorders like Nanometrics Orion or Güralp works in this way (Table 5.3).

The time stamping can be done by the computer or by the digitizer itself whether option 1 or 2 is used. For separate digitizers, the latter is now becoming the standard, which however requires that the GPS is connected to the digitizer, which has its own real time clock. The computer does then not need a real time clock, and the operating system do not have to be real time.

Options 1 and 2 might seem to present important differences, but for the majority of stations it has little importance. The most important issue is how well the time is kept, particularly when the GPS is not available, so the drift of the clock and how smoothly the synchronization is performed will determine overall timing accuracy. It is better that the timing accuracy is always better than 100 ms, that at 1 ms 90 % of the time but uncertain for 10 % of the time. It is therefore important that the recorder continuously monitors the status of the time synchronization and that the data have this information. Typically, a recorder will log when the last successful synchronization was made.

The GPS uses around 1W of power and, in order to save power, many portable recorders have the option of only turning the GPS on for part of the time. This degrades the overall timing accuracy depending on the accuracy of the local clock.

5.7 Storage media and recording in a ring buffer

The data from the digitizer will have a certain buffer in CPU memory and then the complete data or triggered segments have to be stored elsewhere. In early systems, tape was used, but is now only occasionally used for transferring data from the recorder to a central processing facility, like GSN stations with Quanterra data loggers. A few systems use magneto-optical disks (Table 5.3) or CD. However, in recent years hard disks have become the standard recording media. The SCSI disk used to be the standard on high-end recorders; however, 2.5" IDE disks for laptops now have low power consumption and high capacity, so they are fast becoming the standard in portable equipment. Lately, USB devices have become popular. A USB flash disk uses little power and can have a capacity of 1 Gbyte. USB hard disks use more power but can have much larger capacity. Flash cards seem to be quite popular now.

Many recorders are provided with systems that temporarily store continuous signals in ring buffers (usually on disk) for a given period of time, ranging from hours to several weeks. After that time, the system erases the old data and replaces it by new incoming data. However, during the designated time a seismologist can detect, associate, and analyze events far better than any automatic algorithm. Excellent completeness of data is obtained since all data is available and any data can be inspected. In addition, for the most interesting periods, like aftershock sequences or earthquake swarms, the data can be stored in a continuous manner, thus keeping permanently all information contained in the signals. Ring buffered systems are also very useful if the seismic system is accessed by various institutions for different purposes. Everybody can 'browse' for what he is interested in.

Recorders with memory ring buffers are quite common, and now large memory modules can be used (non-volatile memory such as flash type is preferred for this use). Systems with little memory (less than 10 Mb) usually do not have ring buffers and will only record triggered time intervals, see next section.

Ring buffers are written in almost as many ways and formats as there are manufacturers, so it is hard to generalize. The main consequence is that users always have to use recorder specific software and procedures to read the data and convert it into a useful format. One of the reasons for these user-unfriendly systems is the limitation presented by microprocessor-based recorders. This means that there might not be a file system on the disk, so the disk cannot be accessed directly from any other computer. General-purpose computers have been used for a long time for fixed installations (like the Quanterra VME based systems) and on these systems it has been possible to do better.

Ideally, a ring buffer and associated software should provide access to (or possibility to extract) any time segment of interest without reading more than that segment. Unfortunately, some ring buffer systems have to be read sequentially. There have been attempts to unify formats of seismic recorders with the SUDS format (Ward, 1989), but it failed due to its complexity and lack of consensus on a common format for PC and Unix platforms. Fortunately, general purpose computers seem to be the heart of the next generation field stations, so it will be much easier to implement more standard systems. The most standard open format currently available is the Miniseed format used by Quanterra recorders and a few others. It is also an agreed upon international format.

Ring buffers (and event data) are often compressed, which might reduce the size by a factor of 2-3. The most usual compression method is to use the difference between the samples (e.g. the Steim compression, Steim and Wielandt, 1985) instead of the samples themselves. Thus, in a sequence of numbers only the first sample is given and the subsequent numbers are the following differences. If e.g. the samples are 300, 305, 306, 320 and 322, then the difference sequence would be 300, 5, 1, 14, 2. This very simple compression works well for seismic data, which are changing relatively slowly. The disadvantage is that in a ring buffer, elements of constant time length no longer have a constant size. Considering that a standard small disk size is now 20 Gb, there is much less need for compression during recording than before. If a standard computer is used, data can be compressed any time with standard compression tools like gzip for transfer or permanent storage. See also Chapter 8, section on communication, for more on compression.

In a standard operating system, a ring buffer can be implemented in two ways.

Option A: The ring buffer is just a large file with direct access to the segments of data. A simple algorithm can then be used to calculate where a particular time segment (e.g. 1 sec of data) is stored. This type of buffer is simple, fast and effective and requires little computer overhead for writing and extracting data. The drawback is that it is one big file so it is not easy to handle generally (e.g. copy for backup) and taking out data always requires running a program.

Option B: The ring buffer consists of a series of files with names reflecting the time and with a fixed time length, e.g. 20 min. The files can be created either at system installation time or as time goes and older files are deleted. The simplest and most stable is to use a set of pre-created files. The advantage of this system is that it is very flexible. The files can be written in any format, even in a standard processing format, if desired. To extract part of the entire ring buffer is just a question of a simple file copy.

5.8 Seismic triggers

The amount of continuous data is very large and although data storage is cheap and plentiful, we at least want to know which segments of the continuous data contain real events or alternatively just store the data segment with real events. Obviously, in the latter case, the storage requirement is greatly reduced. Whereas one month of continuous data requires a few Gb, the time segments of real interest might only be a few Mb. If e.g. a 3-component station records 10 events/day of 100 s duration, using a 24 bit digitizer at a sample rate of 100 sps, then 30 days of recording would require 10 events * 30 days * 100 samples * 3 bytes * 3 channels * 100 s ≈ 27 Mb or about half that if compressed. This amount can be contained in a small flash chip consuming almost no power, so it is obviously attractive to be able to just store the real events, particularly on recorders for field use, where size and power consumption should be kept low. For most field recorders sold until very recently, the memory size was so limited that only triggered recording was feasible.

A trigger is an algorithm that checks the signal for variations that could indicate an event. There are several trigger algorithms available, some very sophisticated using pattern recognition, spectral analysis, etc. They are sometimes used in seismic network recorders, but rarely in portable data loggers currently on the market. In the hands of an expert, they can significantly improve the event detections/false triggers ratio, particularly for a given type of seismic events. However, the sophisticated adjustments of operational parameters to actual signal and seismic noise conditions, at each seismic site, which these triggers require, have often been proven unwieldy and subject to error. Hence, for practical purposes, only two types of triggers are widely used, the level trigger and the *STA/LTA* trigger (see below) and only those will be described here.

The level trigger or the amplitude threshold trigger simply searches for any amplitude exceeding a preset threshold. Recording starts whenever this threshold is reached and stops when the level is below the threshold or after a given time. This algorithm is often used in strong motion seismic instruments, where high sensitivity is not an issue, and where consequently man-made and natural seismic noise is not critical. This type of trigger can be implemented in very simple instruments and many analog-recording accelerographs had it. The level trigger is now largely replaced by the *STA/LTA* trigger, since computer power no longer is an issue. However, many instruments still keep the level option for historical reasons.

The short-term average - long-term average trigger (STA/LTA) is the most frequently used trigger algorithm. A single channel of seismic signal is typically processed as follows: The signal is band-pass filtered and the absolute average *STA* (short term average) over the STA time window is determined. Typically, the *STA* time window is 0.5 seconds for a short period channel. The same filtered signal is also used to calculate the *LTA* (long term average) over the *LTA* time window, which is typically 50-500 s. Thus, *LTA* will give the long term background signal level while the STA will respond to short term signal variations. The ratio between *STA* and *LTA* is constantly monitored and once it exceeds a given threshold, the trigger level, the start of an event is declared for that trace. Once the event starts, the *LTA* is usually frozen, so that the reference level is not affected by the event signal. The end of the event is declared when the *STA/LTA* ratio reaches the de-trigger level. Trigger levels and de-trigger levels of 4.0 and 2.0, respectively, are typical values. In order to get the complete event, the recording will start some time before the trigger time. This is called pre-event memory time. Likewise, to avoid the truncation of the signal, recording continues some time after the de-triggering. This time is called post-event memory time. The maximum time of recording can usually be limited by a maximum recording time setting and it is also sometimes possible to discard triggers lasting less than a certain time, the minimum trigger time. *STA* and *LTA* must be calculated on signals without a DC component in order to reflect the real signal. The DC is normally removed by filtering, which is desirable also for other reasons (see below). The *STA/LTA* trigger algorithm is well suited to cope with fluctuations of natural seismic noise, which are slow in nature. It is less effective in situations where manmade or natural seismic noise of a bursting or spiky nature (e.g. wind gusts) is present. At sites with high, irregular seismic noise, the *STA/LTA* trigger usually does not function well, meaning there are too many false triggers.

Before discussing the trigger parameters in detail, let us look at a typical implementation. The trigger must be running continuously in the computer, so the *STA* and *LTA* values are calculated as running averages:

$$STA_i = STA_{i-1} + \frac{|x_i| - STA_{i-1}}{NSTA}$$

$$LTA_i = LTA_{i-1} + \frac{|x_i| - LTA_{i-1}}{NLTA} \qquad (5.1)$$

$$R_i = \frac{STA_i}{LTA_i}$$

where x_i is the signal (filtered or unfiltered); *STA*, the short term average; *LTA*, the long term average; *R*, the *STA-LTA* ratio and *NSTA* and *NLTA*, the number of points in the *STA* and LTA windows, respectively. In this case, x is the unfiltered signal. This works if the signal has no DC component, but if x has a large DC component, this will be transferred to the *STA* and *LTA* and they might be so large that *R* never reaches the trigger ratio.

There are two common variations of the trigger algorithm (5.1). One is to use the squared x_i instead of the absolute value. Using the squared x_i, makes the trigger more sensitive to changes. The other is the Allen trigger (Allen, 1978), in which the absolute value of x_i is replaced by the characteristic function c_i . In a simplified form, c_i is calculated as:

$$c_i = x_i^2 + k \cdot (x_i - x_{i-1})^2 \qquad (5.2)$$

where k is a constant. The second term is proportional to the squared first time derivative of the signal. The characteristic function is sensitive to both amplitude variations and frequency changes.

All the above described trigger methods have in common that they are *not-looking-forward* algorithms, i.e., the present value of the trigger parameters depends only on present and past values of the signal. This makes them easy to implement with simple recursive, real time operations. A human analyst works in a quite different way: He detects an event looking at the entire record and evaluates the past, present and subsequent values of the trace (or something like the envelope of it) to decide the presence of an event and its onset time.

Figure 5.7 shows an example of how the algorithm *STA/LTA* works for the standard trigger using the absolute value.

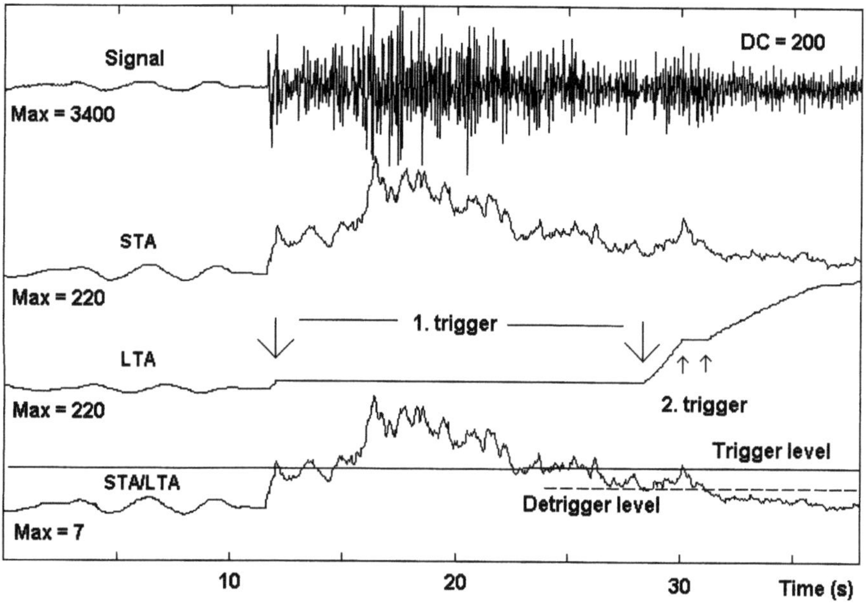

Figure 5.7 The seismic trigger. Top trace is the original signal; the second trace, the *STA*; the third trace *LTA* and the last trace the *STA/LTA* ratio. The *STA* window is 0.4 s and the *LTA* window is 50 s. The trigger ratio is 3 and the de-trigger ratio 2.

The original signal is a small high frequency earthquake with a medium level micro seismic background noise, which is reflected in the slow variations of the *STA* and more smoothed in the *LTA*. At about 12 s, the ratio gets above the threshold of 3 (1. trigger). It immediately gets below again, but since the de-trigger ratio is 2, the trigger remains activated. In the above example, the *LTA* is frozen once an event has been declared, which can be seen from the flat *LTA* level. A few seconds after the de-triggering, the ratio again rises above the trigger level and remains in trigger state for about 1 s (2. trigger). This signal has a good signal to noise ratio, so why has the trigger not been activated much longer? The first reason is simply that the DC has not been removed before calculating *STA* and *LTA*. Since *STA* and *LTA* before triggering are about 220, the event will de-trigger when STA is 440 and the maximum ratio is 7 (Figure 5.7 and Table 5.2). The DC of this signal is only 200 as compared to the maximum value of 3400; yet if it is not removed, it will have a large influence on the trigger performance.

Table 5.2 Effect of DC and filter on STA, LTA and STA/LTA. The units of the numbers are in counts.

	DC not removed	DC removed	Filtered 5 – 10 Hz
STA and LTA before trigger	220	40	3
Maximum STA	1500	1450	600
Maximum STA/LTA	7	36	200
Duration of trigger (s)	15	50	105

When the DC is removed, the figure corresponding to Figure 5.7 will look almost the same, but since both noise, *STA* and *LTA* are smaller due to the removal of the DC, STA/LTA will be much higher for the earthquake signal and the event will remain in trigger state for 50 s instead of 15 s (Table 5.2). The performance can be further improved by filtering. The signal has a typical micro-seismic background noise superimposed, and this gives the main contribution to the *STA* and *LTA* before triggering. By using a band-pass filter from 5-10 Hz, the performance is further improved (Table 5.2) and the trigger lasts for 105 s.

So, the conclusion is that every trigger must have an adjustable band-pass filter in front of the trigger algorithm, which has both the function of removing the DC and making the trigger algorithm sensitive to the frequency band of interest. This is particularly important with broadband seismometers, where small earthquake signals often are buried in dominant 0.2-0.3 Hz seismic noise and a trigger without a filter simply would not work except for the largest events. Some recorders allow several trigger parameter sets to be used simultaneously. This is needed if e.g. a BB station has to trigger on micro earthquakes, teleseismic P-waves and surface waves which each require separate filters, *STA* and *LTA*. The GSN Quanterra station operates in this way.

5.9 Summary of trigger parameters and their settings

STA time. Window length over which *STA* is calculated. Values can typically be from 0.3 to 0.5 s for local earthquakes, 0.5 to 2 s for regional events and longer for teleseismic P-waves. A short *STA* time makes the trigger more sensitive to short term variations in the signal, while a longer *STA* is better at averaging out short term fluctuations and it can be compared to a low pass filter. If there are spikes in the signal, a long *STA* must be used in order to average out the interference. This will of course reduce the sensitivity to short lasting signals, but if a longer lasting signal is the objective, *STA* can be as long as the duration of the main P-wave train like e.g.10 s. The most common mistake in setting the *STA* is to use a too short value.

LTA time. Window length (s) over which *LTA* is calculated. The main objective is to have the *LTA* short enough to adapt to slow changes of the background noise and long enough to avoid reducing the sensitivity to triggering on low amplitude emergent signals. Since the natural background noise usually changes very slowly, it is generally better to use a too long than a too short *LTA*-time, for example, 1000 s could be used. If

LTA is not frozen during triggering (see next section), the *LTA*-time must be at least as long as the event duration to prevent premature de-triggering. Using a too short *LTA*-time combined with a high trigger ratio might even prevent triggering completely since the *LTA* is adjusted upwards so fast that *STA/LTA* never exceeds the trigger ratio. A carefully selected *LTA*-time might prevent triggering on slowly emergent man made signals like a truck passing, which typically can last for 2 minutes. Selecting a *LTA*-time of 30 s might prevent triggering on the truck signal, while still retaining sensitivity to more impulsive signals. However, it might also prevent triggering on other real signals. In general it is recommended to use a long *LTA*-time window in order to avoid reducing sensitivity to low amplitude emergent P-waves. 200 s is a good value.

Freezing LTA. This is the option of keeping the *LTA* at its pre-trigger level during detection. The advantage is that STA/LTA will, during the event detection, reflect the true *STA/LTA* of the earthquake signal and the trigger will terminate at the true de-trigger ratio. If the *LTA* is not frozen, it is continuously increased and the de-triggering will occur too early. Having a very long *LTA*-time can partly prevent this. The disadvantage with a frozen *LTA* is that a sudden increase in background noise that triggers the system will keep it in trigger state forever. This can be prevented by putting a limit on the trigger time after which triggering is terminated and *STA* and *LTA* reinitialized. It is recommended to use a frozen *LTA* in combination with a maximum recording time in order to avoid prematurely truncated recordings.

Minimum trigger ratio or trigger threshold. Triggering starts when *STA/LTA* exceeds this value. The higher the ratio is set, the fewer events are recorded. A too low value results in many false triggers, so the minimum trigger ratio is one of the most important parameters to set for optimum performance. The ratio can be lower at low noise sites (meaning few impulsive disturbances). The most common values to use are between 3 and 6. Values higher than 8 are rarely used, except for strong motion recorders, which often are in noisy places and only need to trigger on strong signals. Note that a high ratio combined with a short *STA*-time can prevent triggering.

Level trigger: The minimum absolute level for triggering. The level trigger and the STA/LTA trigger can, in some systems, both work at the same time.

De-trigger ratio. The *STA/LTA* level at which an event is declared finished. The smaller the de-trigger ratio is, the more complete is the record. A high value will cut the recording short. This can partly be compensated by the setting of the post event memory. A low value might, in combination with a frozen *LTA* and a noise increase, result in very long recordings. This can be prevented by the setting of the maximum recording time. Recommended values are in the range 1.5 to 3.

Pre-event memory time. The number of seconds the recording starts before the trigger time. Since data is digitally recorded, there is always a buffer in memory of past data, so, when the event is written to disk, the recording starts earlier that the trigger time. Even for the most impulsive signals, the trigger time will be delayed relative to the very first onset, so if recording started at the trigger time, the initial onset would be lost. For emergent signals, the initial trigger might be several seconds into the record or as late as the S-phase. There is furthermore a need to have a reasonable noise record in front of the first onset in order to e.g. evaluate signal to noise ratio. So a pre-event memory time must always be set. The value should be large enough to get the P-phase and a noise sample in front if the triggering is on the S-phase. For regional earthquakes, a noise window of 30 sec is reasonable and if typical S-P time is 30 s, pre-event memory time should be 60 s. Since storage space is less of a problem than before, it is advisable to use a longer time window than strictly needed.

Post event memory time. The recording time after de-triggering. This parameter enables the recording to catch the end or the coda of the earthquake signal. Since it is a fixed length, it will be more effective for small earthquakes than large earthquakes where low-level coda waves tend to continue for long with a relatively constant level. Obviously making it very long could catch the end of any signal, but might make too long records. Typical values are in the range 30-60 s.

Maximum recording time. The maximum time a recording is allowed to last. This parameter limits the recording time in case of a continuous noise increase and the use of a frozen *LTA*. It is also useful for limiting recording time if we only are interested in the beginning of the signal. A typical value for small earthquakes could be 50 s and 500 s for regional events.

Minimum trigger time. The minimum time a trigger must last in order for the event to be recorded. This parameter can cut out small events. A typical value to use if we are only interested in regional data is 30 s. A long *STA* can also be used for this purpose, but might prevent detection of signals starting with an impulsive P.

Filters. The filter to use before calculating *STA* and *LTA*. The filter is usually a band-pass filter and it must at least be a high pass filter to remove the DC. The filter can be an effective tool to make the recorder sensitive to the signals of interest. Both FIR and IIR filters are used (see digitizer section). The simplest filter that can be made is the difference filter:

$$y_i = x_i - x_{i-k} \tag{5.3}$$

where y is the filtered signal, x the input signal and k is 1, 2, 3 .... By taking the difference between subsequent samples (k=*1*) or between samples further apart (k=2,3...), we have a very effective way of removing the DC. This very simple FIR filter not only removes the DC but also provides some measure of band pass filtering since it has the maximum gain at frequencies *1·rate/2k, 3·rate/2k, 5·rate/2k* etc. and zero gain at frequencies *0, 1·rate/k, 2·rate/k* etc. where *rate* is sample rate, see Figure 5.8.

Preset recording times. Setting of a particular start time and duration of recording. This is useful for refraction experiments and for collecting noise window samples at predefined times.

Minimum number of triggers. The minimum number of channels triggering within the array propagation window (see next) required for the recording of an event (also called coincidence trigger). If several stations in a network are connected to a recorder, it is to be expected that several will trigger within a certain time window. Therefore, requiring a minimum number of concurrent triggers can drastically reduce the number of false event detections. For a 10-station network, requiring 3-5 stations to trigger might be a reasonable number. Most systems can be set up so only selected stations in a network participate in the triggering, since some stations might be more prone to noise than others.

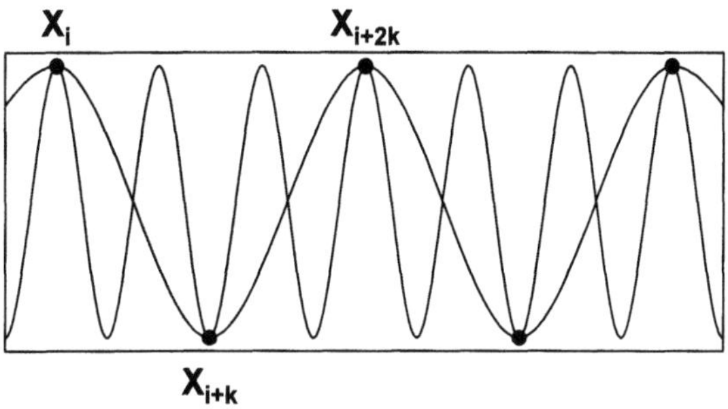

Figure 5.8 The difference filter. A 1 Hz and a 3 Hz signal is shown. Subtracting sample x_{i+k} from x_i gives constructive interference using the 1 Hz signal sampled at 2 samples/second or multiple of this, while subtraction x_{i+2k} from x_i gives destructive interference. The same is true for the 3 Hz signal.

Array propagation window. The maximum time interval between trigger times for different channels in order for the recorder to trigger. If a multi channel recorder is connected to different stations in a network, each channel will trigger at different times due to the propagation of the waves across the network. The array propagation window should then reflect the travel time across the network. Considering that the stations might trigger on S-waves only, the time should be calculated using the average S-wave velocity. For a network with a diameter of 100 km, the array propagation window should then be around $100\text{km}/3.5\text{kms}^{-1} = 29$ s.

Note: Since many recording systems now also have ring-buffers, the exact settings of the trigger parameters is less critical, since missing or cut data can be recovered later, either manually or automatically based on triggers from other stations or known location and origin times. However, even continuously recording systems benefit from a good trigger in order to fish out the real event data.

For more detail of how to set and use STA/LTA triggers, see Trnkoczy, information sheet 8.1 in NMSOP (Bormann (Ed.), 2002).

5.10 Communication and data retrieval

Communication with the recorder has two purposes: (1) Check operation and set parameters, (2) Download data.

Check operations. The most common interface to use is the serial line (RS232) either directly or through a modem, and connects to another computer, mostly a PC. Usually, a particular recorder requires a specific program to communicate with it. The main operation is to check functionality of the recorder and set parameters. Unfortunately, each recorder type has its own interface, so a special communication program is needed for each recorder, and most of these programs only work on a PC. Some systems allow a general login, which can be done both locally and/or remotely with a terminal emulator, so no recorder specific program is needed on the connecting computer. This normally requires that the recorder runs a multi-user operating system. The Quanterra recorders work this way. Newer recorders will also have an ethernet TCP/IP connection, which might be used for general system check.

Some recorders have some kind of built in display to check operation and some also have built in keyboard for parameter setting etc. In the simplest case, only LED's are used to indicate status. If the recorder consists of a software running on a general purpose computer, the console and keyboard is usually the main device for checking and parameter setting.

Download of data. Nearly all recorders have a way of downloading data via a communication port. In smaller, solid-state memory only based recorders, the normal communication is also by the serial line. This is simple and cheap, but since memory sizes have increased, it might also be a slow process. The highest speed on the serial line might be 115 Kbaud or about 11.5 Kbytes of data per second (see 8.4.1). So downloading 10 Mb will take about 15 min. Using an ethernet TCP/IP connection of 100 Mbit/s, it should theoretically take 1 s. However, small microprocessor based recorders usually do not have ethernet. Common to many recorders is that a special program is needed on the connecting PC in order to download the data and manufactures of recorders have shown moderate interest in creating a standard interface. This makes it very hard to automate the download process if more than one type of recorder is used in a network. Data acquisition based on general purpose computers is much easier to use for downloading data, since general purpose software like ftp can be used for copying recorded event files. The trend is to use general purpose software.

A very simple way of downloading data from a recorder consists of displaying, on the screen, the recorded data in some ASCII format and capturing it in a file on the receiver side. This requires no special software. Almost no portable recorders have this ability, but it is built into the GSN system of Quanterra recorders and also available on other public domain recording systems. As the trend is to use more general purpose computers for recorders, more general purpose communication facilities will probably also be available.

The future trend is to use the much faster USB (Universal Serial Bus) for both general communication and download of data and USB will probably replace the RS232 interface. New faster ports and protocols, such as the wireless interface blue-tooth, may be available in the near future as standard.

Download of data via a communication port might not always be practical, because of either the time and amount of data involved or the practicality of connecting to a recorder. Removal storage media are here the solution. Until very recently, tapes were used, but they have now been replaced by removable disks, CD's, hard disks or magneto optical disks (little used) and flash cards (solid state memory). The trend is to use only hard disks or some kind of solid state memory (sometimes called solid state disk and actually acting as a disk). There are many technical solutions to this problem. One of the key issues is whether the disk is hot-swappable, meaning the disk can be removed while the system is running and a new disk put in its place, such that no data is lost. Such systems are available using different technical solutions of which PCMCIA disks and, as the latest possibility, USB connected disks are options. The user might wonder if a few minutes of lost data is really that important but it is clearly used as a selling point for new recorders.

5.11 Public domain data acquisition systems

Many institutions have developed data acquisition systems for their own use using general purpose computers, but few are well documented and tested. Probably the best know is the W. Lee system (Lee, 1989) distributed by IASPEI software libraries. The system works on a PC and can work with both a local ADC card and also with a stream of digital data arriving in real time from remote digitizers. Although limited by the MSDOS operating system, the system has seen widespread use due to being appropriate for many users and it is well documented.

The newer EARTHWORM is now gaining wide popularity (Malone, 1996). EARTHWORM is being developed by several institutions, is very flexible and works equally well on two operating systems, NT and Solaris. It is clearly the most sophisticated public domain system available. The system has data acquisition modules as well as data processing modules. However, EARTHWORM is not as well documented as W. Lee's and requires good computer knowledge to set it up.

The SEISLOG system (Utheim et al, 2001), which was initially developed for the Norwegian National Seismic Network (Havskov et al, 1992), is running on Intel PC's under the real time, commercial Unix-like real-time operating system, QNX, under Windows and under Linux. SEISLOG has been in use for 15 years, and is installed in many countries. The system capabilities and sophistication are probably somewhere between the W. Lee and the EARTHWORM systems. SEISLOG use all the parameters described under trigger parameters and can use about 10 different digitizers and ADC cards. Documentation quality is similar to the W. Lee system.

Figure 5.9 shows an example of the Windows Seislog monitor screen.

The example shows a 3 channel system. The activity monitor shows which channels are entering the system with channel number, name and time of last data received. The In/Disk/Trig LED's indicate if data is received or not, if data is written to ring buffer and whether a channel is in trigger state. If so, one of the 5 network triggers might also be active if e.g. 2 triggers occur within a given time window. The current sample value, *STA* and *LTA* are shown to the right. The log window below gives error messages, start and stop times, GPS status and trigger messages. The window to the lower right shows the signal for one of the channels, several can be shown at the same time. Below is shown the current status of GPS synchronization.

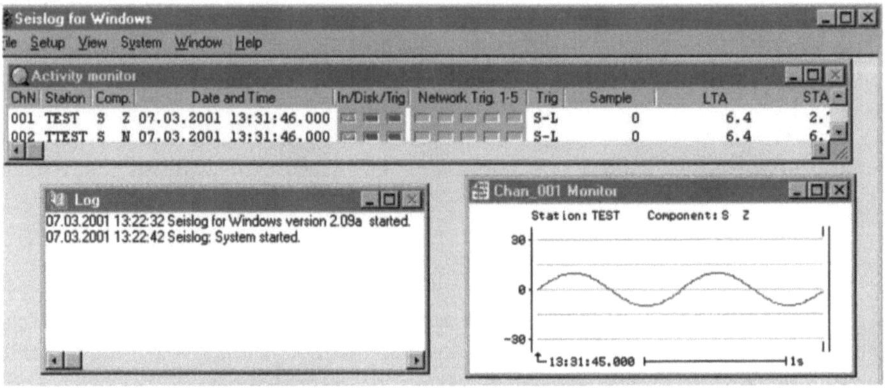

Figure 5.9 The SEISLOG system, see text.

The three public domain systems mentioned above, can all be used for most general purpose data acquisition. Probably the future will see more and more use of public domain systems, since low power general purpose computers become easily available. The role of the manufacturer will then be to provide the hardware, which is a specialist job, particularly with respect to watertight assembly and assured operation under extreme climatic conditions.

5.12 Seismic recorders in use

Recorders are generally divided into stationary and portable and each has different requirements.

Stationary recorders are generally made with general purpose computers and the software used is more often public domain than commercial. There are no special requirements to the environment and the software is the main point making them different. The ADC is often a separate unit although for multi-channel networks with up to 16-bit resolution, a 16-64 channel card might be built in.

A separate class of stationary recorders are accelerographs. Whereas they used to be special products with only level triggers, they now tend to have the same specifications as portable seismographs with solid state memory, complete with GPS timing and communication capability, and due to their high dynamic range and sensitivity they can equally well be considered portable seismographs. An important distinction is that most accelerographs have built-in sensors, which is unusual in seismographs, due to the lower noise requirement for the latter. The trigger system might also be more primitive with only a level trigger (examples in Table 5.3). Another difference is that the sensitivity is often too low to connect a standard velocity transducer directly. E.g. the

Kinemetrics Etna (Table 5.3) has a LSB of 8 μV, which is not enough for a sensor with a sensitivity of typically 200 V/ms^{-1}, while it would be ok with an active sensor with 2000 V/ms^{-1}. Most accelerographs can also have the sensors externally connected.

Portable recorders face tough requirements to environmental protection like being able to work in humid and extreme warm and cold conditions. For prolonged operations, they must be able to work with low power from either solar panels or batteries. Thus, a major part of the cost of producing field equipment is these special requirements. Some lower cost portable recorders are made with less stringent requirements and require more care in deployment. However, not all recorders need to be able to work from –40 to 50° C, so the user should select what is appropriate for local requirements.

A special class of recorders are Ocean Bottom Seismographs (OBS). They essentially have the same requirements as other portable recorders, however there is no need for GPS and data communication at the bottom of the sea! Since no time synchronization can be done, it is important that the built in real time clock is accurate (typically 0.01ppm) and has a linear drift such that accurate timing can be obtained by synchronizing the OBS before deployment and measuring the time error after deployment. Requirements for casing (often a glass sphere), deployment and retrieval are very special. The sensors are often gimbaled so that when the OBS is posed inclined on the seafloor, the sensors will be oriented correctly in the vertical direction. The horizontal direction will be unknown.

A typical scenario for deployment is that the OBS is mounted on a metal frame, which acts as a weight to propel the OBS to the sea bottom and make sure the bottom part is down. Some OBS's have the sensors within the housing while other have a separate sensor housing which is posed separately on the seafloor a couple of meters away in order to reduce the noise. When the OBS is to be recovered, an acoustic ultrasonic signal is sent to the unit, which then releases the weight, and the OBS will rise to the surface. Depending on the water depth, this might take up to one hour since the speed is only around 1m/s. Once at the surface, the OBS will send out radio and/or light signals to help finding it again. On average, 10 % of OBS's are lost in a typical field experiment. Due to the high cost, special requirements and low production number, most OBS' are produced by government institutions. Figure 5.10 shows a typical OBS.

Figure 5.10 Example of an OBS. To the left is shown the key elements of the OBS and to the right is seen an OBS starting its journey to the bottom of the ocean. A flasher light is attached to the left side, a radio beacon to the right and the acoustic release equipment to the rear side. The metal bars on the bottom are the anchor left on the bottom after deployment. This OBS was developed at the Hokkaido University, Institute of Seismology and Volcanology, Japan. Figure from Department of Earth Science home page, www.geo.uib.no.

5.13 The next generation recorder

We will here describe portable recorders since fixed recorders in non harsh environments will probably be based on general purpose computers and public domain software in much the same way as today. In terms of software this might also be true for some portable recorders, but these have in addition other requirements. We will concentrate on the functional requirements, since housing, environment and connector requirements is more or less as made today. The main changes are associated with the development of more powerful low power computers and storage devices.

Based on standard computers and operating systems. This is where the main change will take place and the recorder should use off the shelf computers and operating systems. This will make it much easier to update hardware and software. For the smallest units, more specialized computers might be used. A new trend might be to use a mobile phone with built in CPU, removal storage and TCP/P connectivity.

Modularity. Ideally, the system should be built in a way that different digitizers and computers can be used, so that when a new and lower power computer becomes available, it is just a question of replacement. A problem is that one manufacturer might not be interested in producing units to be used with equipment from other companies. It is also a problem then, that it might be harder to construct some of the very small portable units now available.

Data storage. Solid state memory like flash disks.

Communication and data transfer. There must be a serial line for control and modem communication (will probably be replaced by USB), a 100 MHz ethernet and a removable disk (flash or HD). The communication software is supposed to be standard and part of the operating system. The recorder should be able to send digitized data in real time over TCP/IP or the serial line for central acquisition. Wireless LAN will be used to go to field and download data without connecting a cable or even being near the field recorder.

Sample rate, dynamic range and sensitivity. Sample rates from 1-1000 are desirable, dynamic range should be at least 22 bit and LSB resolution of 0.1 μV.

Data acquisition software. The software should have all the possibilities as described under the seismic trigger and ring buffers. It should be possible to connect to the system and download without stopping acquisition. Recording format should be public and easily readable on PC and UNIX systems.

Power consumption. Should be below 2 W.

We can see from Table 5.3, that some of the recorders on the market now fulfill many of these criteria. The Quanterra has all the flexibility of multi user operating system, but is not modular and low power. The Lennartz system is low power and compact but no communication facility and only 20 bit. The Earth Data recorder has the potential to fulfill most of the criteria. The Nanometrics Taurus is one of the newest generation designs with nearly all desired features and very low power consumption and several other recorders are also near a 1 W power consumption. These are specification values, and the real consumption will depend on sensors used and how much the GPS is needed. On the other hand, some public domain acquisition software has many of the desired options since they were always developed for general purpose computers and this software might potentially be used with future recorders.

5.14 Examples of recorders

In the following, some examples of recorders will be given and an overview of many recorders is shown in Table 5.3. The intention is to give representative examples of different kinds of recorders, like low power, high and low dynamic range etc. All recorders have GPS timing unless otherwise mentioned.

Quanterra, the reference recorder. The Quanterra 24 bit recorders are the backbone of the global seismic network (GSN). They have for long been the standard against which other recorders were measured and no description of recorders is complete without a description of Quanterra. These recorders have always been based on the VME bus, Motorola 680xx type processors (also used in early Sun computers) and are using the commercial real time operating system OS9. This a multi tasking and multi user operating system, which, as described earlier, gives the most flexible system. There has thus always been a simple and consistent way of downloading data from Quanterra recorders without using any special software on the receiving side, and communication can be by either serial line or TCP/IP. This is of course a natural requirement for a station that is open for any seismologist, and an example to follow for other manufacturers. Recording is on hard disk and ring buffers will automatically be dumped on DAT tape when the disk is full or when data has to be transferred. The recorder is not portable and uses a lot of power. The system is not very user friendly by today's standard, but has provided many years of standard operation. The software is comparable to available public domain data acquisition software (see above). The technical solution with 680xx processors and the OS9 operating system is outdated and will probably soon disappear. Kinemetrics has taken over Quanterra and is still selling a newer version of the system, the Quanterra Q4120, which still uses the OS9 operating system and a 68030 processor.

Lennartz-electronic MARSlite/HD, small and very low power. The recorder (Figure 5.11) has a 20 bit dynamic range and 4.3 Gb SCSI recording capacity. With the GPS in 3-hour power cycling mode, the power consumption is around 1 W. The unit was until recently one of the smallest and with the lowest power consumption on the market. Operation can be done without connecting an external computer since the unit has a small keyboard and a display.

Figure 5.11. The Lennartz MARSliteHD recorder. Below the display is seen the 3.5" extractable hard disk. The model with magneto-optical recording is less than half the size (bottom part missing). From Lennartz home page, www.lennarz-electronic.de.

The data can only be downloaded by connecting the disk to another computer and there is no communication facility, so the unit cannot be interrogated remotely. The MARSlite MO with 0.5 Gb magneto-optical disk is less than half the size and one of the smallest on the market.

GeoSIG GBV316 seismograph, small, built in geophones and low price. The GBV316 is very compact complete 16-bit seismograph with built in 4.5 Hz sensors and battery (Figure 5.12). External sensors can also be connected. The recording capacity is up to 64 Mb, which at 100 Hz sample rate gives 5 days of continuous recording, although it is more typically used with a smaller memory and in triggered mode or recording time windows. With GPS it consumes typically 3 W. Communication is by the serial line, can work locally or remotely and is also used for data download. The seismograph is not watertight and uses standard computer connectors. The specifications are thus less stringent than for mainstream recorders, however, the price is correspondingly lower. The combination 4.5 Hz sensor and 16 bit recording maximizes the performance versus price since the lower sensitivity at 1 Hz is partly compensated for with a higher ambient noise level, such that the overall useful dynamic range for the seismic signal is comparable to an 18 bit recorder with a 1 Hz seismometer.

Figure 5.12 The GeoSIG compact seismograph with 3 built in 4.5 Hz sensors. Note the standard computer connectors. In the middle on the lid is located a display giving alphanumerical information and 4 LED's give operational information. This is the most compact complete seismograph on the market. Figure made available by Christoph Kündig, GeoSIG.

Earth Data PR6-24 - new generation. This recorder offers 25 bit resolution at 25 Hz, one of the best resolutions on the market. It is compact (Figure 5.13) but has no internal battery. The PR6-24 belongs to the new generation of recorders based on a standard PC104[4] PC board with a standard operating system (Linux) and therefore has all the standard facilities available in Linux. The user can also put in additional software. No special communication protocols have to be made unlike most other recorders. The low power consumption (1.8 W) has been achieved by stopping the PC while data accumulate in the ADC. The PC then wakes up when the ADC is full or when external communication is initiated.

---

[4] PC104 is a standard BUS (set of connections), with both physical (size and shape) and electrical definitions. It is widely used for PC CPU and peripheral cards for control industry. There exist a number of complete single card PC's with this format, with different performances on speed, power consumption, etc.

Figure 5.13 The Earth Data PR6-24 recorder. The length is 25 cm. Figure made available by Tony Russell, Earth Data.

Kinemetrics Etna – typical accelerograph. This a popular 18 bit unit (Figure 5.14) with a built in Episensor FBA (see Table 2.4 in sensor section). It records on PCMCIA cards (memory or disk) so more than 100 Mb is available. This makes it easy to take out data. A laptop must be connected for configuration. It has a custom made multi tasking operating system making it possible to interrogate the system while acquiring data. The serial line connection makes it possible to download data locally and remotely as well as setting of parameters. Download of data can be automated with Kinemetrics software. Power consumption is 3.5 W with GPS in continuous operation and using the built in sensor.

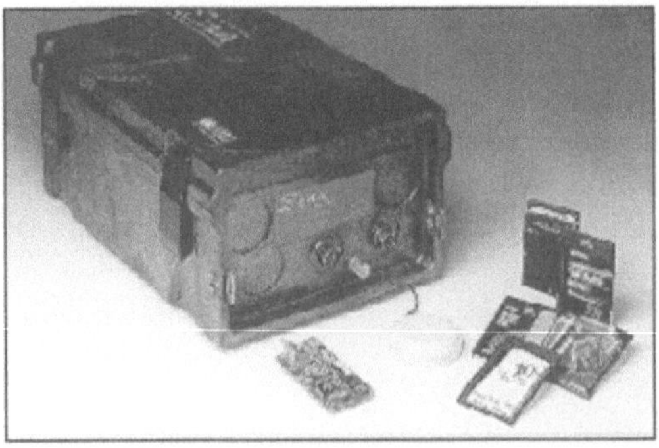

Figure 5.14. Example of a recorder made specifically for strong motion, the Kinemetrics Etna. The round white box in front of the recorder is the GPS antenna and the PCMCIA cards are seen to the right. From Kinemetrics home page www.kinemetrics.com.

RefTek Texan – very small – lowest power- one channel recorder. This one channel, 24 bit recorder (Figure 5.15) with 64 Mb of flash memory is specially made for refraction surveys but can equally well be used for seismological studies. Without GPS the timing accuracy is 0.1 ppm, which e.g. for aftershock studies would be sufficient. The recorder can work with two normal D-cell batteries or external power. A special playback unit is needed to program the recorder and download data. The recorder only has timed or continuous recording.

Figure 5.15 The RefTek Texan one channel recorder. From RefTek home page, www.reftek.com.

Nanometrics Taurus, next generation nearly handheld recorder. This recorder (Figure 5.16) has all desirable inputs and outputs, display and keys so it can be set up in the field without a laptop. The power consumption is only 0.7 W for 3 channels and 100 Hz, it has a web interface and uses Linux.

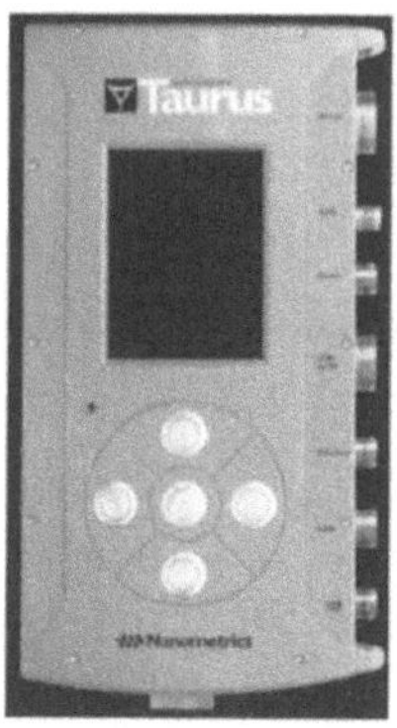

Figure 5.16 Nanometrics Taurus, the next generation handheld recorder (25 x 15 x 6 cm). From Nanomtrics home page, www.nanometrics.ca.

Table 5.3 lists the main characteristics of a number of seismic recorders.

Table 5.3 Overview of seismic recorders. Only the most important parameters have been listed. Some recorders are primarily intended as strong motion recorders and usually have sensors built in. However, they can also be used as a recorder with an external sensor. Abbreviations are:
CHA: Number of channels.
Bit: Number of bits (the true noise-free dynamic range), and an approximate value of noise (μV) with highest gain in standard equipment. This value is rarely given so it has been calculated from other given values and the number are only approximate.
Rate: Sample rate (Hz).
TCO: Free running clock accuracy in ppm.
Storage: Type of storage media, H: Hard disk, M: Memory, usually flash, minimum is usually 64 Mb, T: Tape, P: PCMCIA.
Communication: S: Serial line, T: Ethernet TCP/IP, D: Display, K: Keyboard, U: USB, O: Other type of communication.
Trig: Trigger type: S: STA/LTA, T: Timed windows, C: Continuous, L: Level.
Pow: Power in W without GPS and sensors and using 3 channels if 6 channels is an option.
Env: Environment. P: 100% humidity protection.
Wt: Weight (kg) is with built in batteries if normally delivered and excluding sensors, this data can be a bit uncertain.

Name	CHA	Bit	Rate	TCO	Storage	Communication	Trig	Pow	Env	Wt
Strong motion recorders with built in accelerometers, most can also be used with seismometers										
Geodev. GSMA2400	3	22 5	100-500	0.5	M	S	SLT	1.3	P	10.0
Geotech DS-2400	3-6	22 3.8	10-1000	0.5	P	SD	ST	6.3	P	16.0
Geotech LLC	3	16 14.0	200	0.5	M	S	L	3.3	P	24.0
Kinemetrics QDR (1)	3	11	100	57	M	S	L	1.0		0.1
Kinemetrics Etna	3	18 8.0	100-250		P	S	L	2.2	P	9.0
KinemetricsMtWhit.	18	19 3.5	100-200		P	S	L	18	P	68
GeoSIG IA-1 (2)	3	16	50-300		M	T		5.7		3.3
GeoSIG GSR18	3	18 5.0	100-250	20	M	SD	SL	1.6	P	7.2
General purpose recorders										
Agecodagis Titan	3	20 2.4	20-250	0.5	HM	S	SC	2.5		4.0
Agecodisis Osiris (3)	>3	22 5	1-2000	1.0	HPM	ST	STC	1.0	P	3.0
Datamark LS-000XT	6	23 2.5	1-200		M	ST	SLC	1.2		1.0
Datamark LS-8200SD	1	22	125-1000		M	U	LC	0.5	P	1.5
Earth Data PR6-24 (3)	3-6	24 1.0	1-1000	1.0	H	STU	STC	1.8	P	3.5
ES&S Kelunji D	3	22 0.6	100-200	0.1	P	STR	STCL	2.8	P	9.0
Eentec DAS6102	>3	22 1.2	1-2000		HP	ST	SC	5.0		5.0
Eentec SMR-4000	>3	23 0.6		25	MP	STKD	S	2.0	P	5.0
Geodev. EDAS24M	3-6	22 5	50-500	0.5	M	S	SLT	1.3	P	10.0
Geodev. GSMA2424	24	22 5	100-500	0.5	M	S	SLT	2.2		10.0
GeoSIG GBV316 (4)	3	16 0.08	25-200	10	M	SD	STCL	1.2		5.2
GeoSIG GSR24	3	22 1.0	50-200	20	M	SD	SL	1.7	P	7.2
Geotech SMART24R	3-6	23 2.2	1-2000	0.5	HM	STU	SC	0.8	P	3,9
Geotech DL24	3-6	22 4.8	10-1000	0.5	HP	SD	STC	6.3	P	11.3
Güralp CMG-DM24	3-6	22 4.8	4-200	0.8	M	S	S	1.7		
Güralp CMG-SAM			(5)		H	S	C	1.1		6.0

Kinemetrics, K2	3	19 0.7	20-250		P	S	SLC	4.2	P	10.9
Kinemetrics Makula	3-6	22 5.0	20-250		P	ST	SL	4.2	P	10.9
Leas Hathor 3	3	19	100-300		H	S	SLT	0.8	P	4.5
Lennartz MarsHD	3	20	4-250		H	DK	SLC	1.1		6.5
Lennartz MarsLite	3	20 0.13	4-250		HM(6)	DK	SLC	1.1		6.5
Nanometrics, Orion	3-6	22 0.20	10-1000	0.4	HT	SDK	SLTC	5.2	P	10.9
Nanometrics Taurus (3)	3	24 1.0	10-500		HM	STUD	STC	0.8	P	1.8
Quanterra, Q4120	3-8	24 0.1	1-1000	1.0	HT	ST	SC	30		
Qanterra Q380	3-6	24	1-80		HT	STDK	SC			
RefTek 72 A	3-6	22 0.06	1-1000	0.5	H	S	SLTC	4.6	P	13.0
RefTek, Texan	1	22 0.2	25-1000	0.1	M	O	TC	0.4	P	1.1
RefTek 130-01		23 2.4	1-1000	0.1	MH	ST	SLTC	1.0	P	2.0
Sprengnet. DR3024	3	22 1.2	1-500	0.2	P	S	SLC	4.4	P	12.5
Sprengnet. DR4000	3-6	22 1.2	.1-4000	0.5	H	STDK	SC	5.0	P	4.5

(1) No external timing, internal clock only
(2) Timed by internet
(3) Linux operating system
(4) Built in 4.5 Hz geophones
(5) Data collection unit receiving data from digital sensors
(6) Magneto-optical disk

All recorders except QDR, GeoSIG IA-1 and RefTek Texan have GPS. However, GPS power consumption is not included in total power consumption, since most GPS can be cycled on and off to save power. Depending on the user selectable on-off, the additional power for the GPS can range from 1 W (100 % on) to 0.1 W when turned on less than 10 % of the time. The cycle time depends on the accuracy needed coupled with the accuracy of the free running clock. Example: If the free running clock has an accuracy of 1 ppm and the required maximum deviation is 1 ms, then the clock may drift 3.6 ms an hour. This means that the GPS must be turned on at least four times an hour so the GPS will be on at least 10 % of the time considering turn on time.

The recording formats are mostly company specific; however the trend is that newer recorders use Miniseed. PCMCIA has capacity from 1 Mb to 1 Gb. Not all recorders can accommodate large memory units, but most will be able to use 64 Mb. Memory capacity is fast expanding and some recorders have up to Gb of flash available.

The comparison between recorders shows that many recorders are very similar and different recorders from the same company are nearly identical with maybe one being 18 bit and another 22 bit. The so-called strong motion recorders are in some cases very similar to seismic recorders complete with STA/LTA trigger. In general, seismic recorders have LSB below 1 μV while strong motion recorders are above 1 μV. Most recorders have RS232 communication but only a few have TCP/IP. The trend is that most new recorders have TCP/IP. Note also that there is a large difference in the timing accuracy without GPS ranging from 0.1 to 57 ppm.

5.15 Which recorder to choose

The selection of a recorder is both a question of money and needs, and there is no unique answer to this question, only some general advice can be given. Recorders tend to have a lot of 'bells and whistles' with new models coming out with new features and all might not be needed. The most important issue is to select the right combination to do the job in hand. Some points to consider:

Type of data to record, sensor to use and sensitivity. This is the logical first question. If the intended study is small earthquakes or refraction experiments, there is no need for a broadband sensor and in many cases (down to 0.3 Hz) a 4.5 Hz geophone is good enough. Consider that for the price difference between one 3 component 1 Hz sensor and a 3 component 4.5 Hz sensor, it is possible to buy the cheapest recorder! However, a 4.5 Hz sensor cannot be used with all recorders since the LSB should be below 1μV, ideally less than 0.1 μV in quiet areas (Table 5.3).

Dynamic range. Should be as high as possible, however that comes at a price. The broader the frequency range of the sensor, the larger the dynamic range should be in order to account for the large increase in natural background noise with decreasing frequency (see seismic noise section). For a 4.5 Hz sensor, use at least 16 bit, for 1 Hz at least 20 bit and for a broadband sensor at least 22 bit. It would a waste of money to connect a broadband sensor to a 16 bit digitizer! New recorders on the market will probably not have less than 22 bit dynamic range.

Timing. Normally a GPS is connected, so timing should not be a problem. The free running clock should have as high accuracy as possible within reasonable cost. More than 10 ppm is not acceptable.

Storage. For recording aftershocks and micro earthquakes, 50 Mb goes a long way, but for continuous recording, a hard disk/PCMCIA/flash card must be used. At least 1 Gb is needed.

Communication and data download. As mentioned earlier, download by serial line is not practical for more than 5-10 Mb of memory so for a larger memory other ways of getting out the data must be available. This can be TCP/IP, USB, FireWire or removable disk or memory. There should normally be a possibility to connect a laptop for parameter setup and/or remote communication via modem and/or TCP/IP.

Trigger. Make sure the recorder has STA/LTA trigger with adjustable filters, a level trigger is just not good enough except for very low sensitivity strong motion recorders.

Power. Low power usually means higher cost, so buying a cheaper system might mean changing batteries more often or having large solar cells. A 120 Ah battery, about the largest to carry around, might effectively give 80 Ah, so it would deliver 6 W during one week. This is ok with most recorders, but if the same battery has to last one month (1.5 W), there are fewer general purpose 3 component recorders available. The extra cost of saving a few watts might not be worth the money compared to extra cost and work of changing batteries more often. However, in some cases there is no alternative to using low power. The trend is for more low power recorders to be available.

Environment. If the recording is taking place in an office environment, there might not be any reason to buy a recorder at all, since a good digitizer, a PC and a public domain software might make a better system than most recorders. A good 3 channel digitizer with GPS costs from $4000 so together with a $1000 PC, a complete recorder is obtained for $5000 where the commercial equivalent would be $8-12000. The same can of course be done in the field with a laptop, however laptops consume too much power and are not really suited for being left under field conditions. So, generally, in the field, better protected recorders are needed. It is often a big selling point to be able to put a recorder directly in the open air under any conditions, but rarely done. It is often possible to protect the recorder, so less stringent environmental specifications are needed. However, for long term operation in tropical environments, a good protection is indispensable.

Software. Most recorders come with some software for download and conversion to a suitable format. Some even come with software for limited analysis in the native recorder format. Such analysis software is often more a selling point then really useful, since it can only be used with that format. Therefore, the most important offline software, which must come with the instrument, is a conversion program to a public domain format like SUDS, GSE, MSEED, SAC or SEISAN format.

Price. Complete recorders with sensor cost from $3000 to $15000 with 3-component geophone/short period/acceleration sensors and about $10000 more with broadband sensors. So, one broadband station has the same price as eight of the most economical stations. If money is limited, there can therefore be some important tradeoff between number of instruments and quality. If the objective is to locate earthquakes, a minimum number of stations is needed so instead of e.g. buying two broadband stations at $50000 it might be better to buy six medium quality stations with 1 Hz sensors or one broadband station and eight of the lowest price recorders.

CHAPTER 6

Correction for instrument response

Recorded signals, whether analog or digital, rarely give us the real ground motion, even using a constant factor. The velocity sensors will give output proportional to velocity for frequencies above the instrument natural frequency, however below the natural frequency, there is no such simple relationship (see Chapter 2). Furthermore, seismologists generally want to measure the ground displacement and one very important task is therefore to recover the ground displacement from a given recorded signal. This is also called correction for instrument response. We have seen in the sensor section that, for a given instrument, the amplitude frequency response function can be determined such that for given harmonic ground displacement $U(\omega)$, the output $Z(\omega)$ can be calculated as

$$Z(\omega) = U(\omega)\, A_d(\omega) \tag{6.1}$$

where $Z(\omega)$ can be the amplitude on a mechanical seismograph, voltage out of a seismometer or amplifier or counts from a digital system, and $A_d(\omega)$ is the displacement amplitude response. In order to recover the displacement, $U(\omega)$ can simply be calculated as

$$U(\omega) = Z(\omega)/A_d(\omega) \tag{6.2}$$

$A_d(\omega)$ has traditionally been called the magnification since for an analog recorder, (6.2) gives how many times the signal is magnified. If e.g. the ground displacement is 0.1 mm and the magnification 100, the amplitude on the paper seismogram will be 100 · 0.1 mm = 10 mm. Similarly, if we want to determine the ground velocity, we would get

$$\dot{U}(\omega) = Z(\omega)/A_v(\omega) \tag{6.3}$$

where A_v is the velocity amplitude response. This simple single frequency instrument correction has been widely used for determining maximum ground displacements from analog seismograms in order to determine magnitude. The measure assumes that the signal is nearly monochromatic, which often was the case for the older type narrow-band instruments. Figure 6.1 shows typical magnification curves for the standard WWSSN (World Wide Standard Seismic Network) short and long period (LP) seismographs. For the short period, the maximum gain or magnification near 1 sec is about 70 000 times, while for the long period it is about 3000 times near the 20 sec period. Typical gains used in practice are 50000 and 2000 respectively.

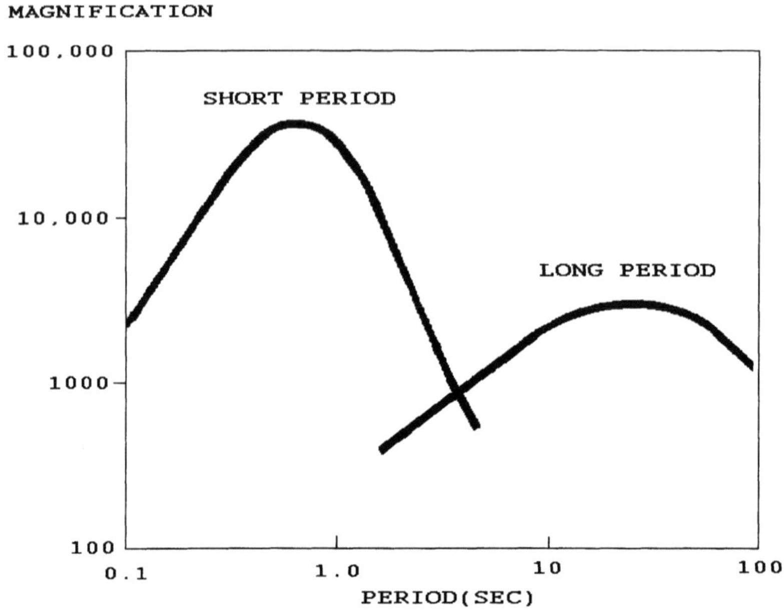

Figure 6.1 Typical magnification curves for the WWSSN seismographs for long period (LP) and short period (SP) instruments.

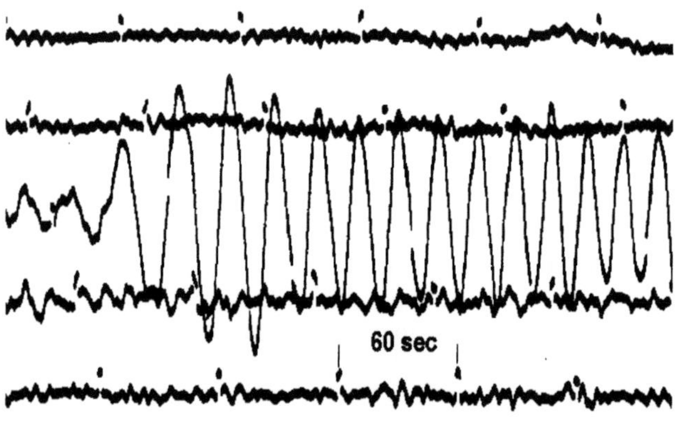

Figure 6.2 Surface waves recorded on a WWSSN long period seismograph. The scale is 15 mm/minute and 10 mm between the traces so a 24 hour seismogram is 90 by 30 cm.

Figure 6.2 shows a copy of an LP record of surface waves from a distant earthquake. The maximum amplitude is about 16 mm at a period of 25 sec. Using the magnification curve in Figure 6.1, we find that the gain is 2900 and the ground displacement is thus 16mm/2900 = 0.0055 mm. In this case, the signal is rather monochromatic, so we obtain nearly the correct ground displacement. For more complex signals, particularly if recorded with broadband sensors, the correction cannot be done so simply and we have to deal with the whole frequency range. It is therefore necessary to briefly describe some of the general concepts of linear systems and spectral analysis.

6.1 Linear systems

In seismology we assume that our instrument behaves as a linear system (6.4). Instruments can here be sensors, amplifiers or complete recording systems. The linearity means that there is a linear relationship between input signal and output signal. If the input signal is $x(t)$ and the output $y(t)$, then multiplying $x(t)$ with a constant will result in an output signal multiplied with the same constant. Like if the ground velocity is doubled, then the output from the seismometer is also doubled. If two signals of different frequency and amplitude are input, then also two signals with the same frequencies (with different amplitude and phase) are output.

Input: $x_1(t)$ Output: $y_1(t)$, (6.4)
Input: $x_2(t)$ Output: $y_2(t)$
Input: $ax_1(t) + bx_2(t)$ Output: $ay_1(t) + by_2(t)$

A linear system is also said to be time invariant if their properties (like filter constants) are time invariant. In practice, in a physical system, the parameters will change with time, but in most cases these changes are negligible in the short interval of a record.

The purpose of this section is to determine how the behavior of the linear system can be described generally so that the complete input signal can be determined from the complete output signal or vice versa, which include the special monochromatic case illustrated above. For a more complete description, see Scherbaum (2001).

Let us first look at a very simple linear system, the RC filter. A RC low-pass filter is seen in Figure 6.3.

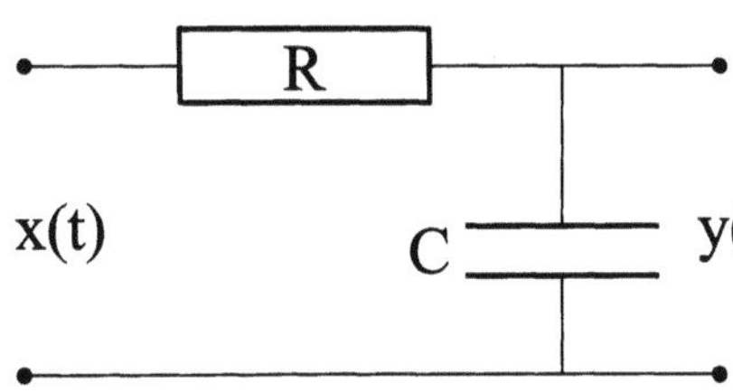

Figure 6.3 RC low pass filter. *R* is the resistor and *C* the capacitor. Input is *x(t)* and output *y(t)*.

It is well known that this circuit lets low frequencies pass while high frequencies are attenuated due to the frequency dependent impedance of the capacitor. The resistance R_c of a capacitor seen by the input (impedance) for a sine wave signal of frequency f is

$$R_c = \frac{1}{2\pi fC} = \frac{1}{\omega C} \tag{6.5}$$

where C is the capacitance (F). Considering that the RC filter is a frequency dependent voltage divider, using a monochromatic signal of angular frequency ω and amplitude $X(\omega)$, $x(t)=X(\omega)cos(\omega t)$, the output signal amplitude $Y(\omega)$ can be written as (see Appendix)

$$Y(\omega) = \frac{1}{\sqrt{1+\omega^2 R^2 C^2}} X(\omega) = \frac{1}{\sqrt{1+\omega^2 / \omega_0^2}} X(\omega) \tag{6.6}$$

where $\omega_0=1/RC$. ω_0 is also called the corner frequency of the filter and for $\omega = \omega_0$ the amplitude has been reduced to $1/\sqrt{2} = 0.707$. If the input signal is a steady state sine wave, the relation between the output and input signal amplitudes $A(\omega)$ can be written as

$$A(\omega) = \frac{1}{\sqrt{1+\omega^2 / \omega_0^2}} = \frac{Y(\omega)}{X(\omega)} \tag{6.7}$$

$A(\omega)$ is called the amplitude frequency response function of the filter, since only amplitudes are considered. If $A(\omega)$ is completely known, then the amplitude of the harmonic input signal $X(\omega)$ can be calculated from the measured signal as

$$X(\omega) = Y(\omega) / A(\omega) \tag{6.8}$$

The amplitude response can be measured very simply as shown in Figure 6.4. By varying the frequency, both input and output amplitudes can be measured at different frequencies to produce the amplitude response function.

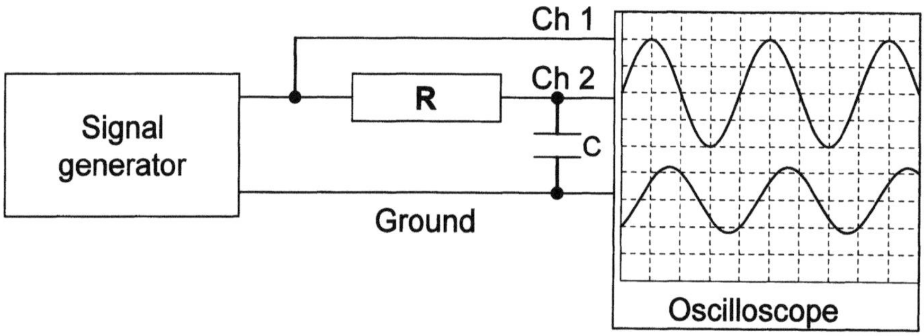

Figure 6.4 Measuring the amplitude response function of an RC filter. The signal from a signal generator goes directly to channel 1 (Ch1) on the oscilloscope and to channel 2 (Ch2) through the filter so both input and output is measured.

From Figure 6.4, it is seen that the output signal not only has been changed in amplitude, but also has been delayed a little relative to the input signal; in other words, there has been a phase shift. Looking at the circuit diagram Figure 6.3, it is obvious why the signal has been delayed. An instantaneous voltage over the resistor will result in a low voltage over the capacitor, since it will take some time to charge it. The charging time (see Appendix) will depend of the size of both R and C and it can be shown that the phase shift Φ is a function of ω and RC. In this example, the phase shift is negative (see definition in (6.9)). The complete frequency response of the filter therefore consists of both the amplitude response function and the phase response function $\Phi(\omega)$, see Figure 6.5. Considering a general input harmonic waveform $x(\omega,t) = x_0(\omega)\cdot\cos(\omega t)$, the output can be written as

$$y(\omega,t) = x_0(\omega)\cdot A(\omega)\cdot\cos(\omega t + \Phi(\omega)) \qquad (6.9)$$

The phase shift is here defined as a quantity being added to the phase as seen above. Thus comparing Figure 6.4 and (6.9), we see that the phase shift is negative. This is the most common way of defining the phase shift, but the opposite sign is sometimes seen and it may then be called phase delay or phase lag. So it is very important to know which definition has been used.

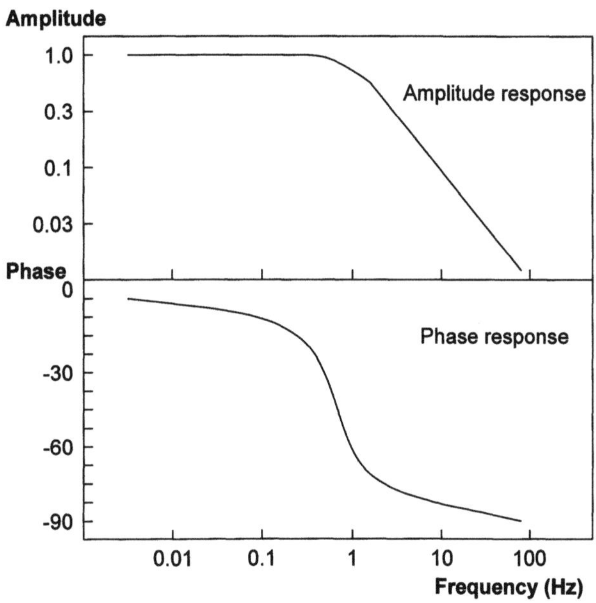

Figure 6.5 Amplitude and phase response of an RC filter with a corner frequency of 1 Hz. The phase response (or phase shift) is given in degrees. See Appendix.

The frequency response function is a bit cumbersome to define and use. This can be considerably simplified if the complex representation of harmonic waves is used. Instead of writing $\cos(\omega t)$, we can use the real part of the exponential function

$$e^{i\omega t} = \cos(\omega t) + i\sin(\omega t) \qquad (6.10)$$

(6.9) can now be written

$$y(\omega,t) = |X(\omega)|A(\omega)e^{i(\omega t+\Phi(\omega))} = |X(\omega)|A(\omega)e^{i\omega t}e^{i\Phi(\omega)} \tag{6.11}$$

y(ω,t) is now a complex number of which the real part is the actual output. This can be further simplified considering that any complex number *Z* can be written as

$$Z = a + ib = \sqrt{a^2 + b^2}\, e^{i\Phi} = |Z|e^{i\Phi} \tag{6.12}$$

where $\Phi = \tan^{-1}(b/a)$ which also follows from (6.10). We can now define the complex frequency response *T(ω)* as

$$T(\omega) = A(\omega)e^{i\Phi(\omega)} = |T(\omega)|e^{i\Phi(\omega)} \tag{6.13}$$

and (6.11) can be written

$$y(\omega,t) = |X(\omega)|T(\omega)e^{i(\omega t)} \tag{6.14}$$

We now have only one complex function that includes the phase shift and therefore completely describes the instrument frequency response. This was also the kind of response obtained when dealing with seismometer theory (see Chapter 2). So far we have only dealt with monochromatic signals and the corrections can easily be made with both the real and complex representations ((6.9) and (6.14)) so it might be hard see why we have to go to complex representation of the response function. Later in this chapter, we are going to see how we correct an observed seismogram, not only for one amplitude at a time, but actually correct the whole signal (containing a range of frequencies), which is possible with digital data. Obviously we cannot deal with one amplitude at a time, but will have to deal with the amplitude and frequency content of the whole signal, in other words, we have to make spectral analysis. It will hopefully then be clear why a complex representation is needed. Before getting into how to do this we have to review basic concepts of spectral analysis.

6.2 Spectral analysis and the Fourier transform

When looking at seismic signals (or other signals), they often appear to consist of a superposition of harmonic signals like Figure 6.2. Let us consider a signal of a finite duration *T* and defined in the time interval *–T/2* to *T/2*, or 0 to *T*. It can be shown that most physical signals can be decomposed into an infinite sum of monochromatic components, the so called Fourier series, which is a sum of sine and cosine functions:

$$x(t) = a_0 + \sum_{n=1}^{\infty} a_n \cos(\omega_n t) + \sum_{n=1}^{\infty} b_n \sin(\omega_n t) \tag{6.15}$$

where $\omega_n=2n\pi/T$. Strictly, the Fourier series expansion is limited to periodic functions. Therefore, this representation implicitly assumes that the signal is of infinite duration and periodic outside this interval 0-T, but the key point is that (6.15), for a continuous signal, is an exact representation of the signal within the interval of interest. If it actually has a finite duration, we are not concerned with the Fourier series result outside this interval. The Fourier coefficients a_n and b_n can be calculated as (e.g. Stein and Wysession, 2003)

$$a_n = \frac{2}{T}\int_0^T \cos(\omega_n t)x(t)dt \tag{6.16}$$

$$b_n = \frac{2}{T}\int_0^T \sin(\omega_n t)x(t)dt$$

with n going from 1 to infinity. The term a_0 is simply the average signal level or the DC level.

What does this mean physically? Imagine a simple signal of just a sine wave with amplitude unity and frequency ω_n. (6.16) would then give

$$b_n = \frac{2}{T}\int_0^T \sin(\omega_n t)\sin(\omega_n t)dt = \frac{2}{T}\cdot\frac{T}{2} = 1 \tag{6.17}$$

and the signal would be characterized by one spectral component with amplitude equal to one. Since the signal can have both cosine and sine waves, the amplitude spectrum is defined as

$$A_n = \sqrt{a_n^2 + b_n^2} \tag{6.18}$$

Note that the spectral amplitudes are normalized such that changing the length of the time window will not change the amplitudes. If the unit of the ground displacement is meter, then the unit of the amplitude spectrum is also meter.

It is conceptually easy to understand that a signal like in Figure 6.2 can be made up of harmonic signals, but what about a short wavelet? Figure 6.6 shows an example of a sum of harmonics that gives a short wavelet by constructive and destructive interference.

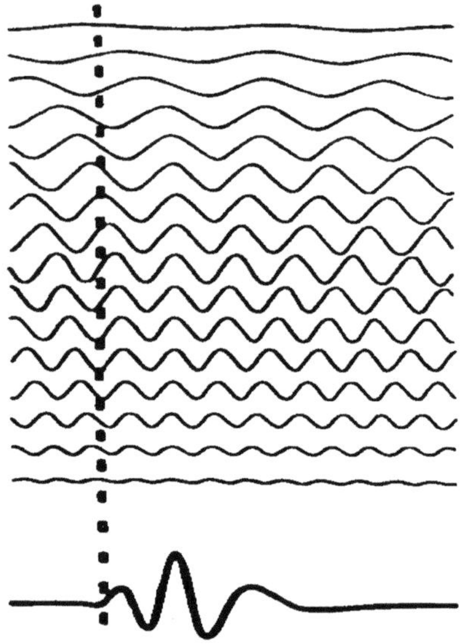

Figure 6.6 Sum of harmonic functions (top) that become the wavelet seen at the bottom. Reprinted from *Modern Global Seismology* by Lay and Wallace (1995), p. 177; copyright (1995), with permission of Elsevier.

(6.15) has both sine and cosine terms, which is a way of being able to sum harmonics with different phase, since the sum of any sine and cosine functions with the same frequency, can be reduced to a single cosine function with a particular amplitude and phase. Considering the trigonometric identity

$$\cos(\omega t_n + \Phi_n) = \cos(\Phi_n)\cos(\omega t_n) - \sin(\Phi_n)\sin(\omega t_n) \tag{6.19}$$

and replacing $a_n=A_n\cos(\Phi_n)$ and $b_n=-A_n\sin(\Phi_n)$, we can rewrite (6.15) as a sum of cosine functions multiplied with the spectral amplitude $A_n = \sqrt{a_n^2 + b_n^2}$

$$x(t) = a_0 + \sum_{n=1}^{\infty} A_n \cos(\omega_n t + \Phi_n) \tag{6.20}$$

where $\frac{-b_n}{a_n} = \frac{\sin(\Phi_n)}{\cos(\Phi_n)}$ and therefore $\Phi_n = \tan^{-1}\frac{-b_n}{a_n}$ and $\Phi_n(\omega)$ is called the phase spectrum. Thus the signal is composed of a sum of cosine terms with different amplitudes and phases. (6.20) can be simplified by writing it in exponential form.

$$x(t) = a_0 + \sum_{n=1}^{\infty} X_n e^{i\omega_n t} \tag{6.21}$$

where $X_n = A_n e^{i\Phi_n} = A_n \cos(\Phi_n) + iA_n \sin(\Phi_n)$. X_n is now the complex spectrum whose magnitude is the true amplitude spectrum and Φ_n is the phase spectrum. x(t) is now formally a complex function of which the real part is the true signal. However, we will continue to call it *x(t)* remembering that only the real part is to be used.

Now consider how a_n and b_n can be calculated with a complex exponential function. Replacing the sine and cosine functions in (6.16) by the complex equivalents

$$\sin(x) = \frac{e^{ix} - e^{-ix}}{2i}$$
$$\cos(x) = \frac{e^{ix} + e^{-ix}}{2} \tag{6.22}$$

it can be shown that (6.15) can be written

$$x(t) = a_0 + \sum_{n=1}^{\infty} F_n e^{i\omega_n t} + \sum_{n=1}^{\infty} F_{-n} e^{-i\omega_n t} \tag{6.23}$$

where

$$F_n = \frac{a_n - ib_n}{2} \text{ and } F_{-n} = \frac{a_n + ib_n}{2} \tag{6.24}$$

Defining ω_{-n} as $-\omega_n = -2n\pi / T = \omega_{-n}$ and F_{-n} as the complex conjugate of F_n , the negative exponentials can be written

$$\sum_{n=1}^{\infty} F_{-n} e^{-i\omega_n t} = \sum_{n=-1}^{-\infty} F_n e^{i\omega_n t} \tag{6.25}$$

Thus by artificially using negative frequencies and setting $F_0 = a_0$, (6.23) can be written simply as

$$x(t) = \sum_{n=-\infty}^{\infty} F_n e^{i\omega_n t} \tag{6.26}$$

such that the spectral coefficients F_n are determined as

$$F_n = \frac{1}{T} \int_0^T x(t) e^{-i\omega_n t} dt \tag{6.27}$$

The amplitude spectrum can now be determined from F_n. Considering (6.24), we have

$$|F_n| = \frac{\sqrt{a_n^2 + b_n^2}}{2} = \frac{A_n}{2} \quad \text{or } A_n = 2|F_n| \tag{6.28}$$

An explanation for the factor 2 is that the energy artificially has been spread out over both positive and negative frequencies, which is just a way of simplifying calculations. Unfortunately, this factor is often forgotten. It has to be included if (6.20) is used for *x(t)*. Nevertheless, it should be pointed out that this factor 2 is sometimes taken as part of the normalization constant and different definitions are used. The various definitions of direct and inverse transforms are matched and the right pair has to be used for the original signal to be recovered from its spectrum. On the other hand, the fact that spectral amplitudes for negative frequencies are the complex conjugate of the positive ones is only true for real signals. So a more general formulation of Fourier series has to use both positive and negative frequencies.

For discrete data we cannot use the integration (6.27), which must be replaced by a summation. We assume an even number of N data points in the time window T and get:

$$F_n \cong \frac{1}{T}\sum_{k=0}^{N-1} x(k\Delta t)e^{-i\omega_n k\Delta t}\Delta t = \frac{\Delta t}{T}\sum_{k=0}^{N-1} x(k\Delta t)e^{-i\omega_n k\Delta t} \tag{6.29}$$

Since the data is sampled, and only available in time window T, we know that aliasing exists (see 4.8) and in reality information is only available up to the Nyquist frequency $\omega_{N/2} = \omega_{NY} = \pi N/T$ or if measured in Hz, half the sample rate $1/2\Delta t$. It can also be shown that the spectral content repeats itself, see Figure 6.7.

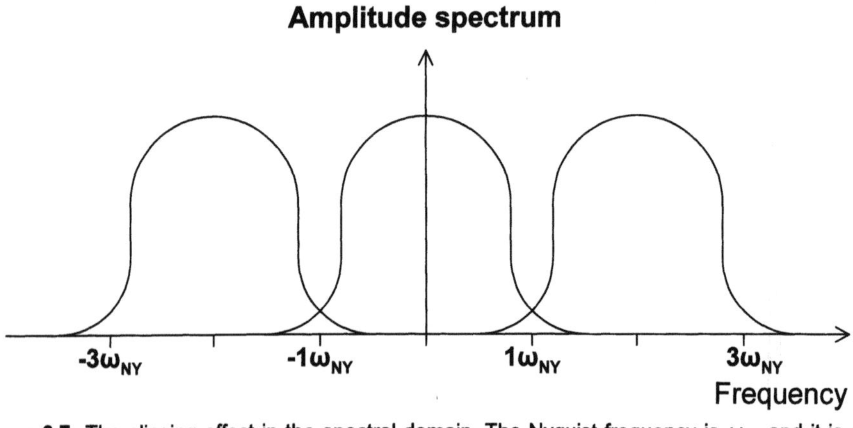

Figure 6.7 The aliasing effect in the spectral domain. The Nyquist frequency is ω_{NY} and it is seen that the spectrum repeats itself for each $2\omega_{NY}$ and that there is overlap when the signal frequency is larger than ω_{NY}.

This means that the spectral values for $n > N/2$ corresponds to the spectral content with the negative frequencies and the first half of the spectrum for negative frequencies is stored in the F_n – values for $n > N/2$, and the N positive frequency spectral values are:

$$\begin{matrix} \omega_0 & \omega_1 & \omega_2 \ldots\ldots & \omega_{N/2} & \omega_{-(N/2-1)} & \omega_{-(N/2-2)} & \ldots\ldots & \omega_{-1} \\ F_0 & F_1 & F_2 \ldots\ldots & F_{N/2} & F_{-(N/2-1)} & F_{-(N/2-2)} & \ldots\ldots & F_{-1} \end{matrix} \tag{6.30}$$

So, when writing the inverse transform, only the summation over the positive frequencies has to be done, since the values for $n > N/2$ in reality are the negative frequencies and it is assumed that there is no energy above the Nyquist frequency:

$$x(k\Delta t) = \sum_{n=0}^{N-1} F_n e^{i\omega k\Delta t} \tag{6.31}$$

Amplitude spectra are not always the most objective way of representing the spectral content, as we shall see in an example. Figure 6.8 shows the spectra of a transient signal recorded in a long and a short time interval respectively. The spectrum from the long time window has a much lower spectral level than the spectrum for the short time window. This is caused by 2 factors: (1) The normalization (*1/T*) over the long window causes any one spectral amplitude estimate to be smaller since a longer window does not mean more energy, (2) The longer window has a more dense spectrum since the frequency spacing $\Delta f = 1/T$ is smaller for the long window, so the energy is smeared out over more discrete amplitude estimates.

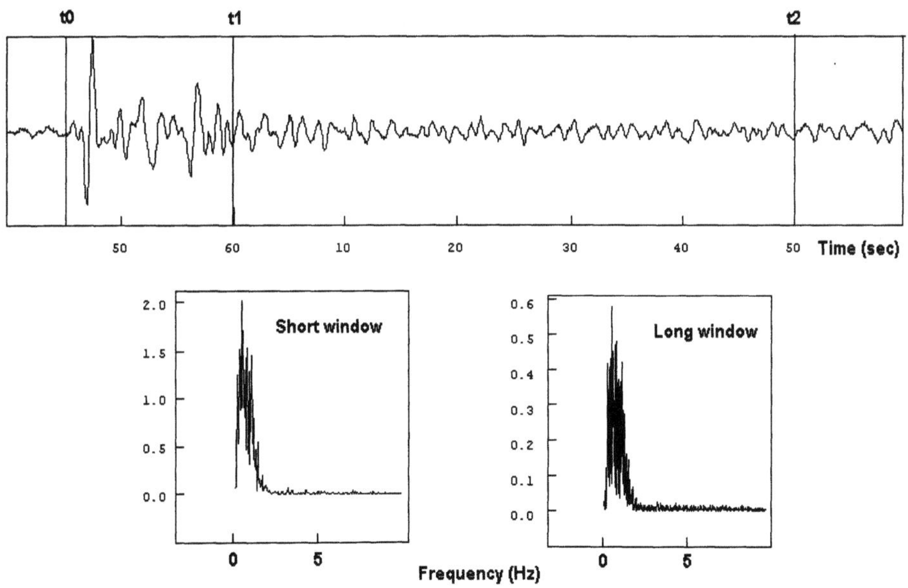

Figure 6.8 The signal (top) is a teleseismic P-phase recorded using a short period seismometer. Below is shown the amplitude spectra of the short window (*t0-t1*) and the long window (*t0-t2*). The scales on the axes are linear. Note that the short window has a higher level than the long window. Also note that the frequency estimates are much closer together for the long window.

Obviously, it is desirable that the spectral amplitude is constant for a given signal irrespective of Δf, so that brings us to the definition of the amplitude density spectrum. If instead of specifying the individual amplitudes, we specify the spectral amplitude multiplied by the number of estimates per Hz, n, then we get an estimate independent of

Δf since the smaller amplitudes correspond to a larger n. The number n of amplitudes per Hz can be calculated as $n = 1/\Delta f = T$. So, multiplying (6.29) by T, we get the amplitude spectral density, $F_n^d = TF_n$. The unit of the spectrum has now changed from amplitude to amplitude/Hz. This is the most usual way of calculating the spectrum:

$$F_n^d = \Delta t \sum_{k=0}^{N-1} x(k\Delta t) e^{-i\omega_n k\Delta t} = \Delta t \sum_{k=0}^{N-1} x_k e^{-i2\pi nk/N} = \Delta t F_n^{DFT} \tag{6.32}$$

where

$$x_k = x(k\Delta t) \text{ and } \omega_n k\Delta t = \frac{2\pi n}{T} k \frac{T}{N} = \frac{2\pi nk}{N} \tag{6.33}$$

and

$$F_n^{DFT} = \sum_{k=0}^{N-1} x_k e^{-i2\pi nk/N} \tag{6.34}$$

F_n^{DFT} is called the Discrete Fourier Transform (DFT) which is the way most Fourier transforms are implemented in computer programs.

The Inverse Discrete Fourier Transform (IDFT) should regenerate the original signal, and since the normalization constant was removed from F_n, it now has to be put back into the inverse transform which then is

$$x(k\Delta t) = \frac{1}{T} \sum_{n=0}^{N-1} F_n^d e^{i\omega k\Delta t} = \frac{1}{N\Delta t} \sum_{n=0}^{N-1} \Delta t F_n^{DFT} e^{i\omega k\Delta t} \tag{6.35}$$

or

$$x_k = \frac{1}{N} \sum_{n=0}^{N-1} F_n^{DFT} e^{i2\pi nk/N} \tag{6.36}$$

It is up to the user to put in the correct normalization constants when using DFT, and as it can be seen from (6.34) and (6.36), no information of time window or sample rate is used when calculating DFT or IDFT. The same computer algorithms can be used by just changing the sign of the exponential.

The expression of the discrete Fourier transform is a functional equivalent to the Fourier series, although conceptually different. Series development is applicable to periodic signals. Therefore, DFT application to a non-periodic signal (e.g. a time window of a seismic record) implicitly assumes that the signal repeats outside the window. If the value at the window start, $t=0$, is different from the end value, $t=T$, we introduce a spurious step (that really does not exist), since the end is implicitly tied to the beginning of the following period. This may produce effects as loss of definition in the spectrum and the so-called "side-lobe contamination". For a discussion of these problems and their treatments, the reader is referred to, e.g., Press et al., (1995).

If the time window goes to infinity, the Fourier density spectrum and inverse transform can be extended into a Fourier transform, which represents the function as an integral over a continuous range of frequencies. It can be shown that (6.36) and (6.32) become (e.g. Stein and Wysession, 2003):

$$x(t) = \frac{1}{2\pi} \int_{-\infty}^{\infty} X(\omega) e^{i\omega t} d\omega \tag{6.37}$$

$$X(\omega) = \int_{-\infty}^{\infty} x(t) e^{-i\omega t} dt \tag{6.38}$$

It is now clear that it is not possible to talk about an amplitude spectrum since the energy has been distributed over an infinite number of frequencies so, as defined above, we can only talk about an amplitude density spectrum.

6.3 Noise power spectrum

One of the important tasks of the spectral analysis in instrumental seismology is to calculate the seismic background noise power spectrum, which is the most standard way of quantifying the noise at a given site. The standard amplitude spectrum, as defined in (6.28), gives an average estimate and we likewise want an average power spectrum independent of window length and in its discrete form also independent of sample rate, particularly since it is assumed that the background noise is stationary. The power spectrum P_n for a periodic function is defined (e.g., Kanasewich, 1973) as

$$P_n = |F_n|^2 = \frac{a_n^2 + b_n^2}{4} = \frac{A_n^2}{4}, -\infty < n < \infty \tag{6.39}$$

Parseval's theorem states that the average power in a Fourier series is the same as in the time series

$$\sum_{n=-\infty}^{\infty} |F_n|^2 = \frac{1}{T} \int_0^T (x(t))^2 dt \tag{6.40}$$

If we have a sine wave with amplitude A_n and frequency ω_n (as defined before) in the time window T, the average power is

$$\frac{1}{T}\int_0^T A_n^2 \sin^2(\omega_n t)dt = \frac{A_n^2 T}{2T} = \frac{A_n^2}{2} \tag{6.41}$$

Using (6.40), we have to sum the two terms F_n and F_{-n} since according to Parseval's theorem, we have to sum over all frequencies, so

$$\text{Total power} = |F_n|^2 = F_{-n}^2 + F_n^2 = 2\frac{A_n^2}{4} = \frac{A_n^2}{2} \tag{6.42}$$

which is the same as (6.41). This result is not surprising. If we have a voltage of amplitude V over a resistor $R = 1$ ohm, then we know that the average power is $V^2/2$, so what is calculated with (6.42) is average power. Since spectra most often are calculated as F_n^{DFT}, using(6.32), the amplitude power spectrum is calculated as

$$P_n = \left|F_n^{DFT}\right|^2 \frac{\Delta t^2}{T^2} \tag{6.43}$$

and if only positive frequencies are considered it will be

$$P_n = 2\left|F_n^{DFT}\right|^2 \frac{\Delta t^2}{T^2} \tag{6.44}$$

Seismic noise is supposed to be stationary, so making the time window longer should give the same result. But if we double the window, the number of power estimates P_n will also double, however since the average energy is the same, the average energy in each P_n will be half so the spectral level will be half. To get a constant value, we must use the power density spectrum (same argument as for the amplitude density spectrum) and (6.44) must be multiplied by T. The power amplitude density spectrum is then defined as

$$\frac{1}{T}X_n^2 = \frac{1}{T}\left|\Delta t F_n^{DFT}\right|^2 \tag{6.45}$$

Considering again only positive frequencies, the seismic power amplitude density spectrum P_n^d must be calculated as

$$P_n^d = \left|F_n^{DFT}\right|^2 \frac{\Delta t^2}{T} 2 \tag{6.46}$$

Note the unit. If the amplitudes are in m, the unit is $m^2s = m^2/Hz$. Usually the amplitude is in acceleration so the unit is $(ms^{-2})^2/Hz$.

6.4 General instrument correction in frequency and time domain

We have now defined a complex frequency response function $T(\omega)$ and we can define the complex spectra of the input and output signals $x(t)$ and $y(t)$ as $X(\omega)$ and $Y(\omega)$, respectively. We use the definition for the Fourier spectrum for simplicity. Knowing the complete complex output spectrum we can determine the complete complex input spectrum as

$$X(\omega)=Y(\omega)/T(\omega) \tag{6.47}$$

If $T(\omega)$ is a seismic instrument response, we can say that we have obtained the instrument corrected ground motion spectrum and since $X(\omega)$ is complex, this also includes the correction for phase. The separate amplitude and phase response can then be obtained as

$$A(\omega)=\sqrt{\mathrm{Re}(X(\omega))^2+\mathrm{Im}(X(\omega))^2} \tag{6.48}$$

$$\Phi(\omega)=\tan^{-1}\left(\frac{\mathrm{Im}(X(\omega))}{\mathrm{Re}(X(\omega))}\right) \tag{6.49}$$

Thus in practice, the ground displacement spectrum would be calculated by taking the complex Fourier spectrum of the output signal and dividing it with the complex displacement response function and finally taking the absolute part of the complex ground displacement spectrum.

All of this could of course easily have been done with the corresponding non-complex equations since, so far, we have not really used the information in the phase directly and usually phase spectra for earthquake signals are not used. However, the next step is to also obtain the instrument corrected ground displacement or, said in another way, get the time domain signal instead of the frequency domain signal. With the knowledge about Fourier transforms, this is now easy, since, when we have all the frequency domain coefficients or the spectrum, we can generate the corresponding signal by the inverse Fourier transform. Since our corrected signal consists of a sum of cosine signals with different phase, each of them delayed differently due to the instrument response, the shape of the output signal will depend on the phase correction. So now life is easy, since the corrected complex displacement spectrum is already corrected for phase and the ground motion can be obtained as

$$x(t)=\frac{1}{2\pi}\int_{-\infty}^{\infty} Y(\omega)/T(\omega)\cdot e^{i\omega t}\,d\omega \tag{6.50}$$

or with real discrete data using (6.32) and (6.34). While it is possible to use only half of the positive frequencies for making the amplitude spectrum, both positive and negative frequencies must be used for the inverse transformation. Normalization constants can be defined in different ways, but if the same routine is used for both forward and inverse transformation (like (6.32) and (6.34)), the normalization constants will cancel out.

Thus in theory, we can recover the ground displacement at any frequency knowing the instrument response. In practice, one has to be careful to only do this in the frequency band where the instrument record real ground motion and not just electronic noise, since the instrument correction then become unstable and the output has nothing to do with the real seismic signal. Figure 6.9 shows an example.

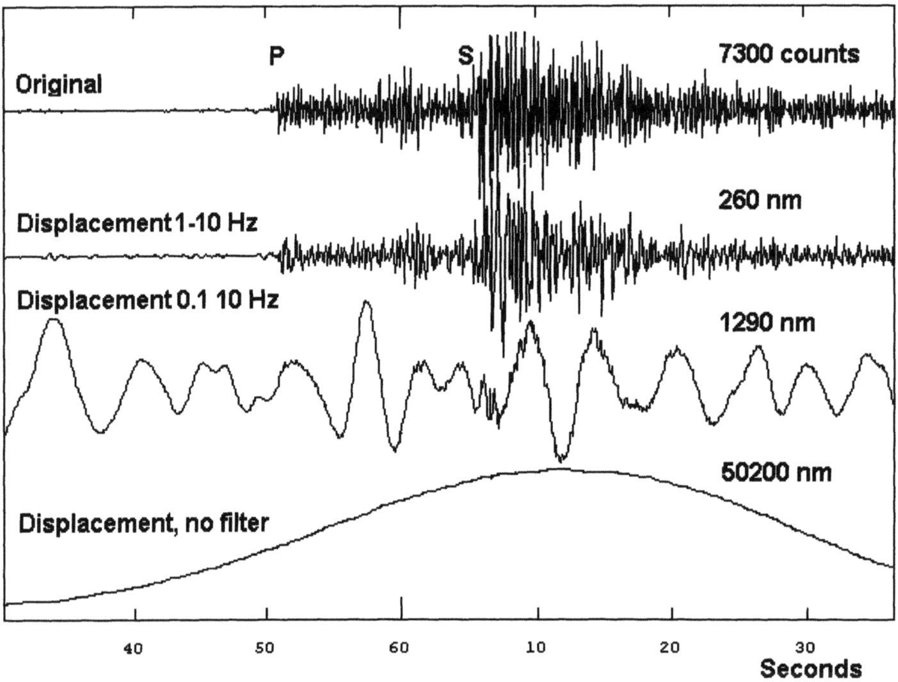

Figure 6.9 Instrument correction in different filter bands. The top trace is the original recording of a small earthquake with a 1 Hz seismometer. The 3 bottom traces have been converted to displacement with different filters. The amplitudes to the right are maximum amplitudes.

The figure shows the influence of filtering, when estimating the ground displacement signal. In the frequency band 1 to 10 Hz, the signal looks very much like the original signal although a bit more low frequency, since it is converted to displacement and can nearly be considered an integration of the original signal. In the 0.1 to 10 Hz range, the earthquake signal almost disappears in the microseismic background noise. Why do we think it is seismic noise and not instrumental generated noise? First, the earthquake signal has about the same amplitude as above, second, it 'looks' like seismic background noise and third, the amplitude at 1290 nm is at a period of 5 sec (peak amplitude of microseismic noise) which looks reasonable compared to world wide observations (see Figure 3.3 in noise section). Note that this is how the earthquake signal would have looked being recorded on a broadband sensor, hardly noticeable. The last trace shows the calculation of the displacement without filtering so the lowest frequency used is *1/T*, where T is the length of the window, here 80 s (only 65 s shown) so f = 0.0125 Hz. The amplitude is now more than 50000 nm and the signal looks 'funny'. The large amplitude obviously cannot be right since the microseismic noise has its largest amplitude around 5-10 s and it was 1290 nm. So we have a clear case of trying to make a displacement signal at frequencies lower than where seismic signals exist in the data. The ratio of the displacement gain for a 1 Hz seismometer at 1 Hz and 0.0125 Hz is $1/0.0125^3 = 5\ 10^5$. In other words, if the gain at 1 Hz is 1.0 we have to multiply by 1.0 to get the displacement, while at 0.0125 Hz we have multiply by $5\ 10^5$. So any tiny amount of instrumental noise present at low frequencies will blow up in the instrument correction. In the above example, it seems that the displacement signal can be recovered down to 0.1 Hz with a 1.0 Hz sensor.

We now have all the elements needed to correct our recorded signals in frequency or time domain, and the main problem is to obtain and describe the frequency response function $T(\omega)$.

6.5 General representation of the frequency response function

We have already seen examples of frequency response functions like the *RC* filter (see Appendix) and the standard inertial seismometer. A seismic station has more elements which might not fit any of these two and a general description of the frequency response function is needed (hereafter called response function for simplicity) which can cover any system used. We will start with some examples.

The response function for the *RC* low pass filter can be written

$$T_{RC}(\omega) = \frac{1}{1 + i\omega RC} \tag{6.51}$$

Similarly, the amplitude response function for a mechanical seismometer (see Chapter 2) is

$$T_d(\omega) = \frac{\omega^2}{\omega_0^2 - \omega^2 + i2\omega\omega_0 h} \tag{6.52}$$

In general, the response function could be any complex function. It turns out that $T(\omega)$ for all systems made from discrete mechanical or electrical components (masses, springs, coils, capacitors, resistors, semiconductors, etc) can be represented exactly by rational functions of $i\omega$ like

$$T(\omega) = \frac{a_0 + a_1(i\omega) + a_2(i\omega)^2 + \ldots\ldots}{b_0 + b_1(i\omega) + b_2(i\omega)^2 + \ldots\ldots} \tag{6.53}$$

where a_i and b_i are constants. The number of terms in the polynomials will depend on the complexity of the system.

It is seen from (6.51) that the RC filter exactly looks like (6.53), while the mechanical seismometer response must be slightly rewritten to

$$T_d(\omega) = \frac{-(i\omega)^2}{\omega_0^2 + 2i\omega\omega_0 h + (i\omega)^2} \tag{6.54}$$

So for a seismometer $a_0=0$, $a_1=0$, $a_2=-1$, $b_0=\omega_0^2$, $b_1 = 2\omega_0 h$ and $b_2=1$. This general representation is sometimes used and is one of the accepted ways of giving response in SEED format (Standard for Exchange of Earthquake Data, defined by Federation of Digital Seismograph Networks (FDSN), IRIS (1993)). However, (6.53) can be written in an alternative and somewhat simpler way. Considering that a polynomial can be factorized, (6.7) can be written

$$T(\omega) = c\frac{(i\omega - z_1)(i\omega - z_2)(i\omega - z_3).....}{(i\omega - p_1)(i\omega - p_2)(i\omega - p_3).....} \tag{6.55}$$

where c is the combined normalization constant for nominator and denominator polynomials, z are the zeros (or roots) of the nominator polynomial while the zeros of the denominator polynomial are p. (6.53) and (6.55) are exactly identical so either might be used to represent $T(\omega)$. Using (6.55) to represent $T(\omega)$ is the so called poles and zeros representation, which has become the most standard way. Representing response curves in terms of poles and zeros is often described in a very complicated way, where it is necessary to understand terms like Laplace transforms and complex s-plane. In reality, as seen above, it is quite simple.

(6.51) for the low pass filter can be rewritten as

$$T_{RC}(\omega) = \frac{1/RC}{i\omega + 1/RC} \tag{6.56}$$

and it is seen that the filter only has one pole at $-1/RC$ and the normalization constant is $1/RC$. For the seismometer, the denominator polynomial must be factorized by finding the roots p_1 and p_2 of the second order in $i\omega$

$$\omega_0^2 + 2i\,\omega\,\omega_0 h + (i\,\omega)^2 = 0 \tag{6.57}$$

It turns out that

$$\begin{aligned} p_1 &= -\omega_0(h + \sqrt{h^2 - 1}) \\ p_2 &= -\omega_0(h - \sqrt{h^2 - 1}) \end{aligned} \tag{6.58}$$

so $T_d(\omega)$ can be written

$$T_d(\omega) = \frac{-(i\omega - 0)(i\omega - 0)}{(i\omega - p_1)(i\omega - p2)} \tag{6.59}$$

So in addition to the poles p_1 and p_2, the seismometer response function has a double zero at $z=0$ and the normalization constant is -1. Note that since h usually is smaller than 1, the poles are usually complex (see example later in this chapter). Complex poles always appear as conjugate pairs.

For the standard velocity transducer, the equation is

$$T_d^v(\omega) = \frac{(i\omega - 0)(i\omega - 0)(i\omega - 0)}{(i\omega - p_1)(i\omega - p2)} \tag{6.60}$$

and there is thus one more zero and the normalization constant is now 1 instead of −1 due to the polarity caused by the velocity transducer (see Chapter 2).

Other filters:

Butterworth filter. The RC filter is simple, but it is not very suitable for higher orders. Ideally, a filter should be very sharp (a square in the frequency domain) but this is not possible in practice. There are many different types of analog filters, but one of the best and most widely used, is the Butterworth filter. This is because the filter has a nice response function, with the minimum ripple (or maximum flatness) for the pass-band, is easy to construct for any order up to 10 and easy to describe. The amplitude response function of the normalized low pass Butterworth filter is (e.g. Kanasewich, 1973)

$$|B(\omega)| = \frac{1}{\sqrt{1 + (\omega / \omega_0)^{2n}}} \tag{6.61}$$

where ω_0 is the corner frequency of the filter (-3 dB point or where amplitude has decreased to 0.707) and n is the order of the filter. The amplitude response function for the high pass filter is

$$|B(\omega)| = \frac{1}{\sqrt{1 + (\omega_0 / \omega)^{-2n}}} \tag{6.62}$$

A nice property of this filter is that the response at the corner frequency remains constant for any order of the filter. The complex response function can be described with a series of poles and zeros. For the first 2 orders, the high pass response can be written on polynomial form as

1. order

$$B(\omega) = \frac{1}{1 - i\frac{\omega_0}{\omega}} \quad (6.63)$$

2. order

$$B(\omega) = \frac{-1}{-1 + i\sqrt{2}\frac{\omega_0}{\omega} + (\frac{\omega_0}{\omega})^2} = \frac{-\omega^2}{\omega_0^2 - \omega^2 + i\sqrt{2}\omega\omega_0} \quad (6.64)$$

Comparing the 2-order Butterworth filter response to the seismometer velocity response (2.28), we see that the response is exactly the same, (except for the sign, see Chapter 2), if the damping constant $h=\sqrt{2}/2=0.707$. This value of damping is frequently used since it gives the most flat response function, so we can describe the seismometer response as a simple high pass 2. order Butterworth filter. Since many modern seismometers shape their output using a Butterworth filter, the equivalent output is described using the normal seismometer response function with a damping of 0.707.

Digital filters. Digital filters operate on sampled signals as described in 4.9 If x_i represents the input series of the filter and y_i the output series, the most general representation of a linear, causal, digital filter is

$$y_i = \sum_{j=0}^{N} a_j x_{i-j} + \sum_{k=1}^{M} b_k y_{i-k} \quad (6.65)$$

In this equation, the first sum is the non-recursive part (it depends only on the present and past input values) and the second sum is the recursive part (depends on the past output values). A FIR filter has only the first part, while an IIR filter has both. If the sum is extended also to future input values (j may also be negative) then the filter will be non-causal. Symmetric ($a_j=a_{-j}$) non-causal filters, often used in seismic digitizers, have the advantage that their phase response is a linear function of frequency and so they do not produce phase distortion, but just a known time delay (see e.g. Scherbaum, 2001).

The shift property of Fourier transform states that a time shift of Δt in the time domain is equivalent to multiplying the transform by $e^{i\omega\Delta t}$. By using this, the frequency response of a filter described by the differences equation given above may be written in a simple way a as rational function of z:

$$\frac{Y(\omega)}{X(\omega)} = \frac{\sum_j a_j z^{-j}}{1 - \sum_k b_k z^{-k}} = \frac{\sum_j a_j e^{-i\omega\Delta t \cdot j}}{1 - \sum_k b_k e^{-i\omega\Delta t \cdot k}} \quad (6.66)$$

where $z = e^{i\omega\Delta t}$ and Δt is the sampling period of the series.

The polynomial $X(z) = \sum_i x_i z^{-i}$ is known as the z-transform of the series x_i. See also Scherbaum (2001).

6.6 Anti alias filters

Analog anti alias filters have been described in the digitizer section. Analog filters are usually Butterworth filters, while the more common digital anti alias filters are FIR or rarely IIR filters. These can only be represented by poles and zeros in the variable $i\omega$ for a limited number of filter coefficients, although a more suitable representation for them is in terms of the variable z, using the z-transform (see above). So, to specify these filters one has to either use the time domain filter coefficients (the most common way) or give discrete amplitude and phase values of the response function. Since FIR filters are very sharp and usually have no phase shift, there is rarely a need to correct for them since the corner frequency is very close to the Nyquist frequency. However, in special cases a correction is made to remove the filter effect of sharp impulsive onsets, where the onset can be masked by a precursor caused by the filter (Figure 6.10).

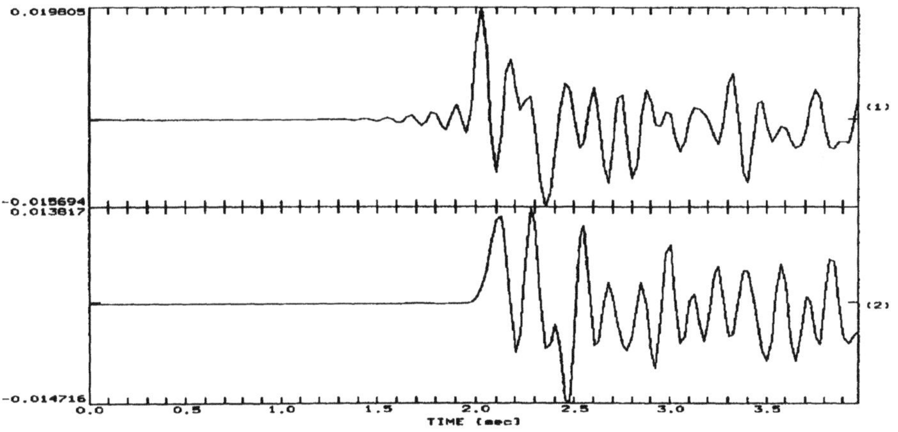

Figure 6.10 . Precursory effect of a FIR filter. The top trace shows the original recorded signal, and the bottom trace the corrected signal. Reprinted from Scherbaum (2001), with kind permission of Kluwer Academic Publishers.

Discussion on how to correct for digital anti alias filters is beyond the scope of this book, see Scherbaum (2001). However, the user of modern digitizers with digital anti alias filters should be aware of the pitfalls and observe precursors to a sharp onset with suspicion. See also 4.9.

6.7 Instrument correction and polarity

The question often comes up whether polarity (first motion is up or down) depends on the type of sensor or more generally on the response function. The answer is NO. If the sensors have been correctly installed, a first motion UP, North or East should always produce a positive signal. However, the response function, particularly digital anti alias filters, might generate precursory signals, so that it might not be easy to observe the first motion as illustrated in Figure 6.10. Figure 6.11 shows an example of a broadband signal (velocity trace) of a teleseismic P-wave which has been converted to displacement and acceleration. Although the signals look different, the same positive polarity is clearly seen on all 3 traces.

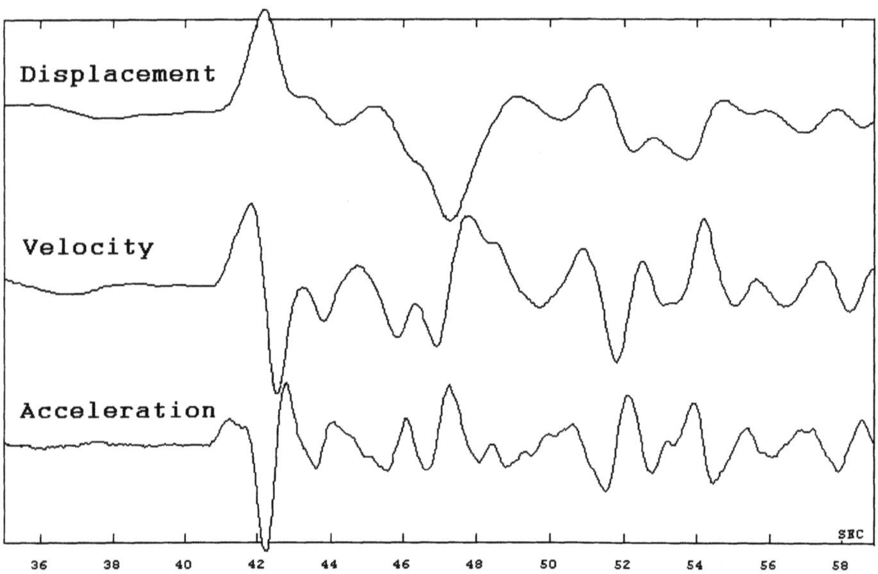

Figure 6.11 A P-wave from a teleseismic event original recorded on the velocity trace. The two other traces show displacement and acceleration. Note that the polarity of the onset is the same on all 3 traces.

Filters can change the onset as seen with the FIR filter example in Figure 6.10. Figure 6.12 shows an example of a synthetic 1 Hz signal with an onset at 2 s. It has been filtered with an 8 pole zero phase shift Butterworth filter (passing the filter forwards and backwards in time domain) and a 4 pole Butterworth filter with a phase shift (passing only forwards). As it can be seen, the filter with the phase shift has preserved the polarity while the zero phase shift filter has shifted some of the energy to before the onset and it is no longer possible to observe the correct polarity and the onset time would be wrong. This is what sometimes happens with digital zero phase shift anti alias filters. An analog anti alias filter would only pass one way and therefore not obscure the onset polarity, but the phase delay may hide the true onset in the background noise if the SNR is not high. High-order analog filters also produce a group delay that may be non-negligible. For more discussion on anti alias filters, see Chapter 4.

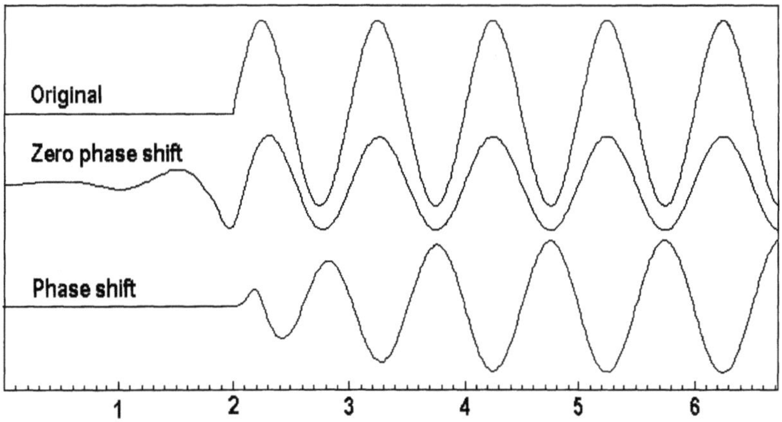

Figure 6.12 The effect of using filters on a signal with a sharp onset. The signal is a 1 Hz sine wave with an onset at 2 s (top trace). The middle trace shows the signal filtered with a zero phase shift Butterworth filter while the bottom trace shows the signal filtered with a Butterworth filter with a phase shift. Both filters are band pass with corners at 1 and 5 Hz.

6.8 Combining response curves

We have now described the response of several individual elements, but a seismic recorder consists of several elements. How do we get the complete response? Simply by multiplying together the response functions of the individual elements. This would also be the case where the response function cannot be represented by a rational function, in which case the values of the function would just have to be evaluated at discrete frequencies. If e.g. a recorder has the following test response parameters:

Seismometer free period:	5.0 sec
Seismometer damping:	0.7
Seismometer loaded generator constant:	200 V/ms^{-1}
Low pass filter for anti aliasing:	25 Hz, 6 poles
High pass filter to cut DC:	0.01 Hz, 1 pole
Amplifier gain:	1000 times or 60 dB
ADC converter sensitivity:	2000 counts/V

We can calculate the response of each element and combine it to the total response T_{tot} as

$$T_{tot} = T_s \bullet T_a \bullet T_{25\,Hz} \bullet T_{0.01Hz} \bullet T_{ADC} \tag{6.67}$$

where T_s is the seismometer response, T_a the amplifier response, $T_{25\,Hz}$ the 25 Hz filter response, $T_{0.01Hz}$ the 0.01 Hz filter response and T_{ADC} is the ADC response. Figure 6.13 shows the response of the elements and the combined response. A similar figure could have been made for the phase response.

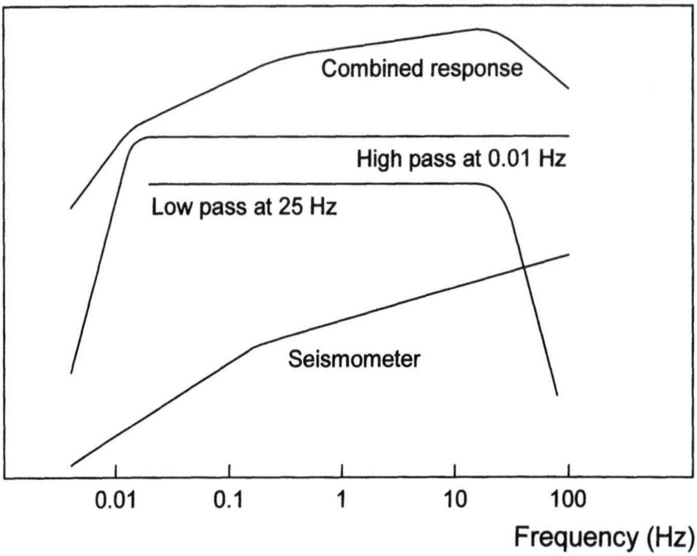

Figure 6.13 Combining response functions. The figure shows the displacement response function of the seismometer and the amplitude response functions of the two filters. Both scales are logarithmic. The y-axis is arbitrary.

In addition to the 3 curves shaping the response function, there is also the constant gain of the seismometer, the amplifier and the ADC. It is assumed that the filters have unity gain, but that is not always the case. The curves were made with SEISAN (Havskov and Ottemöller, 1999) and the total gain at 1 Hz is $2.51 \cdot 10^9$ counts/m, which is a typical value for many sensitive SP seismic stations. This value corresponds to 2.51 counts/nm. It can often be useful to calculate such gain constants manually for checking and to get an idea of their values. We will illustrate this with the above values.

Seismometer: Since the natural frequency is at 0.2 Hz, and the damping is 0.7, we can assume that the response curve is flat for velocity at 1.0 Hz so the gain for displacement is gain for velocity multiplied by the angular frequency:

Seismometer:	$G_s = 200\ \text{V/ms}^{-1} \cdot 2\pi \cdot 1.0\ \text{Hz} = 1257\ \text{V/m}$
Amplifier:	$G_a = 1000$ V/V
Filters:	$G_{0.01Hz} = 1$ V/V
	$G_{25Hz} = 1$ V/V
ADC:	$G_{ADC} = 2000$ V/counts

Combining all, we get a total gain G_{tot}:

$$G_{tot} = G_s \bullet G_a \bullet G_{0.01Hz} \bullet G_{25\ Hz} \bullet G_{ADC} = \tag{6.68}$$

1257 V/m •1000 V/V •1 V/V • 1 V/V • 2000 Counts/V = $2.51 \bullet 10^9$ counts/m at 1.0 Hz.

Note how all the units cancel out to give the correct unit. This is a good way of checking if all numbers have the correct units.

We can now look at how the response function looks like in poles and zero representation. The seismometer has the poles (real and imaginary part):

```
-0.87964594E+00  0.89741838E+00
-0.87964594E+00 -0.89741838E+00
```

and the zeros:

```
 0.00000000E+00  0.00000000E+00
 0.00000000E+00  0.00000000E+00
 0.00000000E+00  0.00000000E+00
```

and a normalization constant of 1.0 ms^{-1}/m. These values can be obtained directly from (6.58). Note the unit: The seismometer input is displacement and the output a measure of velocity. The generator constant then converts this to Volts. For the 25 Hz low pass Butterworth filter, we have the 6 poles:

```
 -40.6552          151.727
 -111.072          111.072
 -151.727          40.6552
 -151.727         -40.6552
 -111.072         -111.072
 -40.6552         -151.727
```

and a normalization constant of $1.50 \cdot 10^{13}$ V/V. Thus to get of gain of 1.0, the response calculated with poles and zeros must be multiplied with the normalization constant.

For the 0.01 Hz high pass Butterworth filter, we have 1 pole and 1 zero:

```
Pole:
 -0.6283E-01 0.5493E-08

Zero:
   0.00000          0.00000
```

Normalization constant = 1.0 V/V

Total normalization constant = 200 $V/ms^{-1} \cdot 1.0\ ms^{-1}/m \cdot 1.5\ 10^{13}$ V/V • 1.0 V/V • 1000 V/V • 2000 counts/V = $6.0 \cdot 10^{21}$ counts/m. NOTE: This is not the gain at any one frequency but the number to multiply with once the total pole and zero response calculation has been made.

This looks very different from the constant we calculated for 1.0 Hz for the total gain $2.51\ 10^{9}$ counts/m. How can we compare one to the other? We again calculate at 1 Hz. The filters used have no effect at 1 Hz (their responses are flat at this frequency) so we can divide by the total filter gain by using the filter normalization constants. The normalized displacement response curve has a gain of 6.28 (= $2\pi \cdot 1$) at 1 Hz, so the total gain at 1 Hz is

$$6.0 \cdot 10^{21} \text{ counts/m} \cdot 6.28 / 1.5 \cdot 10^{13} = 2.51 \cdot 10^{9} \text{ counts/m}$$

Normalization constants used with poles and zeros representation of response curves will then often have little in common with an understandable physical number.

6.9 Common ways of giving response information

Working with response information supplied with some data can be quite frustrating, since it can be given in so many ways and different conventions can be used. One would think that the most natural way of supplying information should be in terms of displacement response with units of counts/m. Unfortunately, life is not so easy. In addition, different processing systems use different ways of specifying the response and also using different formats.

Apart from using different conventions, the response can be given both as velocity and displacement response and sometimes even as acceleration response. The last is of course tempting for accelerometers, since it is so simple, however, *it is strongly recommended to, at least within the same processing system, always use the same type of response curves and preferably displacement*. Doing it differently can create endless frustrations. Like when going from velocity to displacement response, it is not always easy to remember whether we have to multiply or divide by ω, and it is certainly done the wrong way many times.

In summary, the most common ways of representing response information are, in order of importance:

- Poles and zeros (PAZ)
- Discrete amplitude and phase values (FAP)
- Individual parameters (free period, ADC gain etc)
- Polynomials
- Combination of the above
- Time domain filter coefficients

The two most common international waveform formats are SEED and GSE, both created with the help of many experts. However, the response information is not provided in the same way for the two formats.

The most common processing systems are PITSA-GIANT (http://lbutler.geo.uin-potsdam.de/service.htm), SAC (Goldstein, 1999) and SEISAN. PITSA and SEISAN can both use GSE format (in addition to native formats) while SAC has its own format and may use GSE also. SEISAN and GSE formats can include the response with the data, while SAC cannot. Examples of response files will be given in the following as well as examples of how to switch from one format to another, which is not always a trivial matter.

6.9.1 GSE

The combined response, from the test response example above is defined in the GSE2.0 format as follows (Table 6.1)

Table 6.1 An example of a GSE2.0 response file. All numbers are from the test response information in 6.8. First line gives station and component (BERG, S Z), sensitivity at reference period (1 s) in nm/count (0.40), sample rate and date. Second line (PAZ2) gives total normalization factor of seismometer and filter in V/nm (0.3e7) and number of poles (9) and zeros (4). The poles are listed first and then the zeros. Last line gives total gain for amplifiers and AD converter in counts/V (0.2e7) and sample rate (50). Sensitivity and sample rate are not needed. If line DIG2 is not there, it is assumed that DIG2 factors are in PAZ2 line and the normalization constant is more properly called a scaling factor. The first line lists the sensor model (Sensor) and the last line the digitizer model (DigModel).

```
CAL2 BERG   S Z        Sensor    0.40E+00        1.   50.00000 2000/ 1/1 0: 0
PAZ2   1 V   0.30043413E+07                      9    4 Laplace transform

 -0.87964594E+00  0.89741838E+00
 -0.87964594E+00 -0.89741838E+00
 -0.40655220E+02  0.15172728E+03
 -0.11107207E+03  0.11107207E+03
 -0.15172728E+03  0.40655182E+02
 -0.15172728E+03 -0.40655205E+02
 -0.11107205E+03 -0.11107209E+03
 -0.40655193E+02 -0.15172728E+03
 -0.62831856E-01  0.54929354E-08
  0.00000000E+00  0.00000000E+00
  0.00000000E+00  0.00000000E+00
  0.00000000E+00  0.00000000E+00
  0.00000000E+00  0.00000000E+00
DIG2   2  0.20000000E+07     50.00000 DigModel
```

The gain and scaling factors are:

Sensitivity: 10^9nm/m/2.51 • 10^9 counts/m = 0.40 nm/count since at one Hz the total gain is 2.51 • 10^9 counts/m

Total normalization factor of seismometer and filter: 200V/ms^{-1} 1.5 • 10^{13}ms^{-1}/m • 10^{-9} m/nm = 0.3 • 10^7 V/nm

Total gain for amplifiers and AD converter: 2000 V/count • 1000 V/V = 0.2 • 10^7 counts/V

The GSE file has the sum total of the individual elements' poles and zeros as seen above. However, looking at the complete response file in Table 6.1, it is difficult to see which parts originate from which section of the recorder. In SEED, the different parts are usually separated (Table 6.2). This is also possible in GSE format. For a complete description of GSE format, see GSETT-3 (1997). Since the GSE format is an ASCII format, the response information will appear as shown above. In GSE, FAP (frequency, amplitude and phase) can also be used. The GSE format ONLY works with displacement response.

6.9.2 SEED

The SEED response is part of the SEED volumes and is well defined. It can be given in many ways and all stages in the response are described separately. Table 6.2 shows an example from a Quanterra station with a STS2 seismometer. The values are typical for the GSN network. The main parameters are:

Generator constant:	1500 V/ms^{-1}
Period:	120 s
Damping:	0.7
ADC sensitivity:	$0.41 \cdot 10^6$ counts/V

In addition, there are antialias filters and other filters. The response information is only contained in the binary SEED headers and has been dumped with an IRIS read program rdseed. Only a small part of the file is shown giving the main information and some of the filter coefficients.

In SEED, displacement, velocity and acceleration response can be specified. The response has been divided into stages. Stage 1 is the seismometer which has a velocity response described with 2 zeros and 5 poles and a normalization constant A0 of $6.0 \cdot 10^7$. We recognize the 2 zeros and the 2 first poles as coming directly from the seismometer parameters (6.59). What are the last 3 poles? Since a seismometer has a normalization constant of 1.0, A0 must be related to the last 3 poles. Taking out the normalization constant and the last 3 poles and plotting the response, we can get an idea of what the last 3 poles represent (Figure 6.14).

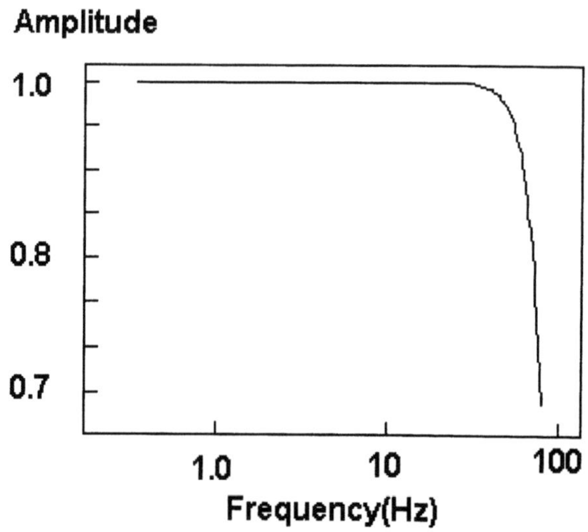

Figure 6.14 Response curve corresponding to the last 3 poles and the normalization constant as given in Table 6.2 for the seismometer in stage 1. Both axis are logarithmic.

Table 6.2 Example of response information for a SEED file. Only a small part of the file is shown, break indicates that a part is missing in between.

```
----------break---------------
#            +           +--------------------------------------------+
#            +           |   Response (Poles & Zeros),   CART ch BHZ  |
#            +           +--------------------------------------------+
#
B053F03      Transfer function type:    A [Laplace Transform (Rad/sec)]
B053F04      Stage sequence number:     1
B053F05      Response in units lookup: M/S - Velocity in Meters per
Second
B053F06      Response out units lookup: V - Volts
B053F07      A0 normalization factor:    6.0077E+07
B053F08      Normalization frequency:    0.02
B053F09      Number of zeroes:           2
B053F14      Number of poles:  5
#               Complex zeroes:
#                 i  real           imag          real_error    imag_error
B053F10-13     0  0.000000E+00  0.000000E+00  0.000000E+00  0.000000E+00
B053F10-13     1  0.000000E+00  0.000000E+00  0.000000E+00  0.000000E+00
#               Complex poles:
#                 i  real           imag          real_error    imag_error
B053F15-18     0 -3.700400E-02 -3.701600E-02  0.000000E+00  0.000000E+00
B053F15-18     1 -3.700400E-02  3.701600E-02  0.000000E+00  0.000000E+00
B053F15-18     2 -2.513300E+02  0.000000E+00  0.000000E+00  0.000000E+00
B053F15-18     3 -1.310400E+02 -4.672900E+02  0.000000E+00  0.000000E+00
B053F15-18     4 -1.310400E+02  4.672900E+02  0.000000E+00  0.000000E+00
----------break----------------
#            +         +----------------------------------------+
#            +         |       Channel Gain,   CART ch BHZ       |
#                      +----------------------------------------+
----------break----------------
B058F03      Stage sequence number:                  1
B058F04      Gain:                                   1.500000E+03
B058F05      Frequency of gain:                      2.000000E-02 HZ
B058F06      Number of calibrations:
----------break----------------
#            +         +----------------------------------------+
#            +         |       Channel Gain,   CART ch BHZ       |
#            +         +----------------------------------------+

B058F03      Stage sequence number:                  2
B058F04      Gain:                                   4.117280E+05
B058F05      Frequency of gain:                      0.000000E+00 HZ
B058F06      Number of calibrations:                 0
----------break-----------------
#                 +----------------------------------------+
#                 |  Channel Sensitivity,   CART ch BHZ    |
#                 +----------------------------------------
#
B058F03      Stage sequence number:                  0
B058F04      Sensitivity:                            6.175900E+08
B058F05      Frequency of sensitivity:               2.000000E-02 HZ
B058F06      Number of calibrations:                 0
----------break------------------------
```

Figure 6.14 shows a filter which turns out to be a 3 pole Butterworth filter. The filter can be considered to be built into the sensor. The stage sequence number 1, second part also gives the seismometer generator constant. Stage sequence 2 gives the ADC sensitivity and stage sequence 3 is the filter coefficients for the first anti alias filter, note it has no gain. Now several filters are given (not shown in Table 6.2) and finally at the end comes sequence number 0 which is all gain coefficients multiplied together, A_{all}, except the poles and zero normalization constant:

$1500\ V/ms^{-1} \bullet 0.41 \bullet 10^6$ counts/V = $6.2 \bullet 10^8$ counts/ms^{-1}

If the data from Table 6.2 are to be used in a GSE response file, one zero has to be added to the poles and zero representation to make it displacement. The other constants are calculated as:

Sensitivity in CAL2 line:
$10^9 /(A_{all} \bullet 2\pi f) = 10^9 nm/m / (6.2 \bullet 10^8\ counts/ms^{-1} \bullet 1 \bullet 2\pi \bullet 1\ Hz) = 0.26$ nm/counts
Total normalization factor of seismometer and filter in PAZ2 line:
$1500 V/ms^{-1} \bullet 6.0 \bullet 10^7 ms^{-1}/m \bullet 10^{-9}\ m/nm = 90\ V/nm$
Total gain for amplifiers and AD converter in DIG2 line: $0.41 \bullet 10^6$ counts/V

6.9.3 SEISAN

SEISAN is using an ASCII format for response files. SEISAN can use GSE, but it also has a native SEISAN format which can use PAZ, FAP or parameters. SEISAN only use displacement response. Assuming a Butterworth filter with 3 poles and a corner frequency of 80 Hz, the SEISAN parameter format with the same information as inTable 6.2 (without the antialias filters) is (Table 6.3):

Table 6.3 SEISAN response file with parameters from Table 6.2. The first line gives station, component and date from which response is valid. The parameters are from left to right: Seismometer period (s), damping, generator constant (V/ms^{-1}), amplifier gain (dB), ADC gain (counts/V), total gain at 1.0 Hz (counts/m), filter frequency and number of poles. The total gain is calculated by the response file program and is therefore not a required parameter.

```
CART BH Z101    1  1  1  1  0  0.000

 120.000  0.7000  1500.0     0.0 .41E+06 .39E+10 80.0        3.00
```

The SEISAN poles and zeros representation is seen in Table 6.4

Table 6.4 SEISAN response file with poles and zeros from Table 6.3. The first 3 values are the number of poles, number of zeros and total gain constant (counts/m) respectively. Following are the first poles and then the zero pairs.

```
CART BH Z101    1  1  1  1  0  0.000

     5     3 0.3685E+17-0.3700E-01-0.3702E-01-0.3700E-01 0.3702E-01
  -251.3        0.000       -131.0       -467.3       -131.0        467.3
   0.000        0.000        0.000        0.000        0.000        0.000
```

The SEISAN poles and zeros representation look almost as the SEED file except that one more zero has been added to make it a displacement response. The total gain constant has been calculated as stage 0 gain times poles and zero normalization constant A0

Total gain = $6.2 \cdot 10^8 \, ms^{-1}/m \quad 6.0 \cdot 10^7 counts/ms^{-1} = 0.37 \cdot 10^{17}$ counts/m

There are thus many ways of representing the response information. SEED is the most complete and best defined but also the most complex. SEED information MUST be part of the SEED data file, while for GSE and SEISAN, the response information can be added later and kept in separate files. The advantage with SEED is that the information always is there, but if it is wrong, it is hard to change. Another disadvantage is that few processing programs use SEED, so the SEED response must somehow be put into another format before processing the data. SEISAN and GSE are less confusing since they always are in displacement.

When operating more than a few stations, it is common to have changes in response information and it is unfortunately not always updated when station parameters are changed. In that case, it is an advantage to be able to add and modify response information afterwards and keep a data base of changing response information with time. If this is too complicated to do, it will often not be done and there are many networks without proper calibration records generally available for processing. It is therefore a help to have a processing system that easily can generate calibration files from any kind of instrumental information.

CHAPTER 7

Seismic stations

In the previous chapters, the equipment used for seismic stations has been described. When putting this equipment out in the field, we have a seismic station. Unfortunately it is not as simple as just putting the sensor and recorder somewhere and start recording. Although we would get data, it is likely that it would be so noise contaminated that it would be useless. So in this chapter we will describe some of the considerations to make when installing a seismic station. In short, the main goal is to make a sensor installation, which is as insensitive to ambient noise sources (human and environmental) as possible, so that the sensitivity for earthquake generated signals is high. In other words, the station should be as far away from oceans and humans as possible. This goal conflicts with practical considerations like access and costs, so a compromise must always be made.

The station can consist of only the sensor and some transmission line with the recorder elsewhere, as it is often the case for seismic networks (see Chapter 8), or the station can be complete with recorder and maybe a communication facility. Therefore individual requirements to the site, communication facilities, housing and power can be widely different. Today, few permanent stations are installed without some kind of communication facility.

Stations need power and putting up a good power system is a more important task than most people imagine, so a rather detailed description of power systems will be given.

In this chapter, we will primarily deal with permanent stations, however stations deployed temporarily will require some of the same considerations as permanent stations.

The most critical aspect of the seismic station is the sensor installation, particularly for broadband sensors. The number of ways of installing sensors seems to be proportional to the number of seismologists and sensor installation can cost more than $100 000, so it is no joking matter. There is much advice available in the literature, enough to fill a whole book. One of the most complete descriptions is found in New Manual of Seismological Observatory Practice – NMSOP (Bormann (Ed.), 2002) and also the old manual (Wilmore, 1979) has useful information. This chapter is to some extent based on information from NMSOP, so the reader is referred to this manual for more detailed descriptions.

Some terms should be defined. Very long period (VLP) means recording signals with periods up to 10 hours while long period (LP) is recording periods as long as 100 s. Correspondingly there are very broadband sensors and broadband sensors although some BB sensors, carefully installed, will record VLP (Trnkoczy et al., Chapter 7 in NMSOP, Bormann (Ed.), 2002).

7.1 Geographical location of a seismic station

When setting up new stations, they can be either part of a local or regional network, or each one can be a single station, which might be part of the global network as the IRIS stations. A station that is part of a local network obviously cannot be located anywhere, since the network must have some optimal configuration in order to best locate events in the given region (see Chapter 8). Studying a map will then give an ideal location, often a remote one, which however could be severely constrained by practical considerations:

Power. The power required by the equipment can vary from 1 W to more than 100 W depending on the type of station. High power stations need access to the power grid while low power stations might get by using solar cells, in which case a careful consideration on the size of batteries and solar cell to use must be made (see 7.7 on power). It is worth compromising other aspects of station installation in order to be able to use a public power line.

Communication. If the station is to transmit by radio, it often has to be located near mountain tops. Satellite communications do not have this limitation, but use more power. Land line communications require the station to be near the public telephone system. Telephone lines are often not available and cost can be high. However costs are rapidly going down and more lines are becoming available, so it is recommended to use public communication.

Noise. The station must be away from noise sources, particularly human made.

Security. Unfortunately, vandalism and theft is not uncommon and some areas are worse than others. There is not much point in putting up a nice low noise station, if the solar cells and batteries disappear within a week. So apart from making a solid construction, the geographical location can be important to avoid these problems.

Access. Ideal locations might require a helicopter to get there. Since few seismologists can afford that in the long run, easy access to a site is essential to ensure long-term maintenance.

Weather. Areas with humid conditions, extreme cold or warm areas and areas with a high probability of lightning should be avoided. Within a given geographical region, there might not be much variation in these parameters.

Topography. It is generally advised to avoid rough topography, which might modify seismic waveforms, however this conflicts with putting stations on hilltops for better communication.

Geology. It is well known that a sensor on soft soil is noisier than a sensor on solid rock, so soft soil should be avoided.

The above requirements are all conflicting since an ideal station is likely to be in a remote area with no power, communication or security. The primary goal should be to ensure low cost and long-term operational stability, since for most organizations running seismic stations it is usually easier to get initial installation funds than operational funds. This often means that the noise level will be higher than ideal and fewer earthquakes will be recorded. However since the operational stability will be higher, the total amount of data might not be reduced compared to a remote station with unstable operation. Considering the exponential growth of the number of seismic stations, it might not be a big loss to record a bit less at a particular site, if a continuous record of good data is obtained.

7.2 Site selection and seismic noise survey

A seismic station located on solid basement rock can always be expected to have a lower noise level than a station on soft sediments, even if there is not a clear noise source nearby. However, there can be unknown noise sources or the geology can be different from what it appears to be on the surface. What appears to be bedrock, might just be a big boulder. Before making a final selection of a site for a seismic station, a noise study must therefore be performed. As we have seen (Chapter 3), noise above 1 Hz is mostly originating in the near field, while lower frequency noise might originate far away. This means that a BB station recording mainly low frequency signals might be operated near a major city if the geological conditions are favorable.

The noise survey consists of recording continuously for e.g. 24 h, the longer the better. It is then possible to take out time intervals of e.g. 10 min, and make average noise spectra (see Chapter 3 and 6), which are then compared to the Peterson curves (Figure 3.4). It is thus possible to evaluate if the average noise level is satisfactory. The recording should ideally be made with the same type of sensor as is intended for installation in order to cover the same frequency band. This might be difficult in practice since the sensor might not be available yet. With sites for broad-band recording, it is also difficult to make a temporary installation with enough temperature stability to render the results useful (see sensor installation).

In any case, the main objective of the noise survey is to investigate near field noise sources, since the low frequency noise is likely to be constant over a large geographical area except for load generated tilt noise generated by e.g. traffic. However, that kind of noise is likely to also generate higher frequency noise, which will be seen in the noise survey. The recording can therefore be made with a short period seismometer, which will, with proper amplification, give good noise resolution down to 0.1 Hz. It is important that the gain is set high enough to get a good resolution of the noise, so the gain should be set substantially higher than for normal recording, where the purpose is to get the best dynamic range for earthquake recordings. With recording down to 0.1 Hz, the microseismic peak (Figure 3.4) can clearly be resolved. Ideally, records should be made day and night and summer and winter, however, that is rarely done. In most cases it should be possible to record continuously for at least 24 h to detect any periodic local noise sources. This means that in addition to the spectral analysis, the continuous record should also be inspected since there might be small impulsive disturbances, which might not show up in the noise spectra or just generally increase the spectral level. Figure 7.1 shows an example from station TRO on the west coast of Norway.

It is seen that the record (Figure 7.1) is not free from occasional spikes and small noise pulses and the record is overlaid with a continuous high frequency noise. The sensor is mounted on a pier on solid rock in the basement of a two stories building in a semi urban environment. The noise spectrum is seen on Figure 7.2.

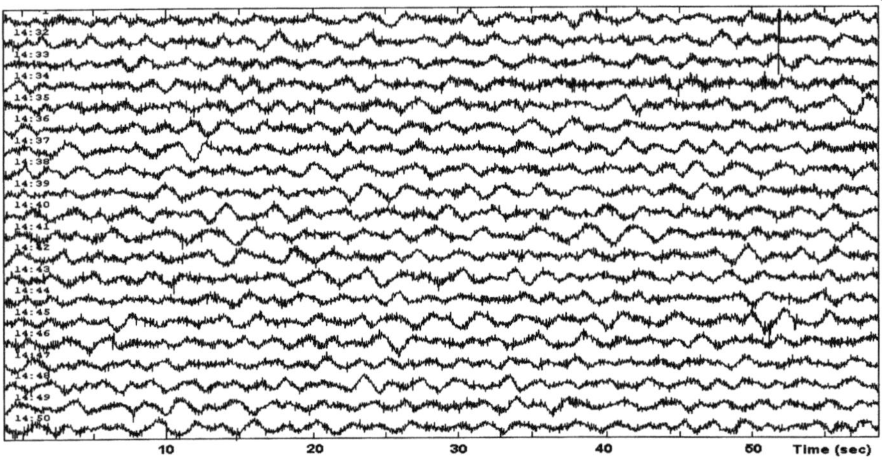

Figure 7.1 Continuous noise record. A window of 20 minutes of noise is seen. The sensor is a short period seismometer. Notice the strong microseismic background noise and several smaller noise spikes.

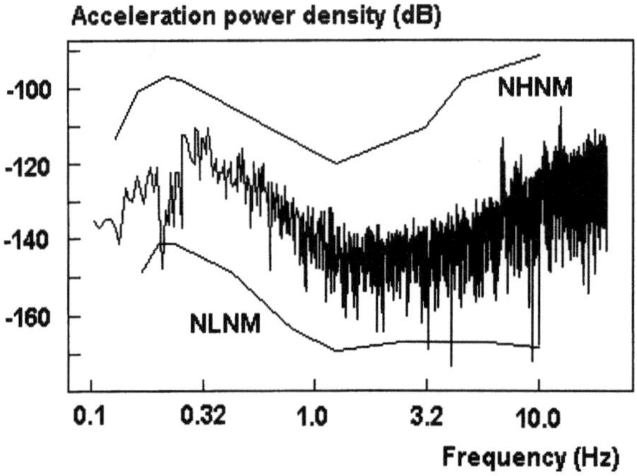

Figure 7.2 Noise spectrum of the first one-minute window from the record in Figure 7.1. The vertical scale is dB relative to 1 $(m/s^2)^2$/Hz. The Peterson noise models are shown (NHNM and NLNM, Chapter 3).

It is seen that noise below 1 Hz is acceptable and as good as for a similar station in a rural environment (Figure 3.5), while above 1 Hz, the city noise clearly shows up as it also does in the record (Figure 7.1). This site is acceptable for teleseismic recording but noisy for local micro earthquakes.

For frequencies below 0.1 Hz (using BB sensors), it is much more difficult to make a good noise assessment without very careful installation. Since the low frequency noise sources generally are far away, there is not much difference in the fundamental long period noise at bedrock level at different nearby sites and most of the low frequency noise recorded originate near the surface due to pressure changes and temperature induced tilt changes. In other words, making a noise measurement with a BB sensor tells us very little more than the noise measurements with a 1 Hz sensor unless an installation can be made on hard bedrock under very stable temperature conditions (1 deg/day change). In any case, the VLP noise is more influenced by the depth and type of installation than by the geographical location.

7.3 Installation of the seismic station

We now assume that a proper site has been chosen where all basic requirements to noise, physical facilities etc. have been met. The installation consists of setting up sensor and possibly recorder and/or transmission equipment. The installation will depend on the type of sensor, and whether local or remote recording is done.

General considerations. Sensors are usually posed on glass, non-porous ceramic or hard plastic plates to avoid stray currents and corrosion of the sensor feet when in contact with cement.

The horizontal sensors are always oriented NS and EW such that a motion N or E gives a positive signal. The most common way to orient the sensors is to use a magnetic compass (not near the powerful magnet of an electromagnetic sensor!), taking magnetic declination into consideration. Inside buildings or vaults this might be unreliable and the direction must be taken outside and transferred inside. At sites with unreliable compass measurements (like far north), the direction can be determined by sighting known land-marks. The orientation of the sensors should be within ±2 degrees.

Short period sensors and accelerometers usually do not require thermal insulation since they are little sensitive to temperature changes. BB sensors require a very good thermal insulation, for more details on BB installation, see 7.4.1 below.

It is important to install all sensors on basement rock, if possible. The depth of the installation is then only important to the extent of eliminating low frequency noise caused mainly by temperature changes. This becomes an important issue for BB sensors, see below. Short period sensors are sometimes installed in concrete over-ground bunkers, which at the same time serve as housing for equipment and a secure installation. If there is no hard rock to be found, it is always an advantage to put the sensor (any kind) a few meters underground.

The most common sensor to install (at least until recently) is a 1 Hz, one or three-component station. The very first choice is where the recording is going to take place. If there are buildings nearby (50-1000 m), the best choice might be to place sensors in the field and put a cable into the nearest building where both power and communication facility might be available. This can make the field installation very simple and all or most electronic equipment is placed indoor in what often might be an office environment. So the indoor equipment can e.g. be a standard PC with ADC, none of which require special environmental protection, thereby saving cost. How far can a sensor be placed away from the digitizer? There is no standard answer to this, since it greatly depends on the electrical noise environment, output from the sensor and type of digitizer. Some examples will be shown:

Case 1. Passive sensor in the field, all equipment indoor: If a cable with individually shielded pairs is used, there are examples of cable lengths of 1 km with a standard 1 Hz sensor (Havskov et al., 1992). However, considering that the signal levels are around 1 μV, this requires very careful grounding of the amplifier/ADC and there is no standard answer on how to do that. Cable lengths of 10-30 m usually pose no problem.

Case 2. Passive sensor and amplifier or active sensor in the field and the rest of the equipment indoor: By providing a higher output, and a lower output impedance, the transmission line is much less prone to electrical noise interference. However, the disadvantage is that there is now electronic equipment in the field requiring better protection and power.

Case 3. Sensor and digitizer in the field, rest of equipment indoor: This is the most stable solution, since now data is sent by a serial line to the recording computer and, in case of transmission errors, data can be resent. Since digitizer and sensor are close together, there is very little chance of electrical noise getting into the system. The RS232 can be used for short cable lengths (up to 20-40 m), while the RS485 can be used for up to 2-4 km cable length (see 8.4.1), depending on the bit rate . Normally, the digitizer will time stamp the data, so the GPS must then also be in the field and the GPS antenna might then be the only visible equipment above ground. Alternatively, some recording software will have the GPS together with the recorder, so time stamping takes place with the indoor equipment (see also Chapters 4 and 5). The digitizer can still be synchronized by sending out time correction signals from the central, e.g. GPS-delivered one-pulse per second (1 pps).

7.4 Sensor installation

The simplest sensors to install are the 4.5 and 1 Hz seismometers or an accelerometer. A typical installation is shown in Figure 7.3.

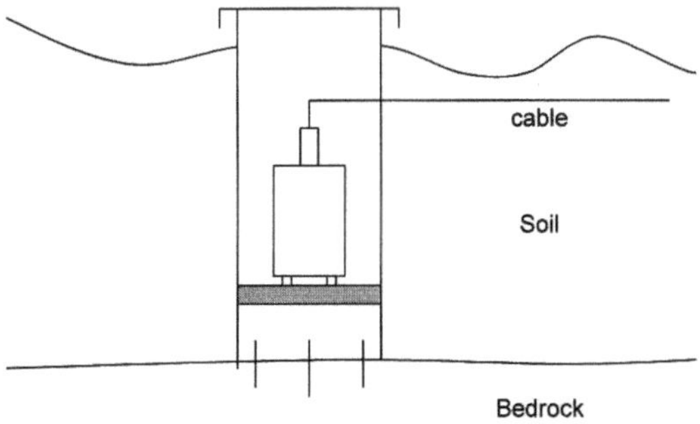

Figure 7.3 Typical installation of a 1 Hz vertical seismometer. The sensor is inside a PVC tube with an airtight screw on the lid (standard sewage tube). There is cement at the bottom, which has been fixed to the bedrock with a few pieces of iron drilled in. The sensor is placed on a glass, ceramic or plastic plate.

The sensor is in a nearly airtight tube only sticking a few centimeters above ground. In this way, there will be little influence from wind. The cable is put out through the side and the hole is sealed with silicone rubber. The lid can also be covered by soil, but usually this is not needed. The cement must be of the waterproof type although as an additional precaution, there is also silicone around the rim of the glass plate in the bottom. By making the tube a bit longer, there could also be room for e.g. a digitizer.

Figure 7.4 shows a simple construction using a commonly available aluminum transport box. This kind of box is available in many sizes so it can fit many types of installations.

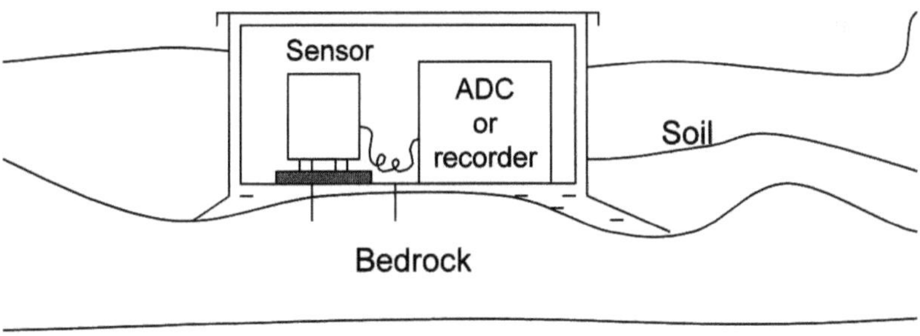

Figure 7.4 Typical installation of a 3 component station and optional equipment. The box is a common aluminum transport box. Part of the bottom has been cut out and is the bottom has been covered by cement, which has been fixed to the bedrock with a few pieces of iron drilled in. Insulation has been put in around the sides and in the lid.

The box has a rubber seal to make it nearly airtight. When putting in cement to fix the box to the ground, it is important to cover the aluminum with a non metallic material like epoxy or plastic tape, since the aluminum will corrode very fast if it is in contact with the cement. The box is normally insulated. For short period instruments, temperature changes have little influence, but insulation, at least in the lid, will reduce condensation. This type of construction can also be used with BB sensors with a moderate requirement to low frequency performance (LP). However in the case of BB sensors, the sensor must be double insulated (see below). The lid of the box is then usually covered with soil to reduce wind noise and temperature variations.

For a telemetered station in a remote area, more security might have to be provided than the simple construction in Figure 7.4 and there may also be a need to install more equipment. A typical installation is shown in Figure 7.5. In this construction, sensor, signal conditioning and battery are placed in the house, while the radio equipment can be in the mast or in the house depending on the distance and type of equipment. Usually, the radio equipment is in the mast. The GPS can be on the roof of the house, but for security reasons it is most likely to be in the mast. Analog stations will not have a GPS.

The mast must be placed away from the house to prevent noise from the swinging mast. A minimum distance is 5 m, however a high mast in a seismic quiet area might have to be 50 m away. This is not very practical, so the mast will often be closer.

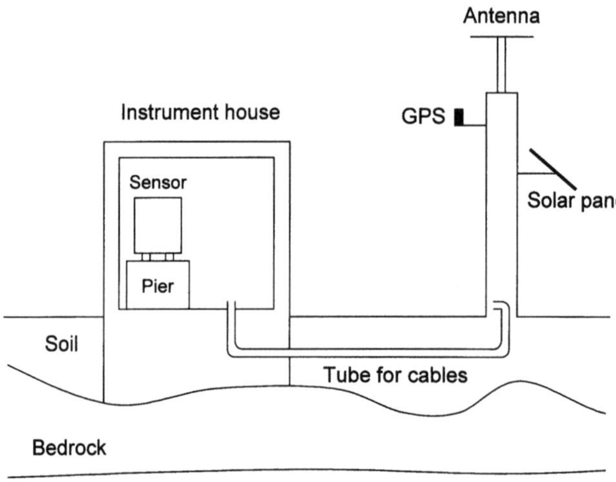

Figure 7.5 Seismic station in a telemetered SP network. The instrument house is made of concrete, and has a typical dimension of 1 to 2 m on the side. Ideally the cement goes down to solid rock. Cables from the instrument house go underground to the communication mast. The house will usually have the sensor and signal conditioning equipment. There might also be a pier for posing the instruments, although that has no influence on seismometer performance.

7.4.1 BROADBAND SENSOR INSTALLATION

Broadband sensors require special care in installation in order to get a good low frequency performance. We should here distinguish between LP and VLP performance. VLP means that frequencies down to earth tides should be resolved while LP performance might be installed to 'only' record regional and global events down to a frequency of 0.05 to 0.01 Hz. The lower the frequency, the more demanding and costly is the installation and the sensor, so it should be carefully considered what the purpose of the station is before starting an elaborate construction.

The main problem with low frequency noise is tilt, which mainly affects the horizontal sensors. A deformation of 1 μm over a distance of 3 km oscillating with a period of 10 min gives a vertical acceleration of 10^{-10} m/s^2 in the vertical direction and 10^{-8} m/s^2 in the horizontal direction (Wielandt, Chapter 5 in NMSOP, Bormann (Ed.), 2002). These values are well above the NLNM. Tilt can be caused by temperature deformations of the ground or atmospheric pressure changes. A good installation should therefore be as deep as possible and on hard rock.

The BB sensor itself will also be affected by temperature fluctuations, so short term temperature changes (with smaller duration than periods to be measured) should be avoided. Different sensors have different sensitivity to temperature changes. Typically, the sensor should be protected from day-night temperature changes. The 24-hour sensor variation in temperature should be less than 1 degree. This stability is mainly achieved by insulation, however, a deep installation will make that easier by having stable surrounding temperatures. Since all BB sensors produce heat, this will produce air drafts that also give noise. The insulation must therefore be installed close to the sensor in order to suppress air drafts.

Air pressure changes will mainly affect the vertical sensors by the effect of variable buoyancy, so most BB sensors are enclosed in airtight containers.

BB seismometers are to some degree sensitive to changes in the magnetic field so nearby electrical disturbances might affect the sensor output, for more details see NMSOP (Bormann (Ed.), 2002).

Installation of different kinds of BB sensors is somehow dependent on the type of instrument and we will here only give general guidelines for BB installation. Since most quality BB sensors are airtight and provide some degree of magnetic shielding, it is assumed that the user only has to worry about temperature shielding. There are many and elaborate ways of doing the thermal insulation. Figure 7.6 shows a relatively simple setup from the GEOFON network (Hanka and Kind, 1994), which has been shown to be very effective down to VLP.

The base plate cannot easily be bent by the influence of pressure variations and gives together with the foam rubber insert, extra thermal stability. If only LP performance is required, a similar installation can be made without the aluminum box. The main thing is to use a double insulation and fill up the box with insulation material to avoid air draft. An aluminum pressure cooker with flat bottom may also make a practical case for thermal and pressure insulation (stainless steel is not recommended because it may influence the magnetic elements inside the seismometer). Anyway, a good thermal contact with the ground provides thermal stabilization, since ground temperature changes are very slow.

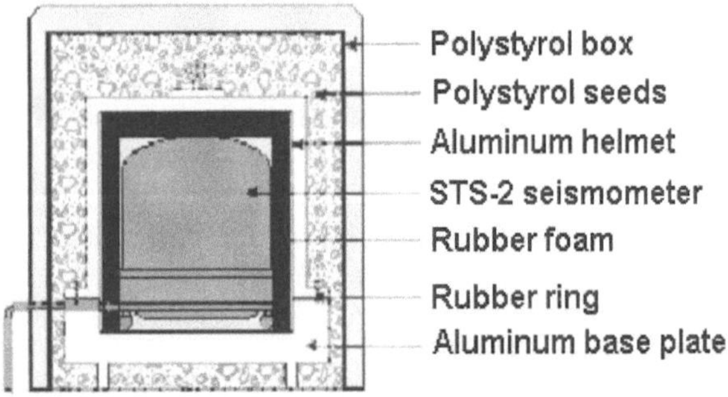

Figure 7.6 GEOFON BB installation. There is a thick (3 cm) aluminum base plate and a thinner aluminum helmet with a cylindrical foam rubber insert (redrawn from Fig. 7.50 by W. Hanka in NMSOP, Vol. 1, Bormann (Ed.), 2002; copyright granted by IASPEI.

Vaults for VLP installations. The vault construction itself has minimal influence on the performance as long as the environmental disturbances can be reduced to a minimum and only in rare cases is it necessary to use boreholes or excavate tunnels, although this often has been used for USGS IRIS stations. Figure 7.7 shows some examples of GEOFON vaults.

The bunkers in Figure 7.7 are made of concrete. Figure 7.7a shows a pier for seismometer installation and the geology is weathered rock. Figure 7.7b shows a solution where a wide borehole is drilled into hard rock and a 1 m wide steel tube has been placed into it. Cement is put at the bottom where the sensor is placed. Figure 7.7c and 7.7d have bunkers placed in sedimentary layers or weathered rock.

A common construction for the USGS IRIS network is shown in Figure 7.8.The figure shows an artificial tunnel with 4 segments, which acts as air locks. The sensors are in the inner compartment.

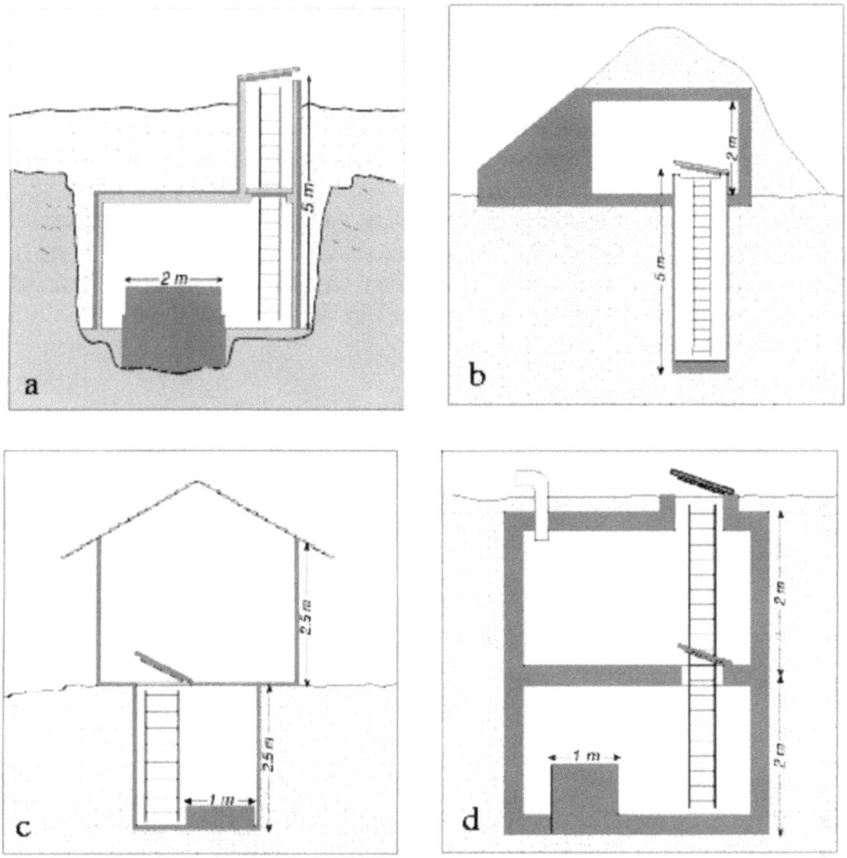

Figure 7.7 Example of BB vaults from the GEOFON network. a) Underground bunker vault for remote recording, b) Wide and shallow borehole type, c) and d) Simple bunker construction. Note that b – d allow onsite recording since there is a separate recording room. Copied from Fig. 7.55 by W. Hanka in NMSOP, Vol.1, Bormann (Ed.), 2002; copyright granted by IASPEI.

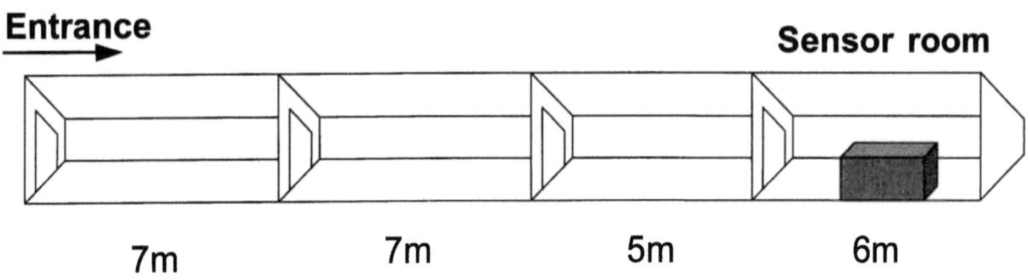

Figure 7.8 Horizontal tunnel construction with different chambers used by USGS IRIS. The total length of the tunnel is 25 m (redrawn from Fig. 7.52 by W. Hanka in NMSOP, Vol.1, Bormann (Ed.), 2002; copyright granted by IASPEI.

7.4.2 BOREHOLE INSTALLATIONS

We have seen that getting a low noise BB installation can be difficult and even installation in mines and tunnels will not always give very good results. An alternative is to install a specially designed sensor in a borehole. This technique has been used for the last 30 years and has shown that the VLP and LP noise level can be substantially reduced. The amount of reduction in the horizontal noise can vary from site to site and the reduction is the largest for the first few meters, which is also why the vaults shown in Figure 7.7 often give good results. The reduction in the noise level is due to the stable pressure and temperature which particularly reduce tilt noise. In a particular case, which might be typical, the reduction was up to 30 dB in noise spectral acceleration density power (Figure 7.9).

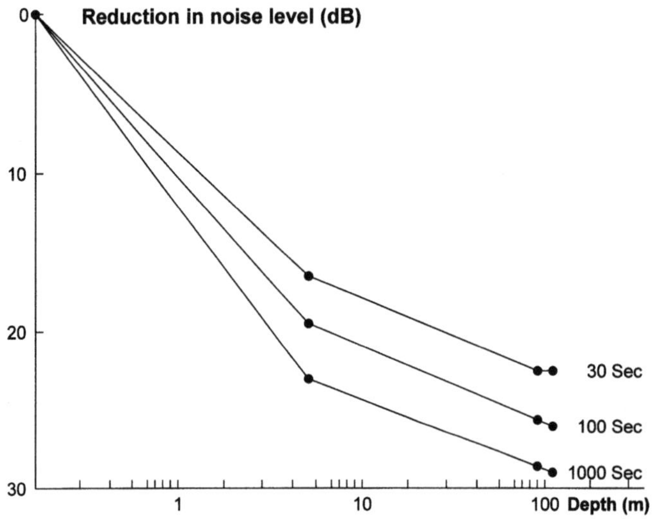

Figure 7.9 Horizontal noise attenuation (dB, spectral acceleration density) as a function of depth and period. Modified from Fig. 7.59 by L.G. Holcomb in NMSOP, Vol.1, Bormann (Ed.), 2002; copyright granted by IASPEI.

The test site consisted of 18 m of unconsolidated overburden overlying fractured granite bedrock (see NMSOP). Note that noise reduction is largest in the first few meters. It is generally considered that a 100 m hole is optimum since little improvement in noise level is obtained below 100 m.

Drilling a borehole and installation of borehole sensors is work for specialists and only a summary description will be given here, for more details see NMSOP. Most borehole sensors will fit in a hole with an internal diameter of 15.2 cm (see examples in Chapter 2). The hole cannot deviate more than 3-10 degrees from vertical depending of sensor type (typical 4 degrees). The steel cased hole is sealed at top and bottom to make it air and watertight. The sensor is lowered into the hole attached to a steel cable, and has a separate cable for electrical connections. The sensor must have good contact with the walls of the hole. Traditionally, this has been done with a clamping device (often motor driven), while lately, fixing of the sensor has been done by pouring sand down and around the sensor. In both cases, the sensor can be removed. The advantage of using sand is that it provides a lower horizontal noise level than clamping (even when additional insulation is used) and it is in addition cheaper.

Borehole installations should not be made within a few km from the coast and on very small islands since the microseismic noise will then be very high just like for any other installation.

The horizontal orientation of the sensor can be determined using a built in compass. An alternative and cheaper method is to temporarily install a well oriented sensor at the surface near the top of the hole. By recording a distant event at both sensors, the orientation of the down-hole sensor can be calculated by correlation techniques to within a few degrees. See also 10.2.4 .

Since the noise level in a borehole can be very low, the sensor self noise must be at least as low, which is not always the case. Figure 2.40 shows the self-noise for one of the best borehole sensors, the Geotech KS54000. At frequencies below 0.01 Hz, the sensor self noise is higher than the NLNM. Considering that the self noise for the KS54000 (Figure 2.40) is a theoretical factory specified level, and real levels are higher, it is clear that available instruments place some limits on achievable noise levels. It is obviously not worth drilling an expensive hole if the sensor cannot achieve the expected low noise level.

A borehole installation is not cheap. A typical borehole costs $ 200000 and the sensors are in the range $30000-60000. In addition it comes recording equipment, housing etc. However, this might still be cheaper than making a tunnel and in some areas with only sediments, there might not be much alternative if a VLP installation has to be made.

Although borehole installations are mostly made to improve VLP noise levels, there are also borehole installations of accelerometers with the purpose of studying ground motion amplification. Since low frequency is not important and high levels of motion are measured, the installation is simpler to make than for BB sensors with respect to temperature and pressure stability. However, in most other aspects, the installation is similar to installation of BB sensors. For more details, see NMSOP.

7.4.3 WHAT IS A GOOD VLP STATION

We now have a description of the possible ways of constructing a good VLP station. Even then, there is no guarantee that a particular installation will be very good since there might be unknown geological or environmental conditions present. A good way of judging the VLP performance is to record earth tides. If that can be done relatively noise free, then the station is a good VLP station.

Figure 7.10 shows records from two VLP stations. Station LVC is built on hard basaltic rock in a desert environment, while KMBO is on soft volcanic conglomerate. Both stations are in identical tunnels (Figure 7.8). This example shows that geological conditions can strongly affect VLP performance. Similar quality records have been obtained by installations as shown in Figure 7.6 and 7.7, so the conclusion is that only in rare cases are boreholes or tunnels needed.

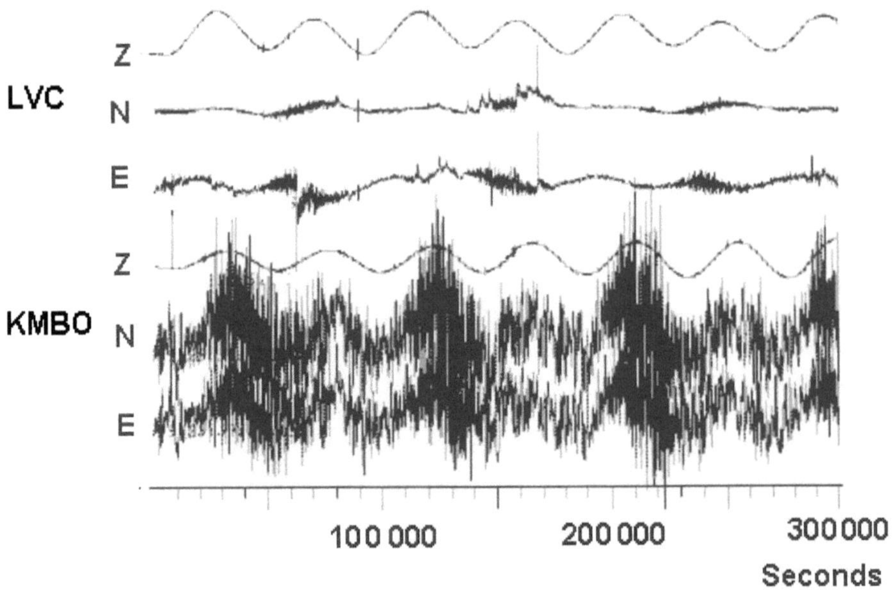

Figure 7.10 Records from two different VLP stations. The earth tides are clearly seen on all records. Note the higher noise on horizontal components than on vertical components (modified from Fig. 7.53 by W. Hanka in NMSOP, Vol.1, Bormann (Ed.), 2002; copyright granted by IASPEI.

7.5 Temporary installation of seismic stations

Temporary seismic stations are often established for field experiments like aftershock recordings or refraction experiments. Obviously, elaborate vaults or cemented piers are not constructed for temporary use; however, the same principles apply for temporary installations as for fixed installations in terms of noise, access and protection of the equipment. The most common sensor to use is the short period, but BB sensors are increasingly used also. It should not be expected to get VLP performance from a temporary installation, but by using a good thermal insulation and installing on bedrock, even a temporary BB installation can give acceptable LP results.

Sensor. If bedrock is not available, which is very likely in temporary installations, a common technique is to bury the sensor 1-2 m and fill up the hole with soil. If in a wet soil, this requires a waterproof sensor and connector. Wrapping the whole system in plastic might provide some protection if the soil is not soaked. This method requires careful leveling of the sensor and might be difficult if the sensor has to be aligned vertically to within a couple of degrees. For surface installation it is important that the sensor is shielded from wind by putting it in a box. The common aluminum transport box is a simple choice, see Figure 7.11.

Recorder. Despite what manufactures might tell you, few recorders will like to sit out in the open, particularly in humid or very hot conditions. They might be completely waterproof when new, but all seals and connectors have a tendency to wear out after repeated use. So it is definitely an advantage to put the recorder in some box for protection for longer duration experiments. Both the recorder and battery might be put in the same box as the sensor if installed as in Figure 7.11. For BB sensors, it is best to have the sensor in its own box and the recorder apart to avoid disturbing the sensor when servicing the equipment. It is important that it is easy to operate the recorder, which might require connecting a laptop or pushing some buttons. So when the rain is pouring down, it is nice to be able to work under the big lid of the transport box. A large plastic box can also be used as shown in the example in Figure 7.12. Using wireless LAN might be a nice way to connect to the recorder without opening lids or using connectors.

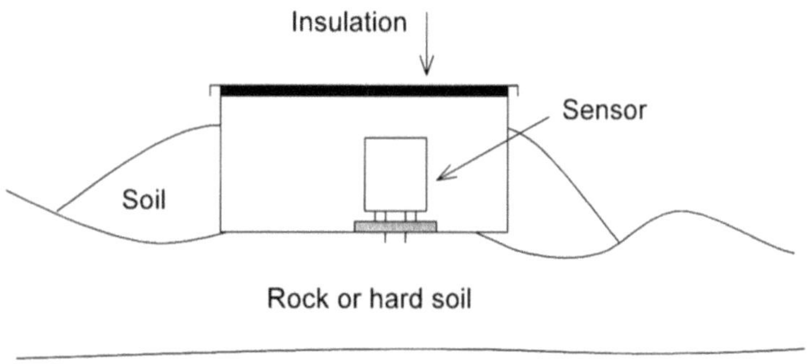

Figure 7.11 Temporary sensor installation. The sensor is placed in an aluminum box. Part of the bottom of the box is cut out so the sensor is placed directly on the ground posed on a glass or plastic plate. Motion of the box will then not directly affect the sensor. The box is almost covered with soil and there is thermal insulation under the lid. For BB sensors, the box can be completely covered with soil for added shielding. Any water getting into the box will drain out of the hole in the bottom. In soil, the box can also be partly dug down.

7.6 Lightning and over-voltage protection

A lot of equipment is damaged due to high voltages, mostly from lightning that causes voltage induction in cables. The cables can be external power cables, seismometer cables or communication cables. There are many ways of protecting the equipment, but common to all is to have a good grounding. If equipment is located in an existing building, ground wires might already be available, otherwise a proper grounding system must be made, which usually consist in digging down enough metal wire to make the resistance to ground low enough (ideally less than 1 ohm). The protection then consists of the following:

- Have all equipment in a grounded metal box, not always easy to do.
- Protect all cables going into the equipment with surge protectors, which means that if the voltage rises above a certain level, the input is shorted to ground.
- Install cables so they are as little exposed as possible, ideally underground.

Figure 7.12 A very simple field installation. A plastic box has been placed directly on the ground. The sensor is placed in the cut out hole. Inside are also the battery and recorder. The cable going out is for the GPS. After installation, soil was put up around the sides and insulation was put on top of the lid. The sensor is a Kinemetrics Ranger 1 Hz seismometer (Table 2.3) and the recorder is a GeoSIG GBV316 (Table 5.3).

Electronic equipment for seismology normally has over-voltage protection built in. The most sensitive equipment is usually computer equipment and particularly communication channels like radio links and modems. A good telephone line protection might cost more than a modem, so it might be as cost effective to change the modem as the over voltage protection. Sensors rarely get hit and digitizers also seem to be quite robust, while incoming power or communication links are highest on the hit list. Lightning protection can be very expensive and complicated with many components to replace after a hit, so there is a limit to how much it is worth doing. On the other hand, ignoring the problem completely is an invitation for problems. As a rule, avoid the top of mountains and, if possible, do not install a mast at a site such that it becomes the higher point in a wide environment. If this is unavoidable, use an even higher mast with lightning protection not too close to the equipment mast, to avoid this to be hit directly by lightning.

7.7 Power

All seismic stations need power and all too frequent do we see stations stopping due to power failure. So it is worth considering the power system carefully.

Nearly all seismic equipment use 12 V. Internally, most electronic systems use ± some voltage, so some manufactures still make a few pieces of equipment requiring ±12V, thereby making life simpler for themselves to the dismay of the users. This is mainly the case with sensors (see Table 2.3).

The power system will strongly depend of the type of installation and can require from 1 W to over 100 W so it is hard to generalize and examples will be given instead. In all cases, there should be some kind of backup power, so if the primary source fails or disappears (like at night for a solar cell), the backup power takes over. The other function of the backup system is to avoid system stop and/or data corruption due to small power failures. In addition, the backup system will smooth out power spikes in the AC system. This requires good batteries, so a detailed description on batteries and battery handling will be given.

7.7.1 STATIONS RUNNING OFF THE MAIN AC SUPPLY

The majority of permanent stations, except local network stations, use the public power grid so there are usually no power limitations. However, if a station uses a lot of power, the battery powered backup systems might be large and expensive. The simplest way to make a permanent station is to buy a self contained recorder with built-in ADC and ready made sensor cables with both signals and power, since many recorders will provide power in the sensor plug. Figure 7.13 shows an example.

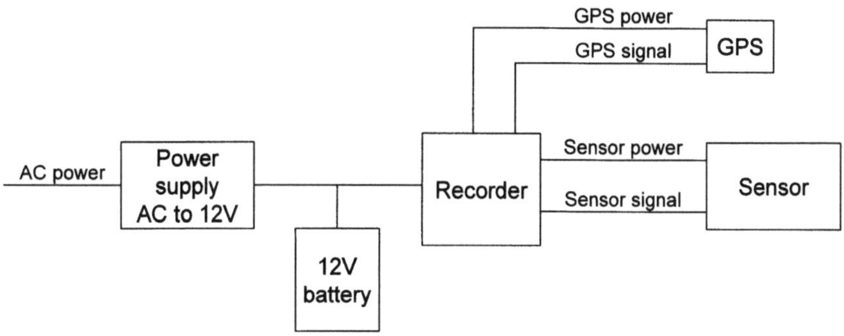

Figure 7.13 Self contained recorder running off main power. The connections are shown as single wires for simplicity.

The power supply is NOT a standard car battery charger, which might destroy both the battery and the recorder by having very high ripple (see Appendix) and very bad voltage regulation, inducing noise into the recorder. The battery charger must be an electronically well regulated power supply. Some recorders might have a built in power supply and backup battery so they can be directly plugged in the AC power. The recorder can operate without the battery, but this would make the system sensitive to irregularities in the AC power supply. For long term operation, it would be an advantage to have the battery outside the recorder, since batteries can fail and leak. Even with a built in power supply, most recorders will also have input for 12 V.

If the user chooses to have a more custom made system, there might be more power circuits to add. A general scenario is to use a standard PC for data acquisition and only buy digitizer, sensor and GPS. A typical connection is seen in Figure 7.14.

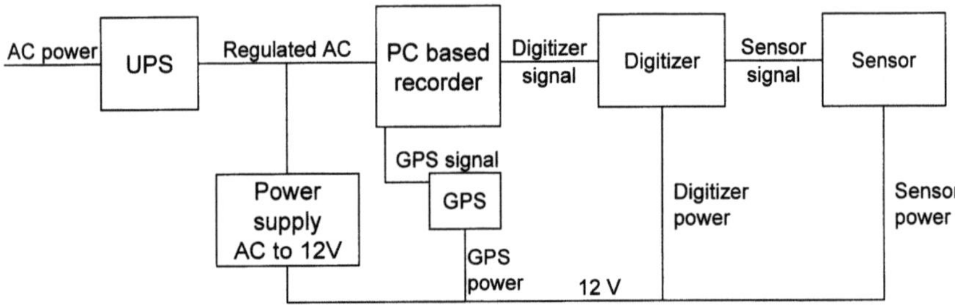

Figure 7.14 Station based on custom made components and AC operation. UPS stands for uninterruptible power supply. The connections are shown as single wires for simplicity.

The uninterruptible power supply (UPS) consists of a battery, a charging circuit and an AC generator. If the outside AC power fails or has variation in power, the UPS will maintain a steady AC output voltage. The autonomy depends on the battery size. UPS units are available everywhere at low prices since they are frequently used to protect computer equipment. A small UPS might only be able to operate a standard PC for a few minutes, but it has an important function in providing a stable power supply. Some UPS's can also communicate with the PC, so when the UPS battery is almost exhausted, the PC can make an orderly shutdown. Some units do not recover if the battery is dead so it is important that the UPS will function with a bad battery if line power is present.

In the example above, the GPS is connected to the PC and must therefore have a separate 12 V wire. The sensor is assumed to work on ±12 V so a DC-DC converter is needed. Some digitizers might be able to provide ±12 V directly and the digitizers could also have the GPS connected directly and the extra converter and cables could be avoided.

7.7.2 BATTERIES

There are several types of batteries available for running seismic stations and by far the most common is the lead-acid type battery. Almost all equipments use a nominal 12 V and usually it can operate in a wider range like 9-16 V. Batteries are composed of chemical cells in series; e.g. 6 cells of nominal voltage 2 V make a 12 V car battery. All batteries share some common properties (to be described in more detail below) like capacity, self-discharge and temperature dependence. Battery failure and overestimation on how long a battery will last are common problems so it is worth dealing a bit with the details. The main types of batteries used in seismic stations are:

Dry cells. These are the common everyday types of batteries, which cannot be charged. They are little used in seismic instruments except for backup purposes, since the capacity is too low and therefore the cost too high. However, with the power consumption continuously being lowered for recorders, dry cells might be more common in the future. At least one recorder (Reftek Texan, Table 5.3) uses dry cells due to its low power consumption. Dry cells are also used in some OBS's. A fully charged dry cell has a voltage of 1.55 V at 20°C.

Lead-acid batteries. This is the most common type used. Lead-acid batteries come in all qualities, shapes and sizes. For internal instrument use, these batteries are available completely sealed, while most batteries are open so they can vent oxygen and hydrogen released during the charging process and consequently they have to be topped up with water. A fully charged cell has a voltage of 2.15 V and a voltage of 2.35 V is needed to fully charge it at 20°C. Standard car batteries are lead-acid.

Lead-calcium batteries. The basic difference from lead-acid batteries is that lead-calcium batteries have little self-discharge and are not designed for supplying high current peaks. These batteries are also called gel-cells since they nearly always come as a sealed battery with a gelled liquid. Sealed lead-acid batteries can also have a gelled liquid. However, since this technology was first used with lead-calcium batteries, reference to a gel-cell normally means a lead-calcium battery. The lead-calcium batteries are commonly used inside instruments. A fully charged cell has a voltage of 2.17 V at 20°C and a voltage of 2.40 V is needed to fully charge the battery.

Nickel-cadmium (Ni-Cd) batteries. This is the most popular consumer type rechargeable battery and it is completely sealed. The battery has a more constant voltage while discharging than the lead-acid battery and can be charged and discharged completely hundreds of times. They present "memory effect", that is a loss of capacity if they are charged before being completely discharged. The disadvantage is high cost compared to a similar size lead-acid battery. A fully charged cell has a voltage of 1.30 V. Usually they are charged at a constant current of 10% of their capacity (see below) per hour: E.g., a battery of capacity 800 mA·h will be charged at 80 mA. The charge time in this way is 12-14 hours. There exist fast Ni-Cd chargers that operate at a non-continuous higher rate, but they need to control the charge parameters to avoid the risk of battery explosion.

NiMH batteries. At present they are substituting Ni-Cd at a similar cost, but without the charge memory effect of nickel-cadmium batteries. Because of their extensive use in cellular phones, a further decrease in cost is expected. The capacity per unit of volume is one of the highest available. The cell voltage is 1.2 V. The battery has an almost constant voltage until 80-90% discharged.

Li-ion batteries. Used also for cellular phones, they have the highest energy density, but are so far more expensive. They do not present charge memory problems. The cell voltage is 1.2 V.

The battery characteristic parameters are:

Capacity. The battery capacity is measured in Ah. This means that e.g. a 60 Ah battery theoretically will be able to give 1 A during 60 hours or 60 A during 1 hour. Unfortunately, the capacity is smaller for larger discharge rates, so a given capacity is only correct at one particular current at a given temperature. For use with seismic instruments, the discharge is usually low, so one can count on the full rated capacity. The capacity is also affected by temperature in different measure for different types of batteries. Typically, the capacity is only half at -10°C compared to 20°C, which is the temperature commonly used when specifying capacity.

Self-discharge. All batteries will discharge if not used; this is particularly the case for all chargeable batteries. Dry cells will often last years before being discharged. The self-discharge depends on type and quality of battery. A standard car battery might completely discharge in 6 months while a higher quality lead acid battery for instrumentation might only discharge 1% per month. The self-discharge is also affected by temperature. It might typically go 5 times faster at 40°C than at 10°C. So batteries should be kept at a cool place and charged regularly when not in use.

Battery life. Rechargeable batteries come in many different qualities. High quality primarily means that the battery can be recharged many times before dying. The second indication of high quality is a low internal discharge. The fastest way of destroying a standard lead-acid battery is to let it completely discharge a few times. A normal car battery might be destroyed after 5-10 discharges while a quality battery might sustain hundreds of discharges. So a lead–acid battery should never completely discharged in order to have a long life.

Battery internal resistance. Batteries have an internal resistance inversely proportional to the state of charge. So when a battery is low in charge, it has a relatively high internal resistance: The open-circuit voltage may be almost normal (as for a fully charged battery), but it drops quickly when a load tries to drain current.

Kind of battery to use. Many recorders come with a built in sealed battery of a typical capacity of 6-12 Ah, so a similar battery should be used for replacement. For permanent stations powered e.g. by solar cells, it is advisable to use a quality sealed battery that can sustain many charge-discharge cycles. A standard car battery will often not last very long and might easily dry up, particularly in hot environments. So the extra cost of the high quality battery (a factor of 2-5) will usually pay off both in battery life and to avoid the extra cost of going unexpectedly to the station to replace the battery. Companies selling solar cells or windmills will usually have a range of these kinds of batteries. The capacity of the battery will depend on the expected power input and the power consumption of the station. The longer the batteries can operate the station without e.g. solar power, the better. However, the battery should have a capacity to operate the station for at least 3 days in order to avoid (e.g. with solar cells) large discharge-charge cycles, and it is hardly worth installing a battery of a capacity less than 100 Ah. These numbers give some bounds on the battery size.

Note that in hot climates, the capacity of a battery will be larger than in cold climates. In subzero temperatures, it is advisable to install the battery in an insulated box so the heat generated in the battery during discharge will help keeping the battery warm.

For a temporary installation it is particularly important that the battery can operate the equipment as long as possible without battery change and less important whether it will last several years, so also cheap car batteries can be used. It is often easier to buy a battery locally, than transporting the battery over long distances. Furthermore, there exists legal regulations that limit the air transport of batteries for safety reasons.

How long will a battery last before having to be recharged. The short answer is: Much less than you think. When charging batteries, it is common to only charge to 80 % capacity (see charging batteries below) and when discharging, it is hardly possible to go below 10 % charge and 20 % would be a better choice to keep the battery from being too much discharged. That leaves a 60-70 % useful battery capacity. So, a recorder using 6 W or 0.5 A requires 12 Ah a day. To operate one week, a $12 \cdot 7 / 0.7 = 120$ Ah battery is needed assuming that 70 % of the battery capacity is used. A 120 Ah battery is a standard 'luggable' size, but much larger than this, and it is hard work to carry more than a few meters. So, recorders using more than 6 W can hardly be called portable recorders. With a 2 W recorder, life of course becomes easier, and these recorders are, at a price, available.

Charging lead-acid batteries. One should think that charging batteries is just a question of connecting a battery charger and waiting until the green light comes on. It *might* be that easy with an intelligent battery charger, but most often batteries are charged with simpler devices.

The first task is to determine the state of charge of the battery. A simple way is to measure the density, which directly gives the state of charge. This is at best messy and cannot be done with sealed batteries. Since the state of charge is nearly linearly related to the *unloaded (open circuit) battery voltage*, a simpler way is to measure the battery voltage. Table 7.1 gives the relation. Note that when charging a battery, the unloaded voltage is higher than 12.9 V right after charging and one has to wait for one hour before the voltage can be used as measure of the battery charge.

Table 7.1 Relation between unloaded battery voltage and percent battery charge for a 6 cell 12 V lead-acid battery at 20°C.

Voltage	% charge
12.90	100
12.80	91
12.70	82
12.60	73
12.50	64
12.40	55
12.30	46
12.20	37
12.10	28
12.00	19
11.90	10
11.80	1

We usually refer to a battery as a 12 V battery, but *a voltage of 12 V means that the battery is nearly discharged* and should be replaced. Most seismic equipment will operate down to 10.0 V. This limit is a bit misleading since a lead-acid battery is completely discharged well before this low voltage and once the battery voltage is below 11.8 V, the equipment will only operate for a very short time.

The simplest way of charging a battery is to use a constant voltage source. If the charging voltage is 14.1 V, the battery will be fully charged eventually. The charging current will initially be high, however it should not be very high (less than 0.2-0.3 times its capacity in Ah), which will lower the life of the battery. This is usually not a problem with standard chargers and large batteries since the charger itself will have a limited current like 2-5 A. As the battery gets charged, the charging current becomes smaller and smaller. Reaching 80 % charge is relatively fast, while charging the last 20 % might take a couple of days with a constant voltage charger. This is the main reason that 'newly charged' batteries rarely are charged above 80 % capacity. There are intelligent chargers available that will charge with a higher voltage than 14.1 V, sense the state of charge and automatically turn down the voltage to 14.1 V when the battery is fully charged. However, most chargers, including the chargers built into recorders, are of the constant voltage type. If the voltage is kept above 14.1 V, the battery will 'boil', excess energy is converted into hydrogen and oxygen and the battery will soon dry out. For open batteries, some gas is also released during the normal charging process, so ventilation is important to avoid collecting this explosive gas. When the battery is fully charged, a voltage of 14 V (better with a limited current) can be applied to trickle charge the battery to compensate for self discharge.

Charging lead-calcium batteries. These batteries have similar characteristics as the lead-acid batteries with a similar relation between battery charge and unloaded voltage as shown in Table 7.1. However, charging has to be done with a limited current. For a 12 V battery, the charging voltage is 14.70 V and the charging current in A is a maximum of 0.2 times capacity in Ah. So for a 10 Ah battery, maximum charging current is 2 A. When the charging current reaches 0.01 times the capacity, the battery is fully charged and a voltage of 13.5 V can be applied to keep the battery charged (trickle charging).

7.8 Power sources

Many seismic stations are powered by AC, so either an instrument built in power supply or an external unit is necessary (Figures 7.13 and 7.14). For remote field stations without AC power, the most common power sources are solar cells or more rarely wind generators.

7.8.1 SOLAR CELLS

Photovoltaic panels make a direct conversion from sunlight to electric energy. The efficiency of this process is around ten percent with the present state of technology, which is high enough if we take into account that about 1 kW/m^2 of solar energy arrives to the surface of the Earth at average latitudes.

Solar panels may yield some electric energy even on a cloudy day. In high latitudes a relevant proportion of the solar energy arrives as diffuse radiation, while in lower latitudes the major contribution is due to direct sunlight.

Two fundamental types of panels exist: Crystalline and amorphous. Amorphous panels are about 4 to 6% efficient. They seem to be the future trend, but at present they are not yet as reliable as the crystalline type. Crystalline panels are either poly-crystalline or mono-crystalline, both with the same efficiency of 10 to 11%. So they need less space than amorphous for the same power. If maximum efficiency (about 13%) is needed, Laser-Grooved, Buried-Grid panels (LGBG) may be the choice.

Mono-crystalline panels may supply some 110-130 W per square meter, at a little higher cost. LGBG are especially suitable for low light or cloudy conditions, as are amorphous panels. Figure 7.15 shows an example of mono-crystalline solar panel.

Figure 7.15 A small 12 W mono-crystalline solar panel mounted on top of the case of a temporary seismic station. Both sides of the panel are light sensitive, so the roof of the case is white-painted to reflect the diffuse radiation to the rear side of the panel.

As the sun travel through its daily path in the sky, the rays strike on the panel with a varying angle, so what is the best orientation for the panel to obtain the optimum efficiency? Ideally, a system following the sun's apparent orientation in such a way that incident rays are always normal to the panel (panel tracking). This kind of installation is used in large solar power facilities, but is usually cost-ineffective for the low power required in a seismic station. The moving parts and the control system have a higher cost than the extra solar panels required to compensate for the smaller efficiency of a static installation. Of course, the best horizontal orientation is toward South on the northern hemisphere and toward North on the southern hemisphere. Optimum tilt for summer time is different than in winter, as apparent elevation of sun on the horizon changes. Most installations have fixed tilt angle, nevertheless, and in this case, an optimum tilt from horizontal to maximize the annual energy output may be found in Table 7.2.

Table 7.2 Optimum tilt angle of solar panels as a function of site latitude

Latitude of station	Tilt angle from horizontal
0-4°	10°
5-20°	Latitude + 5°
21-45°	Latitude + 10°
45-65°	Latitude + 15°
65-75°	80°

The power that a panel supplies depends on the radiation intensity of sunlight and on the length of atmospheric path (the so-called AM – air mass) that the sun rays have to travel, since air attenuates light and modifies the light spectrum. This length depends on the elevation of sun on the horizon. For a sun directly overhead (at zenith), AM is 1. For an incidence angle θ from vertical, the AM is given as AM=1/cos θ. Nominal power of solar panels is specified for standard test conditions (STC), that means a light intensity of 1 kW/m^2 (equivalent of a full sun) and a spectral distribution corresponding to a value of air mass AM=1.5, considered as representative of an average terrestrial condition.

Under direct sunlight, a solar panel of nominal 12 volts may supply some 20 volts with an open circuit. This is needed to assure that even in less favorable light conditions, some current flows to the battery. Actually, open circuit voltage is almost constant until light intensity drops under about eight percent of nominal, but the current supplied depends on irradiation, so the panel internal resistance decreases with increasing sun intensity.

Charge regulator. A solar panel is usually connected to the battery by means of a charge regulator. A panel not illuminated behaves like a diode (or a series of diodes) in the wrong direction. So, the most simple charge regulator consists of a simple series diode to avoid discharging the battery through the panel if it does not have enough illumination. Another important function of the regulator is to avoid overcharge. For most simple devices, this task is performed by switching off the charge, when the battery reaches a given maximum voltage, and connecting it again if the battery voltage drops under a given threshold. Ideally, these voltages should depend on temperature, to match the battery characteristics. This On-Off technique is the simplest way, but not very efficient. It also makes the battery lose water since it has to overcharge a bit, requiring frequent maintenance. If the average power supplied by the panel is not much higher that the power consumption of the load and the battery has a high capacity, there is little danger of overload and a regulator may not be necessary. A simple series diode is sufficient to avoid nighttime discharge.

In more sophisticated regulators, the load (i.e. the seismic station) is disconnected from the battery if its voltage drops under a certain minimum, to save battery life, which is lowered by deep discharges. This function can of course also be present without a solar cell regulator.

Maximum power output of panels is rated at a voltage of about 17 V. When connected to a charging battery, the output voltage will be at most 14V. If it is charging at a current of e.g. 4 A, this means that 3 V · 4 A = 12 W are wasted as heat. The power losses are even higher if the temperature is low, since the panel increases its output voltage and power as temperature decreases. The worst case is when the battery is almost discharged, since the charging voltage then is much lower than the panel voltage. This power loss can be avoided by using an intelligent DC-DC converter (maximum power point tracking), which converts the solar cell voltage to the correct voltage for charging. This particularly improves efficiency in winter time, when it is more needed.

All high efficiency DC-DC regulators operate as switched-mode power supplies (pulse-width modulation, PWM). Spikes generated may introduce noise in the seismic equipment unless they are well filtered out.

Power needed. The nominal power that is needed to install depends on the average radiation received at a given site. This is a function of latitude, local weather (average number of daily hours of sun), season and of course average power consumption. A conservative estimate for average latitudes may be to consider that a panel will give its nominal power for three hours a day as an average. So, for example, if an average power drain of 6 W is needed, at 12 V, this is a current of 0.5 A, or 0.5·24 = 12 Ah per day. The panel nominal power must supply this in three hours, that is 12/3 = 4 A. At a voltage of 17 V (typical voltage at which power is specified), this means 4·17 = 68 W of nominal power to install. Or, in terms of current, the peak current should be six to eight times the average current consumption of the equipment.
As a rule, battery capacity for solar powered stations must be calculated for at least one week of operation without charge, although this may depend very much on the weather conditions at the site.

At low latitude sites, panel elevation relative to horizontal is low. This facilitates the accumulation of dust on them, which leads to loss of efficiency. The coating surface of the panel must be dust-repellent in this case. Snow accumulation may also be a problem in cold sites, but most panels have a surface suitable for avoiding the snow to be kept for long time.

Solar panels may be mounted in a simple mast if they are not too large. But if there is a small house for the instruments, it is better to fix them on the roof by a suitable frame. The frame and anchors strength must be enough to stand against possible strong winds on the panels. They are a preferred candidate for vandalism and theft, as they are exposed, so they should be difficult to dismount (height, security screws). The best protection, nevertheless, is to keep a friendly relation with the nearby inhabitants.

Figure 7.16 The Air 403 wind generator. To the left is shown the monthly energy output as a function of average wind speed. To the right is shown the power output as a function of the instantaneous wind speed (Figure modified from Southwest Windpower, www.windenergy.com).

7.8.2 WIND GENERATORS

Wind generators have not been very common, however recently many very robust models have become available. The generators have mostly been developed for the marine market and for use in remote secondary residences, so they are built to withstand harsh conditions over several years without maintenance. Due to a large production, they are also relatively cheap and a windmill station will cost approximately the same as a solar powered station ($1000). Windmills are particularly useful in high latitudes where there is often too little light in the winter to use solar cells. The only problem seems to be that in a climate with frequent change between freezing and thawing combined with snow and little wind, the wings might freeze to be locked in a fixed position. Since wind is less predictable than light, wind generator stations might require a larger battery than the solar cell powered stations to cover large periods of no wind. A one week backup capacity is considered a minimum in general, but this may be strongly dependent on the wind regime at the site.

For most geographical areas, average wind speed maps or even average wind power potential maps are available. So it should be possible to make a reasonable prediction at least of the average wind power potential at a given site. Figure 7.16 shows one of the smaller generators, which has been on the market for several years.

This small wind generator (produced by Southwest Power) can produce up to 600 W and will give out 20 W at a wind speed of 5 m/s. So even in light wind, the generator gives more than enough power, and it is therefore essential that the generator has power regulation like the solar cell. The weight of this generator is only 6 kg so it does not require any special tower for installation

Some seismic stations have both wind generators and solar cells using the argument that when the wind blows, there is likely to be little sun and vice versa.

7.8.3 FUEL CELLS

The fuel cell technology is rapidly evolving and smaller units are now available for the domestic market. A fuel cell produces electricity directly from fuel by an electrochemical process. They can, according to manufacturers, operate in harsh conditions. An example is the SFC Smart Fuel Cell (www.smartfuelcell.de), which will give about 150 Ah using 2.5 l of methanol.

CHAPTER 8

Seismic networks

The very first and basic problem to solve in seismology is to locate of the seismic events. For that purpose we generally need at least 3 stations (Figure 8.1). We therefore define a seismic network as being a group of stations working together jointly for data collection and analysis.

Before 1960, there were in general only individual seismic stations operating independently. Each station made its observations, which were usually sent to some central location, if several stations were operating in a country or region. In that sense it was possible to talk about networks, however the time lag between recording and manual processing was so long that such networks are not considered as seismic networks in the modern sense. In the 1960's, 'real' seismic networks started operating. These were mainly networks made for micro earthquake recording and the distance between stations was a few kilometers to a couple of hundred kilometers. The key feature, used to define them as networks, was that the signals were transmitted in real time by wire or radio link to a central recording station where all data was recorded with central timing. This enabled very accurate relative timing between stations and therefore also made it possible to make accurate locations of local earthquakes. Recording was initially analog and has over the years evolved to be nearly exclusively digital. Lee and Stewart (1981) made a good general description of local networks. With the evolution of communication capabilities to cover the whole world, seismic networks are now not limited to local networks but can be either regional or global too. The distinction between them is primarily no longer the differences in data transfer, accuracy of timing, time lag between data acquisition and analysis, etc, but the scope of investigation, spatial resolution, and quality of data in terms of frequency content and dynamic range.

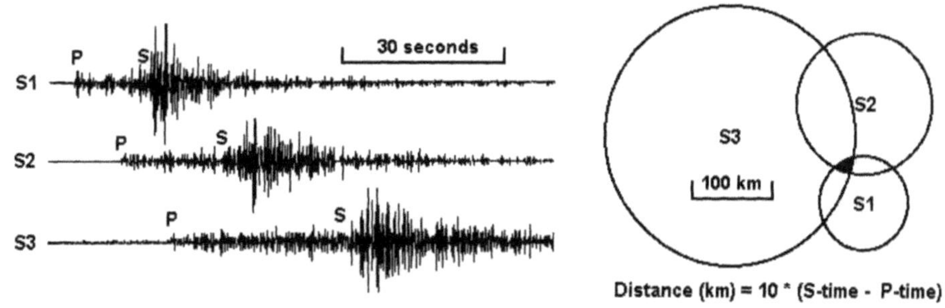

Figure 8.1 Location by the circle method. To the left is shown the seismograms at 3 stations S1, S2 and S3 recording a local earthquake. The amplitude scale is different on the 3 traces. The stations are located at S1, S2 and S3 (left). Hypocentral distance is proportional to the S-P time difference by a factor approximately $V_p/0.73$ (in this example the average P velocity V_p is taken as 7.3 km/s). The epicenter is found within the black area where the circles cross. These circles will rarely cross in one point, which indicates errors in the observations and/or model. With only 2 stations, we see that there are 2 possible locations. With more than 3 stations, the uncertainty in location decreases.

Since communication ability defines a network, this chapter will to a large extent deal with communication systems.

Temporary seismic networks, like for aftershock monitoring, usually do not have communication capability, although this might change in the future.

This chapter follows in some parts of Trnkoczy et al. in Chapter 8 in NMSOP, (Bormann (Ed.), 2002), with a few paragraphs taken literally from NMSOP.

8.1 Purpose of seismic networks

We have seen that the primary goal of all networks is to be able to locate seismic events, whether local or global. Further research done with the data can be magnitude and source parameter determination, hazard studies and knowledge of the earth's interior. While earthquake location only requires a SP capability, other research tasks might require a LP or VLP capability. Equipment and layout is therefore to some degree different for the different types of networks, although the most high quality stations can be used in most networks. However, in some cases, that would be overkill.

Sensor selection has been discussed is Chapter 2. A brief summary of network types/purposes is given in Table 8.1 together with possible sensor selection.

Table 8.1 Types of networks and purposes. For all networks where events with magnitudes larger than about 4 are expected, it will always be an advantage to have at least one sensor with LP or BB capability.

Type of network/purpose	Distance, km	Frequency range, typical and extreme	Sensor
Mining earthquakes	10	5-2000	Geophone, accelerometer
Dam induced seismicity	50-100	0.1-100	Geophone, SP, accelerometer
Volcano monitoring	30	0.1-50	Geophone, SP, BB
Strong motion monitoring	1000	0-100	Accelerometer
Local seismicity	100	0.1-100	Geophone, SP, accelerometer
Regional seismicity	1000	0.01-50	All types
Global seismicity	All	0.0001-20	All types, depends on specific purpose
Refraction surveys	2000	1-20	SP, geophone
Global earth structure	All	All	BB,
Test Ban Treaty monitoring	All	All	SP, BB

One- and three-component seismic stations. Historically, many seismic stations and networks used single component sensors, usually vertical seismometers. Many of them still operate. This has been the case because the equipment was analog and required paper recording for seismic data. If three components were used, then exactly three times more equipment was required but three times more valuable information was not gathered. It was also very difficult, if not impossible, to generate a vector of ground motion from three separate paper seismograms. Furthermore, analog radio or telephone telemetry of three signals required analog multiplexing in a narrow-band channel, which degrades its dynamic range.

Today, in the era of digital recording and processing of seismic data, the situation is different. The price/performance ratio is much more favorable for three component stations. Most data recorders and data transmission links are capable of accepting at least three channels of seismic data. The costs for upgrading the central processing facilities for an increased number of channels due to three-component sensors are relatively small and generating vectors of ground motion is easy with computers.

Since ground motion essentially is a vector containing all seismic information, and considering the fact that many modern seismological analyses require this vector as input information, one-component stations are no longer a desirable choice for new installations (note that seismic arrays are not discussed here, see Chapter 9). One component seismic stations are still a choice where communication capability and economy are limiting factors and also where the main purpose is to locate events.

8.2 Network geometry

Since networks as a primary goal has to locate seismic events, network geometry becomes an important factor in addition to the site selection factors discussed in Chapter 7. We have seen that at least three stations are needed to locate an earthquake, although a single 3-component station might be enough to make a preliminary epicenter, see Figure 8.3. Ideally, the epicenter should be inside the network since the locations become less and less reliable as the epicenter moves outside the network. The "azimuth gap" (the largest of all angles among the lines connecting a potential epicenter with all the stations in the network, which recorded the event) should ideally be less than 200 degrees.

Station density, under equal noise conditions, essentially determines the detection threshold (smallest magnitude always detected, see Figure 8.2) so more stations are needed for high noise installations than for low noise installations to get the same detection threshold. Alternatively, stations have to be placed closer to the area of interest. A rule of thumb is that for an average station (e.g. Figure 3.5), a magnitude 2 event will be 'well recorded' at a distance of 100 km.

The hypocentral depth is even more difficult to calculate accurately. A rule of thumb is that to get an accurate depth estimate, the nearest station should be no further away than 1-2 times the hypocentral depth. For many areas, typical depths are 5- 20 km, so to get good depths might require a very dense network, much denser than most often is possible from practical or economical considerations. It is therefore important to put stations near active areas where hypocentral depth determination is important. Station distribution might then resemble earthquake distribution (Figures 8.25 and 8.29).

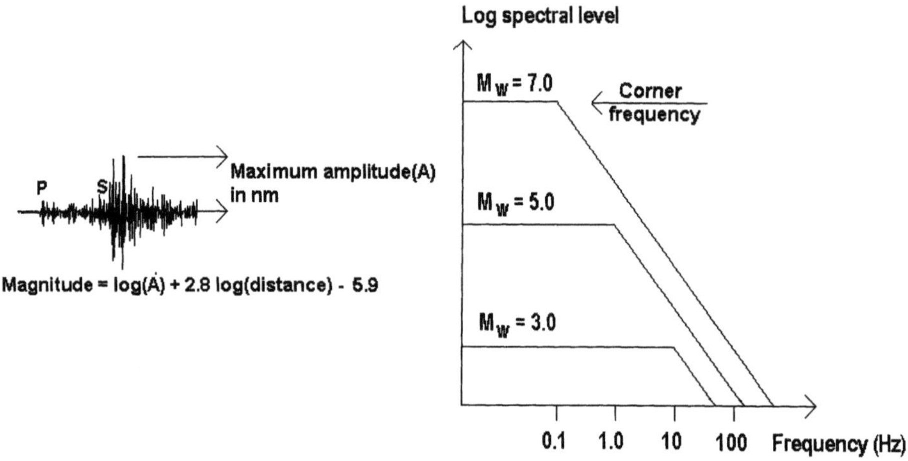

Figure 8.2 Magnitude determination. The simplest way of determining magnitude of a local earthquake is to measure the maximum amplitude (nm) on a displacement seismogram and use the formula above (figure left), where distance is measured in km. This gives the so-called local Richter magnitude (Lay and Wallace,1995). A more general way is to make the displacement density spectrum (see Chapter 6) of the P or S-waves. From the spectral level (flat part of spectrum), the magnitude can be determined with a simple formula (Lay and Wallace,1995). The figure (right) shows how the spectrum increases in level with the magnitude (*M*). It also shows that the spectrum corner frequency becomes smaller with higher magnitudes, which explains why it is important to use BB sensors when recording large earthquakes.

The geometry of a network will determine the accuracy of location in different directions, and a reasonably regular grid will give most uniform location accuracy. The worst configuration is a network with aligned stations, see Figure 8.3 for an example.

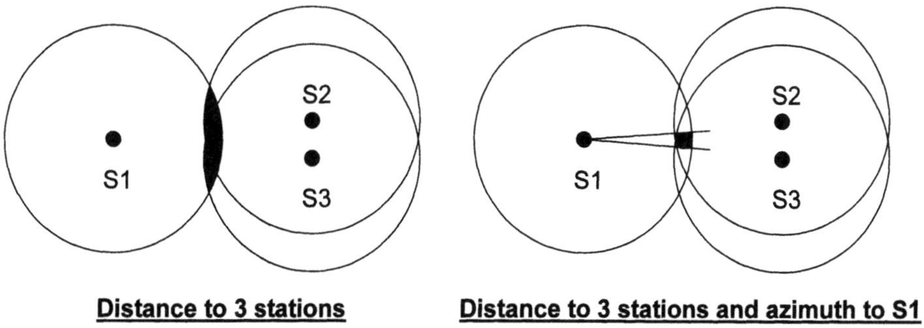

Figure 8.3 Network geometry of aligned stations. The figure to the left shows 3 stations (S1, S2 and S3) almost aligned in the x-direction (left - right). The event has been located by using the distances to the 3 stations and the shaded area in the middle gives an indication of the area within which the epicenter can be found. The figure to the right shows the same situation except that an azimuth determination has been made with station S1 which limits the direction to which the epicenter can be located to within the angle shown and thereby makes the estimate of epicenter error smaller in the y-direction.

Figure 8.3 (left) shows a network with 3 stations almost on a line. It is seen that the location accuracy is good in the x-direction while very bad in the y-direction. With the stations exactly on a line, it would be impossible to determine if the event was above or below the line of stations. Since many stations now have 3 component sensors, one can use standard techniques to determine, to within a few degrees, the direction of approach (azimuth) by correlating the P-signals of the two horizontal components. Figure 8.3 (right) shows how the location accuracy has been improved by using the limitation imposed by the azimuth observation. In principle, the epicenter can be determined with just one station when 3 component data is available. An alternative way of determining the azimuth is by using a seismic array, which in addition also improves the signal to noise ratio, see Chapter 9. The lesson is that, when one has to use unfavorable station geometry, 3-component station or small arrays might significantly improve location accuracy. Traditionally this has not been considered when designing seismic networks, partly because easy-to-use analysis software has not been readily available.

How many stations should a network consist of?

Three is the smallest number and five would be more reasonable, since the network will not be so vulnerable if one station stops working. In principle, the more, the better since it improves locations and make it easier to do other kinds of research as well, like determination of fault plane solution, attenuation and crustal structure studies. However, there will always be a tradeoff between cost of installation, and not least operation, and size of the network. There are several computer programs which are capable of calculating optimal network configuration given number of stations, preferred source areas and size of network (see Chapter 8 in NMSOP, Bormann (Ed.), 2002). Alternatively, the programs can give likely location accuracy for a given network configuration. Figure 8.4 shows an example.

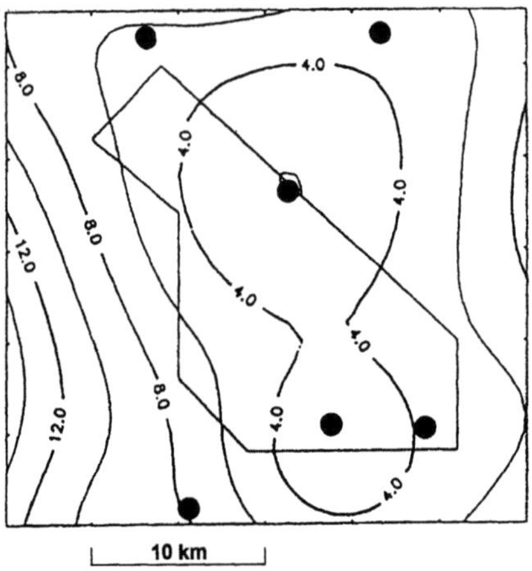

Figure 8.4 Isolines of the uncertainty of epicenter determination in km for a small local network, figure modified from Trnkoczy & Zivzic (1992), by permission of E.C.G.S.

Note that one should not rely too much on proposed algorithms for overall optimization of seismic network configurations (e.g. Kijko 1977; Rabinovitz and Steinberg 1990). The theoretically recommended optimal configurations can hardly ever be realized or their predicted theoretical potential information gain be exploited under real conditions simply because the stations cannot be put on the recommended locations due to inaccessibility of sites, very bad underground conditions, proximity of strong noise sources, non-availability of required power or data communication links or other reasons. On the other hand, some of these programs may be of help in quantifying the task dependent performance of various options of realistic station siting (e.g. Trnkoczy and Zivcic 1992; Hardt and Scherbaum 1994; Steinberg et al. 1995; Bartal et al. 2000), of assessing the degree of achievable up-grading of network performance for given event distributions by adding some stations to an existing network or to assess where reduction or closing of stations (e.g. for cost saving) can be made with the least negative effect on overall network performance. Also note that optimal configurations for event location are often not optimal for source mechanism determinations, tomographic studies or other tasks (Hardt and Scherbaum 1994).

If a new network is a radio frequency (RF) telemetry system, one has to correlate RF data transmission requirements with seismological requirements. If one plans the use of phone lines for data transmission, their availability and the distances to which new phone lines would have to be laid have to be checked. If one plans to power future stations by mains power, the availability of power lines and the distances to which new lines would have to be laid must be checked - or one must decide for solar panels. In practice, it is often hard to get the best geometry since other deciding factors, like access limitations, communication and security of equipment, will limit the choice. There is a strong tradeoff between the possibility of constructing and servicing the stations and the quality and location of the sites.

8.3 Network configuration: physical and virtual networks

We now assume a network geometry has been decided, type of sensors selected, so now it 'only' remains to set up how the network will actually work, which can be called network configuration or network functionality. This consists of two parts: Communication and interaction of the central recording system with the field stations. The communication will be dealt with in a later section.

In the days of only micro-earthquake networks and one-way data transmission (from stations to central recording site), it was quite clear how a seismic network was defined. Today the situation is more complex. Nowadays, more and more seismic stations are connected to the Internet or to the public phone system. Such stations usually have a local seismic signal recording capability and sometimes there might not be any real time data transmission to a given site. However, these stations still can be defined to be in a network since they are all connected to the global communication network. Any networked computer can in principle be used to collect data from a number of stations in what functionally is a seismic network. By defining a seismic network in this way, the distinction between local, regional, and global networks does not exist anymore in terms of hardware, data transmission and acquisition, but is merely a question of how the data collection software is set up to handle communication, data collection and processing.

This means that two types of seismic networks can be defined: **Physical and virtual**.

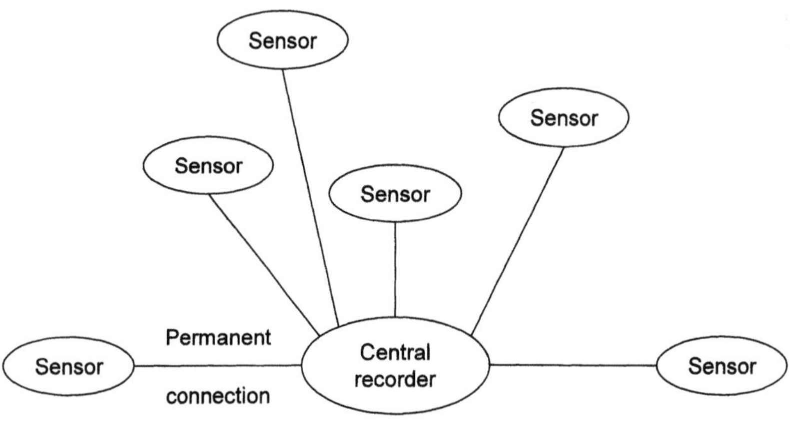

Figure 8.5 A physical seismic network. The sensors are connected to a central recorder through a permanent physical connection like a wire or radio link. In this example, transmission is analog and digitization takes place centrally, but the ADC could also have been placed with the sensor and transmission would then be digital.

A physical seismic network (usually local) consists of closely linked remote seismic stations (Figure 8.5). The remote stations detect the ground motion and usually send data in real time to a central recording station for event detection and recording. This type of network covers both the old analog systems and the current digital systems.

A virtual seismic network consists of stations selected among many stations connected to the global communication network or a public phone system (Figure 8.6). A station can also be a central recording station of a physical network. The remote stations must be capable of event detection and local recording and data is normally not sent to a central recording system in real time. The remote stations must have 2-way communication capability (duplex or at least semi-duplex). The central recording station can manually or automatically connect to selected remote stations and download triggered and/or continuous data and determine if the individual station triggers can be combined to an event.

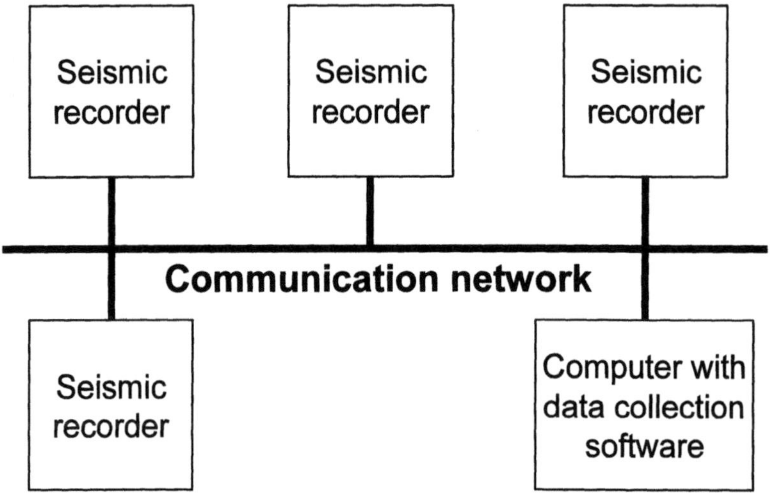

Figure 8.6 A virtual seismic network. The thick line is the communication network, which can have many physical solutions. The data collection computer can collect data from some or all of the recorders connected to the network provided that it knows the protocol used by the recorder. Both types of seismic networks might result in the same kind of output although the virtual network will deliver data with a larger time delay than most of the advanced physical networks.

8.4 Physical seismic networks

The physical network, according to the definition above, is characterized by central recording and transmission of data in real time from the field stations to the central recorder. The network can be made in different ways:

Analog transmission and analog recording. The data comes by cable, phone or radio and is recorded on drum recorders or other analog media like film or tape. Film and tape is not used anymore, but large quantities of data are available in this form. These networks were common around 1965-80 and many networks are still in operation, where backup recording is taking place on drum recorders, but rarely as the primary recording media (Figure 8.7). There is obviously no event detector so processing is slow. As a curiosity, we can mention that there were analog detection systems that performed analog event detection (as described for digital systems in Chapter 5), and started a multi channel recorder so that all traces of the event were recorded on one piece of chart recorder in a much higher resolution than available on the drum recorder. There was even an analog pre-event memory built in.

Analog transmission and digital recording. This kind of network is a logical extension of the pure analog network. Since analog data is readily available, it is quite easy to connect a multi channel digital recorder to a stream of analog data. Since this is not a field operation, there is no need for a sophisticated low power recorder (Chapter 5) and a standard PC with a multi-line ADC card and appropriate software can be used (Figure 8.7). Since the data already has been amplified, there is no need for a sensitive ADC, ±5 V is a typical output of an analog transmission system (see transmission section). Many such data acquisition programs have been written for PC (e.g. Camelbeck et al, 1990, Lee and Stewart 1981, Utheim et al, 2001, Chapter 5), and just as many different recording formats have been used! A few commercial programs were also sold, but are hardly available anymore. As mentioned in Chapter 5, the most common systems used now employ public domain software.

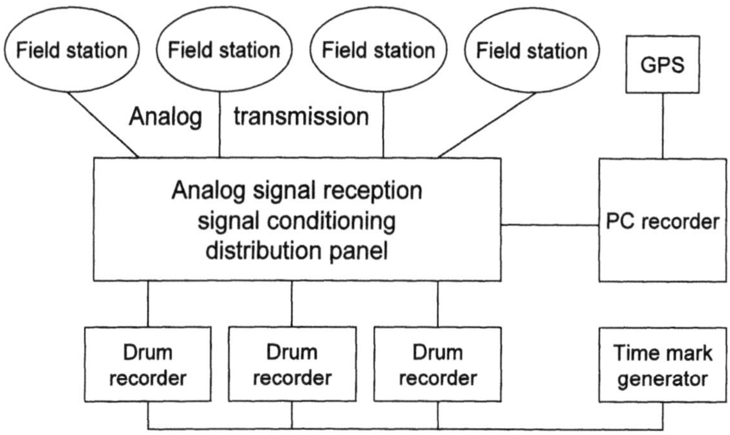

Figure 8.7 Typical analog-digital network. The analog data are transmitted to the central site over fixed analog communication channels. At reception, the signals are put into a distribution panel where some filtering and/or amplification might take place before recording in analog and/or digital form. In this example, 3 of the 4 stations are recorded on paper. Timing is done with the PC recorder. The time mark generator for the drums can use the recorder GPS or it has its own timing reference.

Due to limitations in dynamic range of analog transmission, it is enough to use a 12 bit ADC, so such a recording system only has the cost of an inexpensive 12 bit ADC (< \$1000), a PC and a GPS. However, the ADC card has no antialias filter and most systems have relied on anti aliasing filtering in the analog amplifiers and/or the analog transmission system. Since these generally are only 2-4 pole filters, there are many network digital recorders without appropriate antialias filters. Also many old records from analog tapes, that were digitized without suitable filters, are aliased (Anderson, 1978).

Continuous recording is straight forward, since all the data arrives to the central recorder, so recording is only a question of disk space. A 10 station 3-component network with a 12 bit digitizer and sampling at 100 Hz gives 10 stations x 3 channels x 2 bytes x 100 samples/s x 3600 s x 24 hours = 0.5 GB per day, hence few high frequency networks store the continuous data.

This kind of network can accommodate off-line as well as automatic near-real time computer analysis. One can use most modern analysis methods, except those which require very high-resolution raw data. Such systems are still useful for some applications when the high dynamic range of recorded data is not of prime importance and the purpose of the seismic network is limited to a specific goal. An advantage of these systems is low cost and easy maintenance. The power consumption is also low, around 2 W for a radio transmission station, making this type of network easy to power with solar panels.

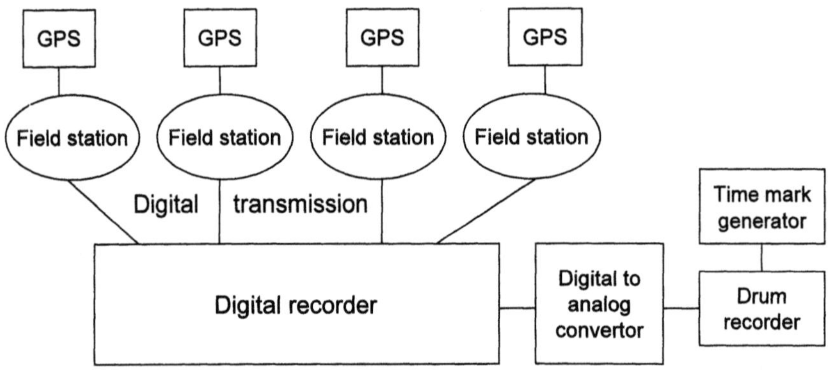

Figure 8.8 Typical digital network. The digital data are transmitted to the central site over fixed digital communication channels. At reception, the signals enter the recorder directly. Timing is normally taking place at the field station although some systems also time the signal on arrival. In this example, one station is recorded on paper and the data therefore has to be converted from digital to analog. The time mark generator for the drum can use the recorder GPS, if it has one, or it has its own timing reference.

Digital transmission and digital recording. This is now the most common way in operating a seismic network. In principle the system works just like the analog network except the digital conversion takes place at the field station and real time digital data are sent to the central recorder (Figure 8.8). Like for the analog network, data must be continuously received so any transmission break will usually mean lost data (not so with virtual networks, see next section). However, some digitizers have memory so shorter transmission breaks might not mean loss of data, if a duplex connection is used, see transmission section. Let us for the time assume that the transmission is continuous and error free, which is quite possible. The big advantage of the digital network is that the dynamic range is as high as the dynamic range of the ADC and the data is error free. The recorder can still be an inexpensive PC, which permits entry of multiple streams of digital data. There are several public domain systems available supporting different digitizers (see Chapter 5). Each equipment manufacturer of course has his own system, which only works with that company's type of digitizer although there is now a trend, also for private companies, to use public domain software.

There is sometimes a demand for additional paper drum recorders in a purely digital system because there is a wish to continuously monitor incoming signals. However, there are a number of problems with paper drum recorders in digital systems. In terms of hardware, they are fundamentally incompatible with digital systems. This requires additional digital to analog converters. Being mechanical devices, they are and will continue to be expensive (they often cost more than a multi channel digital recorder) and they require continuous and specialized maintenance and consumables. On the other hand, nearly all modern observatory seismic software packages allow continuous observation of the incoming signals in (near) real time and some even simulate the traditional appearance of paper seismograms. The experience is that once the user becomes familiar with the digital system, expensive paper drum recorders soon prove to be of little use.

A real time digitally recording network will usually have a trigger system as described in Chapter 5 and, due to the real time transmission, particularly the coincidence trigger makes it possible to cut down the rate of false triggers without using very sophisticated trigger algorithms.

Physical digital networks are obviously the state of the art for real time networks, however, the requirement for continuous transmission can also be costly in installation and maintenance, so a virtual network might be an alternative, see 8.5.

8.4.1 COMMUNICATION STANDARDS

Since nearly all new networks are digital, it is useful first to review the most common communication standards.

RS-232-C is an interface standard for serial data transmission. The protocol is not specified, and in almost all cases, it is asynchronous. That is, the bits are transmitted at a given rate and the reception end has to extract the clock from the data sequence. Actually, the receiver has its own clock and it locks the phase using the "start bit" of each data frame, usually a byte. The signal is single-ended. The transmitter electric levels may be ±5 V to ±15 V. At reception, a level above +3 V is a logical 0 and a level under –3 V is a logical 1. A level between –3 and +3 V is undefined. RS-232 is a point-to-point interface and does not permit multiple receivers or transmitters. The maximum allowed cable length depends on the data rate, but cannot be longer than 20 m, except for very low rates. RS-232 needs a minimum of two wires for one-way transmission and three wires for duplex communication.

RS-422 interface uses a differential signal and usually a twisted-pair cable. The interface permits one driver and up to 10 receivers on the same line, and cable lengths of more than 1 km. The maximum theoretical data rate is 10 Mbit/s, although this is not possible with long cables. A rate of 19200 bits/s is achievable in practice with a cable of 2 km. Two twisted pairs are needed for duplex transmission. The transmission level is ±2 V minimum and at reception the level must be at least ±0.2 V.

RS-485 is a superset of RS-422, which allows up to 32 drivers and 32 receivers in the same line, although the drivers have to switch to a high impedance state when not talking, so that only one driver is transmitting data at a given time. The cable length, data rate and receivers sensitivity are the same as for 422 and the transmission levels may be ±1.5 V.

The common way to specify the data rate or *transmission rate* is in **bit/s**, which is the number of bits of information transmitted per second. The **baud rate** is different and refers to the *modulation rate*. There may be two reasons to distinguish between them: a) The modulation may not be binary, but multi-state (e.g. a phase modulation with 4 possible phases) and in this case the baud rate would be lower than the bit rate, and b) some synchronization signals are used, which do not carry information (e.g. in RS-232 each character transmitted is contained in a frame, typically with a start mark of one bit length and a stop mark of at least the same length, thus making a frame of 10 bits). Standard baud rates are, among others, 4800, 9600, 14400, 19200, 28800, 38400, 57600 and 115200 bauds. If we use, for example, 9600 baud with 8 bit characters, one start bit, one stop bit and one parity bit, the maximum number of information bytes transmitted per second will be 9600/11, and the bit rate would be 8·9600/11=6.98 kbit/s. The user is concerned with the bit rate, while the telecommunication engineer is concerned with baud rate, related to the required bandwidth of the channel.

Other serial interface standards are *USB* (Universal Serial Bus), used for communication between PC's and peripherals or devices; *IEEE 1394* (Firewire), mainly for digital video, *IrDA* (infrared) and *Bluetooth* (short range radio). The first two allow for very fast data transfer and IrDA runs up to 115 kbaud. The short range of all these interfaces, however, makes them of little use for communications within a seismic network, therefore they will not be treated here.

Ethernet is a standard for LAN (Local Area Network) at a rate of 10 Mbit/s (later extended to 100 Mbits/s as Fast Ethernet), that uses a protocol Carrier Sense Multiple Access/Collision Detection (CSMA/CD) as the access control to avoid conflicts between nodes trying to transmit simultaneously. On an upper level, a protocol such as TCP/IP (see below) is implemented for data transfer. The physical interface may be of several types:

- 10BASE2 Thin coaxial cable (RG-58), with maximum segment length of 185 m (a repeater is needed for larger length).
- 10BASE5 Thick coaxial cable (RG-213), with maximum segment length 500 m.
- 10BASE-T Unshielded twisted-pair (UTP) cable, with maximum segment length 100 m, node-to-node connection. This may be used for star-shaped networks in which nodes

are all connected to a common HUB. Valid also for Fast Ethernet. It uses 4 wires and a connector RJ-45.
- 10BASE-F Fiber optics cable, with connector ST, maximum segment length of 2 km.

TCP/IP (Transmission Control Protocol/Internet Protocol) is a standard protocol, which runs on a higher level than the data exchange level, and is used for network interconnection, independently of the net topologies. Initially it was adopted by the U.S Defense Department. Later it was incorporated in the Berkeley Unix and is now the most used protocol for open networks. It includes the utilities for virtual terminals (Telnet) and for file transfer (FTP).

PPP (Peer to Peer Protocol) is used for connections point-to-point such as dial-up phone access via modem. TCP/IP can be mounted on top of PPP.

8.5 Virtual seismic networks

For virtual seismic networks, the setup is dependent on the mode of communication. In the general case, all field stations are connected to the Internet and/or, the public telephone system and there might not be any a priori defined network, since general protocols are used. In the more commercially set up systems, the stations can only be reached by communication from a dedicated central computer using proprietary software. In both cases, the systems do not operate in real time. The network operation usually follows the same principles as for physical networks with some additional capabilities. A common scenario is:

The central computer copies detection lists and/or automatic phase picks from remote stations. Based on the detection lists and trigger parameters, events are declared (Figure 8.9). Here two options exist. Either recorded event waveforms are copied from the field stations to the central computer and no waveforms are copied from stations that do not trigger, or (assuming the field stations have ring buffers with continuous data), the same time interval of waveform data is extracted from all remote stations. In this way, waveform data from all stations in the network (like for the physical network) are gathered at the central station.

There is no requirement of continuous connection to the stations, the connection is set up when data has to be transferred by e.g. ftp. The simplest scenario for a virtual network is seen in Figure 8.10. The field station only has a waveform file for each event and no ring buffer or other parametric data. The virtual data logger simply connects to each station in the network at regular intervals and copies all the waveform files to some directory on the central computer. This system is a very simple automation of what is often done manually. The virtual data logger will check that files are not copied more than once and might delete files on the field station after they are copied. This simple data logger has the disadvantage that all triggered data are transferred, not only the network declared events. The simplicity is attractive and for low seismicity areas it does not make much of a difference and enables detection of small events only seen on one station.

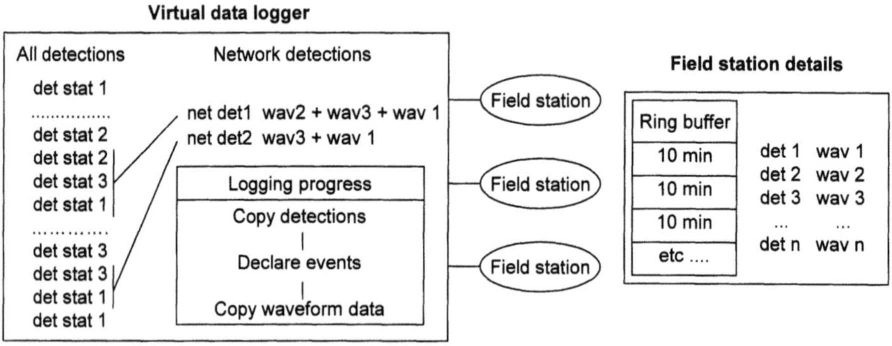

Figure 8.9 Typical virtual data logger. The field station (right) has a ring buffer with files or segments 10 min long. It also has a list of detection times with associated parameters (det 1, det 2, etc.) and corresponding waveform files (wav 1, wav 2, etc.). The virtual data logger (left) has the following logging process: First get a copy of the detection times from all stations (det stat 1, det stat 2, etc.; ... indicates a longer time window), which are time ordered. Based on these, network detections are made if at least 2 detections occur within a short time window (net det 1 and net det 2). Finally, the waveform files are copied. In this example, the ring buffer is not used (see text).

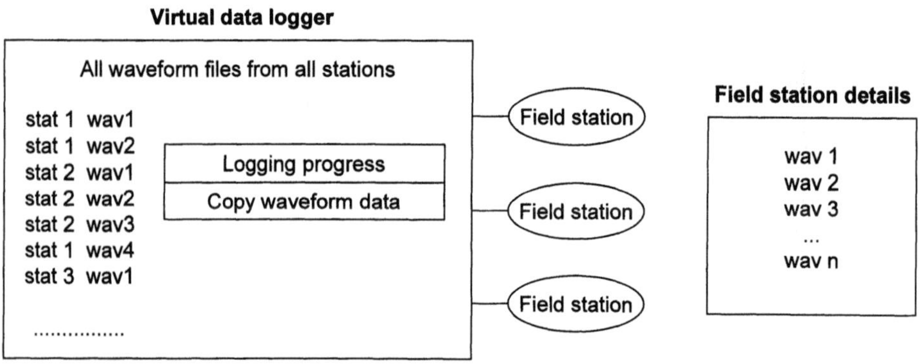

Figure 8.10 Simple virtual data logger. The field station (right) has a waveform file for each detected event (wav 1, wav 2 etc). The virtual data logger (left) simply copies all the waveform files from all stations to a directory on the central computer (stat1 wave1 etc). Further processing is done manually.

The speed of data collection depends on the communication system and the way the data collection system is set up. In a typical scenario, all data collection is controlled from the central computer and data is collected at the time interval set up in the central computer.

Transmission speed example: Suppose that, on average, 3 MB of data per day is generated at each remote seismic station (2500 s of uncompressed, 4 byte data at a sample rate of 100 Hz from 3 channels). A network consisting of a central computer, 10 remote stations, and a single modem at the central recording site, having 9600 baud data transfer, needs about 10 hours/day to transmit this data. This means that the maximum delay in getting the data will be 10 hours and the data transfer would typically be started once or twice a day. The same network connected to Internet, having a speed of 128 kbaud and a multi-line ISDN port would need less than 10 minutes for the same task. In this last example, transmission of data takes place at the same time from all nodes. If the data collection software is set to operate more frequently, less data is to be transferred at once, and an even shorter time delay can be achieved. Thus it can be said that the system operates in semi real time.

Continuous central recording is also possible with virtual networks if the connection is fast enough. The data collection software could e.g. initiate a copy of station ring buffer files once an hour. Therefore, like for detections, continuous data would not be instantly available.

The above systems are based on the traditional idea that the central computer controls the network. However, with some equipment it is also possible to set up the remote station to send parametric data to the central computer immediately after an event is detected. The central computer would then request waveform data, if sufficient detections within a given time window arrive. In this way, events are declared immediately after the event occurrence and the process does not have to wait for the polling. The data might be available to the central computer faster then if the central computer has to initiate polling at regular intervals. The problem with this solution is that it is not easy to develop reliable software which controls the data flow in case the remote stations trigger wildly. This situation may block the system in the worst case. Currently most systems are based on the central computer keeping control.

For the virtual networks, the main challenge to the network operator is to get hold of the right software that links together the stations reachable by computer, into a network. The main difficulty today is lack of standards. There are many ways of accessing different types of seismic stations and many different formats of parameter and waveform data. When setting up a virtual network, the user is then limited to use only a certain kind of stations for which data collection software exists. In the following, some examples of virtual networks will be given.

Example 1. The IRIS/Global Seismic Network (GSN) is a typical example of a virtual seismic network. This global system consists of more than 120 seismic broadband stations, which can be reached by modem and/or Internet. At the IRIS data management center in Washington, a public domain software SPYDER running on Unix automatically retrieves data from selected GSN stations based on preliminary determination of epicenters by NEIC (National Earthquake Information Center) in Boulder, USA. Thus event detection is not part of the Spyder system. The Spyder system has been installed in several places for local or global use. It only works with GSN type stations. Figure 8.11 shows the GSN network and Figure 8.23, the type of communication used.

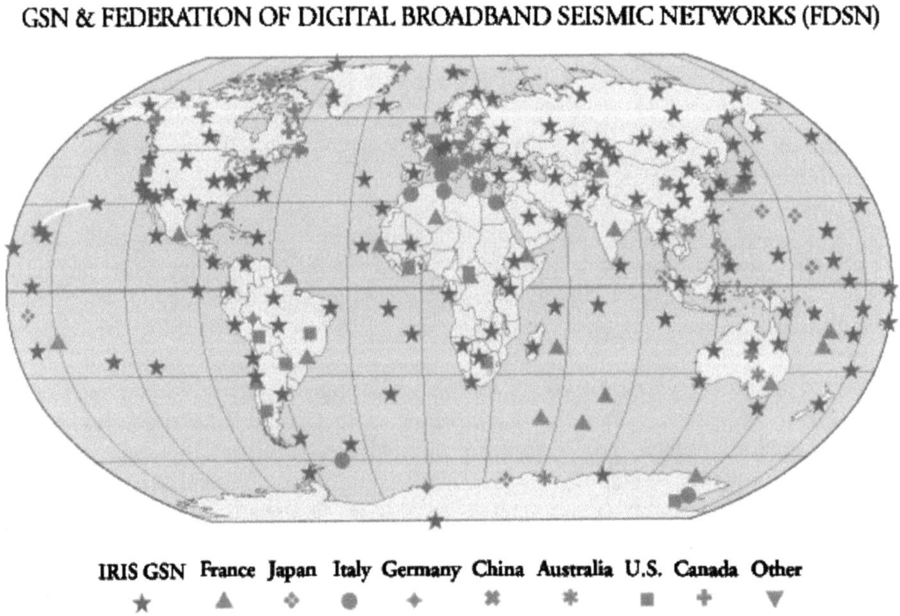

Figure 8.11 The Global Seismic Network (GSN) and other global broadband band stations being member of the Federation of Digital Seismic Networks (FDSN). Figure from www.iris.washington.edu.

Example 2. The public domain SEISNET system running on Unix is another software enabling establishment of virtual seismic networks. SEISNET is similar to Spyder, however, it operates with other types of stations in addition to the GSN stations and also performs network detection and preliminary location (Ottemöller and Havskov, 1999). It was developed for the Norwegian National Seismic Network (Figure 8.27) and is in addition also used in several other places. SEISNET is very flexible and can be adopted for virtually any type of field station, the main capabilities are:

- Retrieval of detection information from seismic stations (GSN, SEISLOG and ftp servers)
- Retrieval of epicentral information provided by seismic centers
- Retrieval of waveform data from seismic stations
- Retrieval of waveform data using AutoDRM (Kradolfer, 1996)
- Network event detection
- Automatic phase identification, hypocenter location and magnitude determination
- Transfer of waveform data from selected field stations based on a given hypocenter location and origin time

Example 3. ANTELOPE is yet another virtual seismic network software package in the market and is free for IRIS members. It supports a wide range of seismic stations as well as other environment monitoring equipment. Its open-architecture, modular, UNIX based real-time acquisition, analysis, and network management software supports telemetry using either standard duplex serial interfaces or TCP/IP protocol over multiple physical interfaces. In addition to data acquisition, it includes real time automated event detection, seismic event association and location. ANTELOPE runs on Sun Microsystem' Solaris OS using SPARC workstations and Intel architectures, and it also works with Linux. It was developed by the BRRT Company and Kinemetrics and is currently used by IRIS networks and several other networks. It is probably the most advanced software of its kind available and requires good computer skills to set up.

8.6 The choice between physical and virtual seismic systems

The decision on which type of network is optimal depends mainly on two factors: Cost and requirement for real time data. For seismic networks with important alarm functionality, the main requirement is to locate events and determine magnitude as fast as possible. For this purpose one has to have raw data available in real time. This usually means that most of the virtual seismic networks are ruled out and a physical network must be used. Two exceptions exist: Virtual seismic networks which can handle real time data transfer via Internet (like the 3 examples above) or if accurate automatic event location and magnitude determination can take place at the field station, and this information is immediately sent to the central station. Remote stations must initiate data transfer. The drawback is that remote automatic locations based on single station data are less reliable and that the results cannot be verified before the complete raw data arrives.

For seismic networks with the exclusive purpose to monitor general seismicity and/or to serve research purposes, there is no need for real time data. The main factor in deciding which network is the most appropriate is cost of operation and quality of data. For research purposes, flexibility is also a very important issue. If phone lines or coverage by a cellular phone system are available at reasonably seismically quiet station sites, it might be the cheapest way to construct a virtual network. For very large networks, where dedicated radio links are difficult to use, virtual networks will probably always be the cheapest alternative.

Since, in general, communication costs are drastically going down and services are getting available everywhere, it is likely that more and more networks will operate as virtual networks in the future.

8.7 Seismic data transmission

We have now described physical and virtual networks. Both types need some physical medium for the communication. A physical network will usually have some very direct means of communication, like a radio network, while virtual networks can be a complicated mixture of different kinds of point to point digital connections. However, the user still has to connect some physical equipment like modems, although in some places it is as simple as going to the local telecommunication company and ordering a plug in the wall, where the station is to be connected, and the rest is just software set-up.

While data transmission may seem like a less important technical task of a seismic network, poorly selected or designed data transmission capacities are the most frequent cause for disappointments and technical failures. The technical quality of a seismic network operation rests largely on the reliability and the quality of data transmission.

Another very important, but frequently overlooked factor, is the cost of data transmission. Note that these costs may largely determine the budget for long-term seismic network operation.

The three key technical parameters in physical data transmission links related decision-making are:
-The required information flow (channel bandwidth with analogue links or data transfer rate with digital links).
-The distances to which data must be transmitted (becomes unimportant with computer network based virtual seismic networks).
-The desired reliability (acceptable time unavailability of the links - that is the maximum time period per year where signal-to-noise ratio is lower than required (analog links) or bit error rate (BER) is higher than allowed (digital links).

In virtual seismic networks, two decisions are the most important:
-The physical network which will be used in establishment of a virtual seismic network (Internet, proprietary WANs (Wide Area Networks), analogue public phone network, ISDN, ADSL or other types of digital data networks.
-The protocol that will be used.

In seismology there are several different kinds of physical data transmission links in use, from simple short wire lines to satellite links at global distances. They differ significantly with respect to data throughput, reliability of operation, maximal applicable distances, robustness against damaging earthquakes, and in cost of establishment, operation, and required maintenance.

Note that strong-motion seismic networks generate far less data than weak-motion networks and their designs might therefore differ significantly. Seismic data transmission links that are fully acceptable for strong-motion data may be inadequate for weak-motion data and data transmission links used in weak-motion networks may be an absolute overkill and too expensive for strong motion networks.

8.8 Analog data transmission

Analog transmission links, whether radio or telephone lines, have a limited frequency band like 200-4000 Hz. Typical frequencies of seismic signals are DC to 100 Hz, so seismic signals cannot be transmitted directly through standard analog communication channels. The seismic signal is therefore transformed to a frequency modulated signal using a voltage controlled oscillator (VCO). The VCO has an input of e.g. ±1 V and the output is e.g. 1700 ±125 Hz (Figure 8.12). Zero volt in, gives out the frequency of 1700 Hz and 0.1 V in would give 1712.5 Hz out. At the receiver end, there is a demodulator turning the FM signal into the original seismic signal. In the example with the 1700 Hz VCO signal, the seismic channels only uses the band 1675-1825 Hz to transmit the signal and there is therefore room for several seismic channels in a standard voice grad link. Fortunately, different manufacturers have agreed on a standard set of frequencies to use and all channels therefore also have the same frequency deviation of ±125 Hz. Up to 9 channels can be put into one analog voice grad line, but often the quality of the line is not good enough for the lowest and highest frequencies (340 and 3060 Hz, respectively) so 7 channels is more realistic.

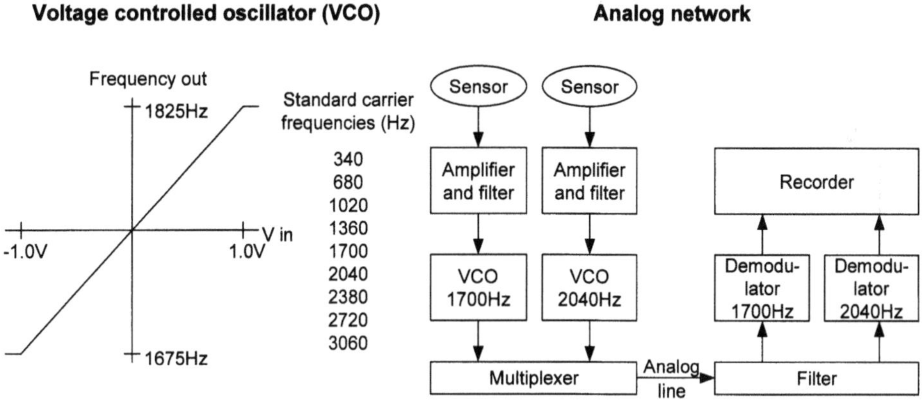

Figure 8.12 Analog transmission using frequency modulation. To the left is shown the VCO and standard frequencies used. The figure shows the linear relationship between the voltage input and frequency out for a carrier frequency of 1700 Hz. To the left is shown a network of two stations using two different carrier frequencies which are summed in the multiplexer, separated in the filter and demodulated in the demodulator.

Figure 8.12, right, shows an example of a transmission of two channels through the same analog line. The multiplexer (see also below) simply sums the two signals, which then consists of the two audible tones 1700 and 2040 Hz. On the receiver side, the two signals are separated by means of band pass filters and converted back to the original signals with the demodulators (sometimes called discriminators). The filter is usually built into the demodulator, but has been shown separately on the figure for clarity. The big disadvantage of analog transmission is that the dynamic range is only 60 dB. That can be improved by using a larger frequency deviation than 125 Hz at the expense of having fewer channels in the analog line. An example of this is using a narrow-band radio channel for a single seismic component. In this case, an audio FM center frequency of 1000 Hz is used with a deviation of ±400 Hz and a dynamic range better than 80 dB (>13 bit) may be achieved. For a phone line, this use of one line per seismic channel would be too expensive, particularly when a high rate can be obtained with a digital phone line.

8.9 Radio links

Local seismic networks or microearthquake networks have traditionally used radio links and this is still a popular system for local networks. For analog data transmission there already exist many hundreds of such links, for details see Lee and Stewart (1981). The current trend is to install radio links with digital transmission. Essentially the same technology and hardware are used as for analog transmission with the addition of a modem at each end (now usually built into the radios and transmitters). Most country regulations for radio use allow some VHF and UHF bands to be used for data transmission within licensed narrow-band channels. These bands are saturated in some areas and it may not be easy to obtain a channel license for continuous transmission. The narrow-band channels are separated by either 12.5 kHz or 25 kHz, with useful bandwidths of 8.5 and 16.5 kHz, respectively. Point-to-point, highly directive, fixed, low-power links may in some cases share the use of a channel with other users within the same zone. An interesting alternative to narrow-band channels is now the spread-spectrum radio, as described later, which permits higher data rates and does not need a license for limited power transmissions.

8.9.1 SIMPLEX-DUPLEX

Digital simplex transmission. In simplex transmission, data is sent only one way, from a field site to a central site. This is the simplest and cheapest and only requires one transmitter in one end and one receiver in the other end. This can be used with both analog and digital transmission.

Radio interference or fading may corrupt data during transmission and there is no way of recovering data, unless digital transmission and forward error correction (FEC) methods are used (see error correction section). However, the FEC methods are rarely used except with satellite links. They require a significant bandwidth overhead, which is hard to provide using standard, low cost 8.5 kHz bandwidth RF channels. The one-way links usually use different types of error-checking methods that allow recognition of corrupted data (see below) but not its correction. The error recognition methods used range from simple parity check and check-sum (CS) error detection, to cyclic redundancy check (CRC) methods.

Digital half duplex transmission. In this system, both ends of the communication link have both a transmitter and a receiver operating on the same frequency with only one antenna. They cannot operate both at the same time, which means that the communication is half duplex. Since most of the data is coming from the field stations to the central site, this is often not a problem. Many companies now sell so called transceivers, a box containing both the transmitter and receiver, so the overall cost is less than buying each unit separately.

Digital duplex transmission: This system has separate receivers, transmitters and antennas in each end and operates on 2 different frequencies. The link is therefore duplex and allows for communication both ways at the same time, like a telephone line. This is the most expensive option for radio communication.

With two-way transmission (half duplex or duplex) and error detection, bad blocks of data can be resent until they are received correctly. In this way a very significant increase of reliability of data transmission is achieved. Another very important benefit of the two-way links is that they allow remote access to data acquisition parameters of the remote seismic stations and their control as well as remote use of various diagnostic commands at the remote stations from central facilities for checking vital operational parameters.

The highest practical speed used in standard links is currently 19200 baud although much equipments only use 9600 baud. Special links can be set up for higher speeds.

8.9.2 POINT TO POINT OR POINT TO MULTI POINT RADIO NETWORKS

So far we have only talked about point to point radio links. However, it is also possible to have a network where the center transceiver communicates to one of the stations at a time (Figure 8.13).

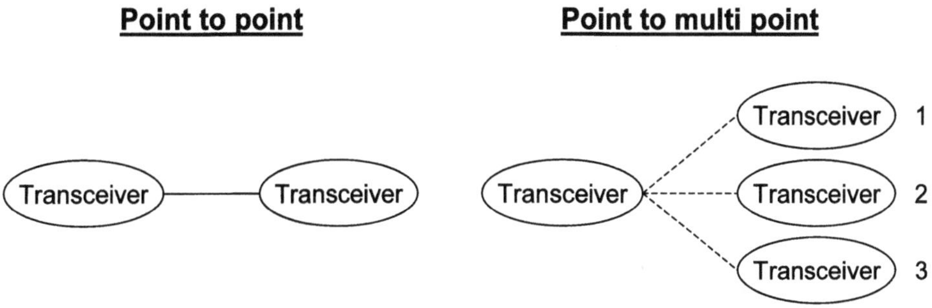

Figure 8.13 Point to point or point to multi point system. To the left is shown the traditionally point to point system while to the right is shown a point to multipoint system. Here only one frequency is used to communicate to 3 stations, one at a time. A solid line shows that there is a continuous communication.

Each station will listen to all messages from the central station but only react to messages coded for itself, so the central station can collect data from any station, one at a time. The software for operating the central station is usually built into the transceiver (example in Table 8.2). A point to multipoint system will have less data rate capacity than a point to point system and is usually not suitable for real time transmission. However, for a virtual seismic network where only a fraction of the total data is downloaded, a point to multipoint system might be an economical solution.

8.9.3 SPREAD SPECTRUM

A new technology for public use, called Spread Spectrum transmission, is now becoming available. Spread Spectrum transmission uses wide-band noise-like signals. The systems efficiently uses available communication space and can communicate at more than 100 kbaud. However, if the airspace becomes congested, the effective speed might go down to unpredictable levels and a continuous throughput cannot be guarantied. Spread Spectrum transmitters use similar transmission power levels as narrow band transmitters. Because Spread Spectrum bandwidth are so wide, they transmit at a much lower spectral power density, measured in Watts per Hertz, than narrowband transmitters. This lower transmitted power density characteristic gives Spread Spectrum signals a big plus. Spread Spectrum and narrow band signals can occupy the same band, with little or no interference. This capability is the main reason for all the interest in Spread Spectrum transmission today.

Spread spectrum systems are easy to use in connection with repeater stations. The communication is corrected for errors and can use both RS232 and Ethernet for communication. Spread Spectrum equipment is usually built to work with either point to point or point to multipoint. Spread Spectrum transmission is probably one of the coming technologies for seismic networks due to the ease of establishing a computer network. Spread Spectrum transmission is operating in the 900 MHz band or higher, e.g. 2.4 GHz. An example is shown in Figure 8.14. Wireless LAN may easily be implemented between laptops using PCMCIA cards with Spread Spectrum technology, covering distances of kilometers with suitable antennas, if there is line-of-sight.

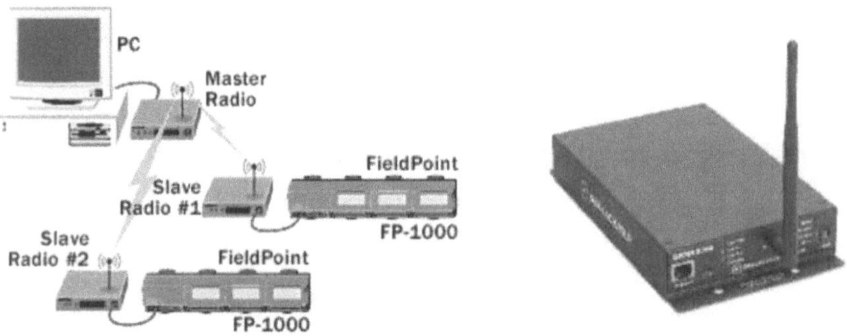

Figure 8.14 A Spread Spectrum central station communicates to two data acquisition systems (left). Right is shown the Data-Linc Spread Spectrum radio modem model SRM6000 (RS232) or model SMR6000E (Ethernet).The range is 40 km with whip antenna and 55 km with Yagi antenna (40 and 50 km respectively for model E). For more details, see Table 8.2. Figure from www.data-linc.com, by permission of Linc Technology Corporation/dba Data-Linc Group.

8.9.4 RADIO LINK CONSTRUCTION AND EQUIPMENT

Fixed radio links normally operate in the VHF (~ 100 MHz), UHF (~400 MHz) or 900 MHz bands. Initially, all links were in the VHF band, but now the higher frequency bands are used almost exclusively due to fewer frequencies available and more disturbances on the VHF bands than on the higher frequency bands. Normally, to operate a radio link, local authorities will assign a specific frequency and a license must be obtained. The Spread Spectrum system has not yet been approved for use everywhere, but in general operates without a license and also has the advantage that no fixed frequency is assigned.

Radio links have the disadvantage that, for any one link, line of sight must be available. This is seldom possible for more than 100 km and typically links are 30 km long. For long links or where line of sight is not available, repeater stations must be used. A repeater station receives the signal on one frequency and sends on another. Reception and transmission frequencies have to be separated enough (usually several MHz) to avoid the output signal to be picked-up by the input and special high-Q resonant filters (cavities) are often needed both at input and output. An analog repeater station will often have the capability of receiving data from several stations, multiplexing the data and resending several incoming data streams in one (see repeater section).

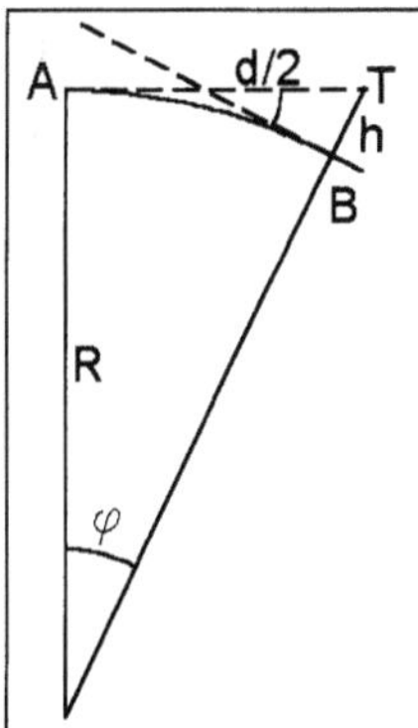

$$\sin(\varphi) = \frac{h}{d/2} = \frac{d}{R}$$

$$h = \frac{d^2}{2R}$$

d (km)	*h*(m)
10	8
20	31
30	71
50	196
70	385
100	785

Figure 8.15 Theoretical calculation of the height h required at a distance *d* over a flat topography to get line of sight. From the geometric scheme on the left, the formula for *h* is obtained (center) as a function of *R* (Earth radius) and the distance. The table on the right list the heights needed for some distances.

In practice, the most frequent technical problems with radio-frequency (RF) telemetry seismic networks originate in inadequately designed data transmission links. The design of RF telemetry links in a seismic network is a specialized technical matter, therefore guessing and "common sense" approaches can cause problems. The issues in the RF link design and link reliability calculations are: The frequency of operation, obstructions by topographic obstacles, the curvature of the Earth, the gradient of air reflectivity in the region, expected fading, potential wave diffraction and/or reflections, time dispersions of the RF carrier with digital links, etc. A simple line of sight test is therefore not a guarantee. For more details, see Chapter 7 in NMSOP (Bormann (ed.) (2002)). Figure 8.15 explains the required height of the antenna to obtain line of sight between two points at the same height above sea level. It is clear that for distances above 20 km, a link is not possible unless either the transmitter or the receiver (or both) is situated on a hill. Often this is a strong restriction for the station site selection. But if other conditions permit the installation on a mountain, links of about 60 km are possible with a (directional) radiated power of a quarter of a W, and a radio consumption of less than 1 W. Directional antennas are not only preferable, but often required by the telecommunication authorities in order to avoid interference. For VHF signals, there will always be some bending of the raypath around the surface of the earth, so it is often possible to obtain communication with a longer distance than theoretically predicted. For UHF and higher frequency signals, there is little bending.

Table 8.2 Digital radio link equipment. This table gives a few example of digital radio equipment used in seismology. All equipment use 12 V. The switching time between transmitting and receiving is usually less than 10 ms. Abbreviations are T: Transmitter, R: Receiver, SS: Spread spectrum, V: VHF, U: UHF, 900 is the 900 MHz band. Power consumption for transmission is given for maximum rated output.

Model and manufacturer	Band	Channel Spacing (kHz)	Baud rate	Power out (W)	Power used (W)
Earth Data TR2400 TR	U	12.5 25.0	4800 9600	0.1-0.5	5.5 T 1.4 R
Nanometrics TX-2 T	V U	25.0	9600	2.0	6.6 T
Nanometrics RX-2 R	V U	25.0	9600		0.9 R
Wood and Douglas	V U	12.5 25.0	4800 9600	0.1-0.5	3.2T 0.9R
Electronic systems technology, Esteem, Model 192C TR (1)	U	12.5 25.0	9600 19200	2	18 T 4.8 R
Adaptive Broadband, MDS4710 TR	V U	12.5	9600	0.1-5.0	24 T 1.5 R
Dataradio T-96SR	V U 900	12.5 25.0	9600 19200	1.0-5.0	
Data-Linc SRM 6000 SS	900	----------	115k (2)	0.1 1.0	7.8 T 1.2 R
Dataradio Integra H SS	900	----------	19200	0.1-1.0	7.8 T 2.6 R
FreeWave DGR-115 SS	900	---------	115k (2)	1.0	7.2 T 1.2 R
WiLan vip 110-24 SS	2400	---------	10 M (2)	0.001-0.1	5.0

(1) Modem has error corrections and can be used for point to multi-point communication
(2) Theoretical maximum speed. At long distances and 'normal' conditions, lower rates are more likely.

The simplest way of testing a potential site, where at least line of sight is available, is to put up the equipment intended for use and let it operate for some time; that takes the guesswork out of the office. Note that it is no good to test with a higher power or different frequency equipment, since that might work, while the equipment intended for use might not.

Also be aware that in many areas, there might be a lot of RF interference from other radio equipment, even if there is no other equipment operating on the same frequency since harmonic distortion from other transmitters might enter the frequency band used. Radio links can usually be constructed with low power consumption, so that they can be operated by solar panels. A typical transmitter output is 0.1-2 W.

Table 8.2 shows examples of radio equipment sold by seismological equipment companies (Nanometrics, Earth Data and Integra do not produce the radios inside). These units tend to use lower power than general purpose systems, where power consumption is less critical. These models will work particularly well with other equipment from the same manufacturers, but might not work with equipment from other companies. The above overview is limited and there might be many other units on the market suitable for seismology. The table mainly has the purpose to illustrate some typical types of equipment. Also note that Spread Spectrum radios have a power consumption comparable to low power 'seismic' transceivers. The channel separation for standard radio channels is either 12.5 kHz or 25 kHz. This is not a strictly technical matter since the bandwidth is linked to the permission to use a particular frequency and in a congested area, it might be hard to get permission for the 25 kHz channels. As it can be seen, the maximum data rate for the 12.5 kHz band is 9600 baud.

Fixed frequency transmission requires a license and the authorities will give a particular frequency for a particular link. In early radio models, only one frequency could be used, so the radios could not be bought before the frequency was known since a particular crystal had to be installed and the radio tuned. The models shown above all have programmable frequencies, so it is much easier, but the radios also contain more components that can go wrong compared to the simpler models. There are examples of some of the older single frequency analog radios working for 10 years without a single revision (like Monitron, www.monitron.com).

The radio technology is changing very fast and what has been presented here is just a fraction of available equipment, which might be outdated very soon.

8.9.5 RADIO LINKS WITH REPEATER STATIONS

When line of sight is not possible, the data must be received at one high point and retransmitted. In some cases, several signals are received at one repeater station, put together and then retransmitted (multiplexing). Figure 8.16 shows an example where simplex transmission is used.

In the case of analog transmission, one voice channel can accommodate up to 9 VCO modulated channels (see section above). The process of multiplexing is simply a question of summing the different VCO signals and then re-transmitting them. There will be some loss of signal quality and more so if several repeaters are used. In repeater station 2 (Figure 8.16), no multiplexer is needed, since it is the same voice grade signal being retransmitted.

In case of digital simplex transmission, we assume that the radios have built in modems and they are therefore sometimes called radio-modems. The process of multiplexing is now more complicated. The protocol used is usually RS232, so the signals cannot simply be summed and retransmitted, that would completely ruin the communication. The multiplexer must be a unit, which can receive RS232 from 3 units, buffer the data and retransmit without mixing the data. A scenario could be that each station sends a one-second buffer of data, properly identified with station code (Figure 8.17).

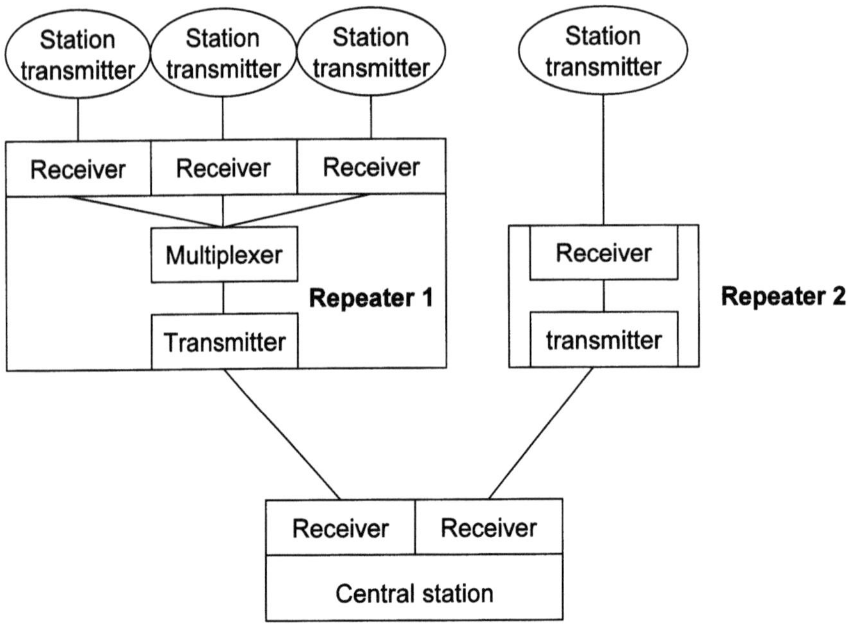

Figure 8.16 Network of 4 stations with 2 repeater stations using simplex transmission of analog or digital signals. The signals from 3 stations are combined at repeater 1 and retransmitted. At repeater 2, the signal is simply received and retransmitted. Note that the receiver and transmitter at the repeater stations must have different frequencies. Communication is simplex since the flow of data only goes from the field stations to the central station so for the case of digital transmission, there is no error correction, only error check.

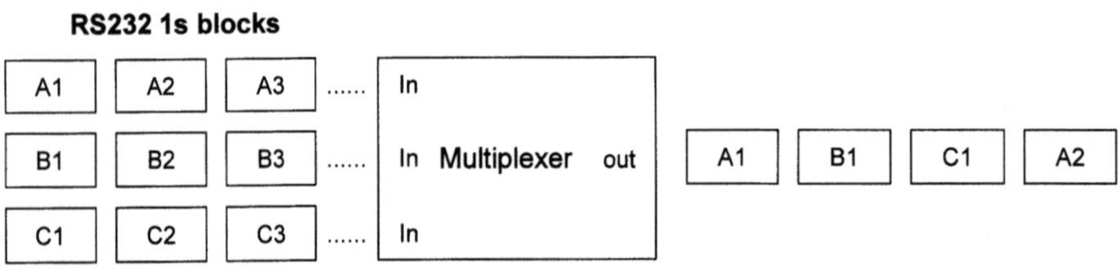

Figure 8.17 Multiplexing of digital data. Digitizers A, B and C send out 1 s buffers which might (as used in this example) or might not arrive at the same time to the multiplexer. The multiplexer stores the buffers and send them out in order A, B, C, A..... on one line. If the buffers are properly identified, they do not have to be received or resent in any particular order.

The multiplexer stores each buffer separately and sends out one at a time. The multiplexer therefore is a specialized product that has to fit the digitizer used and it will therefore be made by the digitizer company. Few such products are available since most instrument companies do not generally produce communication equipment.

In many cases, a multiplexer cannot be used due to the limitation in the baud rate of the radios. A 3 component station with a sample rate of 100 Hz and a 24 bit digitizer will require a baud rate of at least 3 x 100 x 3 x 10 = 9000 so most radio channels will only have enough bandwidth for one 3-component station. However, if 3 single-component stations are installed, 3 channels could be multiplexed into one with a 9600 baud line. Another possibility is using data compression, however, the amount of data would then depend on the amplitude of the signal, so the repeater must have a large buffer capacity to deal with periods of large amplitude data..

In case of digital half duplex or full duplex transmission, more radio equipment is needed. The setup is as in Figure 8.16 and 8.18, except that all receivers and transmitters are replaced by transceivers for half duplex (examples in Table 8.2) using the same frequency, or pairs of receivers and transmitters in case of full duplex. Communication for half duplex can only go one way or the other and the better radios have built in intelligence, so that they automatically switch from transmit to receive depending on data input. In case of multiplexing, the difference from the simplex solution is that errors that might occur on any part of the link can be corrected by requesting retransmission. This requires a special ability of the multiplexer, so half duplex might be difficult to use in connection with multiplexers. In this case, most companies only work with full duplex systems.

Multiplexing with digital data links is not a very attractive solution and most digital network with real time transmission use repeater stations without multiplexers, which in the case of Figure 8.16 would now look as shown in Figure 8.18.

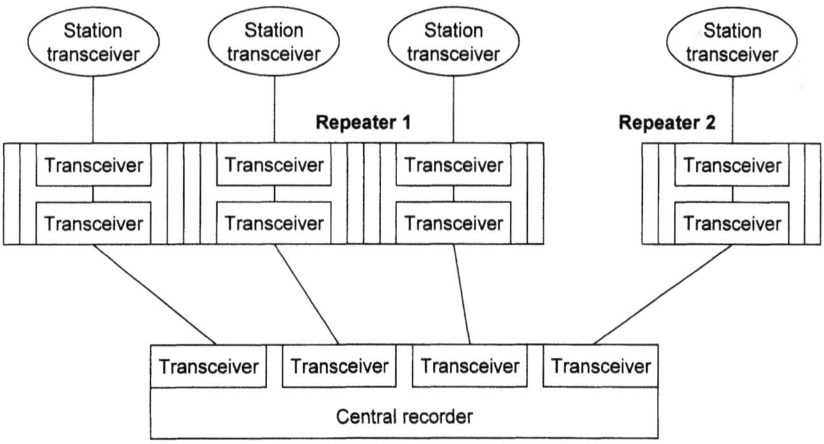

Figure 8.18 Network of 4 stations with 2 repeater stations using duplex or half duplex transmission. The transceiver can be understood as a transceiver (half duplex) or a receiver and transmitter using different frequencies. The signals from all stations are simply received and retransmitted. Note that the receiver and transmitter at the repeater stations must have different frequencies. If full duplex is required, there will be 12 antennas at repeater 1 and 4 at repeater 2 and half that much for half duplex.

The simplest way of making a repeater is using Spread Spectrum radios, since they usually can be set up to work as repeaters and secondary masters without any additional equipment. So the Spread Spectrum radio can be used as both multiplexer and repeater. A careful evaluation (or testing) on amount of data to be sent considering local radio traffic must be made. Spread Spectrum transmission might not be suitable for real time transmission, since a constant throughput cannot be guaranteed.

8.10 Telephone and satellite

Telephone lines. The telephone lines are ordinary dialup, leased analog or leased digital lines.

Dial up lines. The common telephone line globally available is analog and a standard modem is used for communication. In general such dial up systems are only used for short term communication to connect to a field station to get specific data. Continuous operation would be cost prohibitive. The maximum speed generally available is 56 kbaud, however this depends to a large extent on the quality of the phone lines and also to some degree on distance. In practice it might be hard to get above 19.2 kbaud. The modem development is continuing rapidly and higher speeds might be expected in the future.

In many countries, the telephone systems are becoming digital, using the ISDN (Integrated Service Digital Network) standard, where 128 kbaud is the standard speed. While modem connection can be unstable and take up to a minute to establish, the ISDN system connects within a second and provides very stable communication.

A newer approach to digital lines is ADSL (Asymmetric Digital Subscriber Line), which uses the standard telephone cable pair and offers a permanent (no dial-up) connection at a speed of up to 2 Mbit/s. The term “asymmetric” refers to the different speeds for download and upload from the subscriber. The basic system usually permits 256 kbit/s in the direction from network to subscriber and 128 kbit/s from subscriber to network (in this case, this is the important for data transfer). The distance from the subscriber to the nearest active ADSL central cannot exceed about 5 km, and the present (2003) cost per month is about $ 40.

In practice, dial-up phone lines based networks cannot 'digest' earthquake swarms and the numerous aftershocks after strong events. With large capacity local recorders no data is lost, but the delay in receiving the data could be large. In addition, they often do not function at all for several hours after strong events, due to either especially high usage of public phone system or technical difficulties. They are not the right choice for networks with the predominant purpose of giving seismic alarms.

Leased analog lines. These lines function like analog radio transmission and have about the same bandwidth. Although several are in operation, they are now being phased out due to low quality and high cost.

Leased digital lines. These lines are standard 2 way digital serial links. Usually the user has the choice between several transmission speeds, which is cost dependent. On long distances, this type of line is usually very expensive. There is usually no error correction built in.

Cellular phones. For many areas of the world, this might be the simplest way of setting up communication for a seismic network. The operating price is still too high for continuous operation, but for periodic short-term communication to a field station, it is viable today. The installation price will be much cheaper than any radio network. While any voice terminal with data facilities may be used, special data terminals exist, such as Siemens S-1, with more transmission power and an external antenna connection. The maximum present speed is 9600 baud for the GSM system. This may soon change with the new GPRS (General Packet Radio Service) technology, which is expected to improve the effective speed by a factor of 3 to 6. It is based on packets, like the IP protocol. In fact an IP address is assigned to the terminal and the cost depends on the amount of data transferred, instead of the connection duration (an earthquake swarm may be quite expensive!).

Satellite links. Satellite links function like radio links with the advantage that no line of sight is required. Several different companies offer different types of technical solutions. In general, satellite links use more power and are more expensive to set up and operate than radio links. However, there is such a rapid development in the satellite technology, that it can be expected that telephone and satellite communication will soon be similarly priced. The first global satellite telephone system using handheld phones is already in operation.

The most common system to use is the Very Small Aperture Terminal (VSAT) system, which has been in operation since 1985. Several companies make equipment and several service providers are available. Generally, it is only cost-effective to use VSAT if a certain number of stations are used, in which case monthly charges can be less than $ 100 per station. This is a lot more than a radio network, but less than renting a permanent digital phone line for a few hundred kilometers. Currently, the only seismological instrument company marketing their 'own' system (inside produced by others) is Nanometrics. Their Libra system is fully integrated with the rest of Nanometrics equipment with the same advantages and disadvantages as described under radio links. The VSAT system itself consumes 18 W so it can be operated on remote sites. Other systems are as low as 11 W (*www.gilat.com*).

There are generally two ways getting VSAT, (1) to get a complete private system with satellite systems at both ends, or (2) to get just the remote transceiver which then connects to the public internet. Option 2 is by far the cheapest solution, and technically the simpler, but provides a bit less data security than option 1. Option 2 is often provided to users who have no other means of access to internet and is in some cases intended for home installation, therefore relatively cheap (< $3000) and simple to install.

8.11 Digital data transmission protocols and some examples of their use

When the hardware connection between two sites can be established, the next question is how the data is sent along the established connection or what protocols are used for the two units to communicate. In the following, some of the main types of communication will be discussed.

8.11.1 SERIAL DATA COMMUNICATION

Most seismic digitizers will send out a stream of data in serial format and all computers have hardware and software to communicate with serial data (8.4.1). The serial lines use either RS232 or the RS422 interfaces (8.4.1) (USB might soon become also standard). The first one can run on up to 25-50 m long cables and the second one over up to 2 km long cables. A serial line RS-232 requires at least 3 lines: One for sending data, one for receiving data, and ground. If data is only to be sent or received, 2 lines suffice. Serial line communication is used by modems, by radio links, fixed telephone lines, cellular phones, and satellite links. Let us give some examples of how serial data transmission is used in practice.

Example 1, One-way continuous communication. (Figure 8.19). A remote station has a digitizer sending out RS232 data, which enters a radio link to a PC, which reads the data and processes it. The communication is governed by the RS232 protocols. The software on the PC can run a continuous or triggered mode data acquisition.

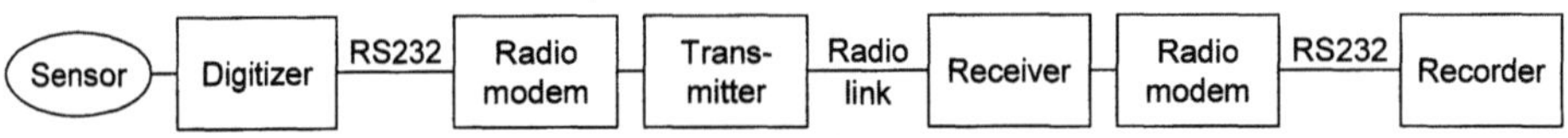

Figure 8.19: One way communication from a remotely installed digitizer via a digital radio link to a centrally located PC. The radio modem and transmitter /receiver might be one unit.

Example 2. Interactive communication with a remote seismic station (Figure 8.20). A user calls up a terminal emulator on his PC, connects to a modem with one of the PC's serial lines, dials the number of the modem connected to the remote station, and logs into the field station. Once logged in, several options are usually available. One is to browse a log file containing all triggered events in the local memory of the station. Another option is to initiate a download of event data. A way to do this over a serial line connection is to list the event file in ASCII form and set up the terminal emulator at the local PC to capture the data. This is one way of getting data from the standard GSN seismic stations (see below). The advantage of this type of communication is that the system requires only very simple software. In this way it is easy to get access to many different seismic stations. This process can be easily automated. This type of connection can also be made with Telnet over any TCP/IP connection.

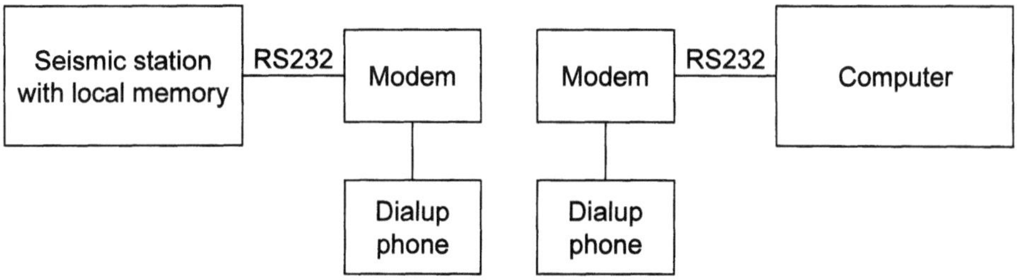

Figure 8.20 Manual dial up to a seismic station for data inspection and/or download. The computer dialing can be any type of computer with a terminal emulator program like Hyperterminal in Windows.

The following shows an example of logging into a GSN station and downloading data. Comments are in parenthesis:

```
OS-9/68K V2.4   Motorola VME147 - 68030   99/08/26 22:11:51
```

`User name?:` **`seed`** *(Enter **seed** for the User Name)*
`Password:` *(Enter **data** – the Password will not be displayed)*

```
Process #39 logged on    99/08/26 22:11:59
Welcome!


type ? for help
```

`Command?` **`t`** (Enter **t** to select the transmission format)

```
Possible transmission formats are:

         "C" - Compressed hexadecimal ascii
         "CS" - Compressed hexadecimal ascii SEED
         "V" - Expanded variable record length ascii
         "F" - Expanded fixed record length ascii
         "S" - SAC ascii digital counts
         "X" - Exit to main command menu
```

(At this point, the user can decide which type of data to transfer. To choose SAC ASCII digital counts, use option **S**)

`Transfer mode?` **`s`** (Enter **s** to select SAC ASCII digital counts)

```
Current transmission mode is SAC ascii
```

`Transmit card numbers with each line of data? (y/n):` **`n`** *`(no Card numbers)`*

`Command?` **`s 00-bhz 99/8/26 1:00:00`**
(Select the channel and start time for data retrieval)

```
   Search requested starting at 1999/08/26 01:00:00

Time window begins in segment 71 at buffer record 2419
```

`Maximum number of samples to transfer?` **`25`** (Number of samples, limit = 10000 samples)

```
Buffer server is processing your request

skipping first 5165 samples...
transmission will begin at requested starting time within 0.014161 sec

Use the "G" command to begin transmission
or to re-transmit data received incorrectly.
```

`Command?` **`g`** (Enter **g** to transmit the data)

```
GUMO   H99238010000-BHZ
       1565.0      1565.0      1565.0      1565.0      1565.0
       1565.0      1565.0      1565.0      1566.0      1565.0
       1565.0      1565.0      1565.0      1566.0      1566.0
       1565.0      1566.0      1565.0      1565.0      1565.0
```

Example 3. Interactive communication with a remote seismic station using proprietary software. The setup is as in Figure 8.20. The user starts up a specific program on his local PC made by the manufacturer of a particular field station. The program handles all the communication through the serial line (dialing, connecting, login, etc) and the user will be in connection with the field station as if sitting next door. Data download, acquisition parameter settings, system state-of-health verification, and diagnostic commands (if applicable) are managed through simple menus and the event files will be automatically transferred to the users local PC. Figure 8.21 shows an example.

The process can be run manually or automatically, unattended at given times, or, with some systems, remote stations can initiate it after triggering on seismic events. The advantage with this setup is that communication with a particular remote station is very easy. Unfortunately, a typical software like the one shown in Figure 8.21 is completely useless with other types of seismic stations than from that particular recorder.

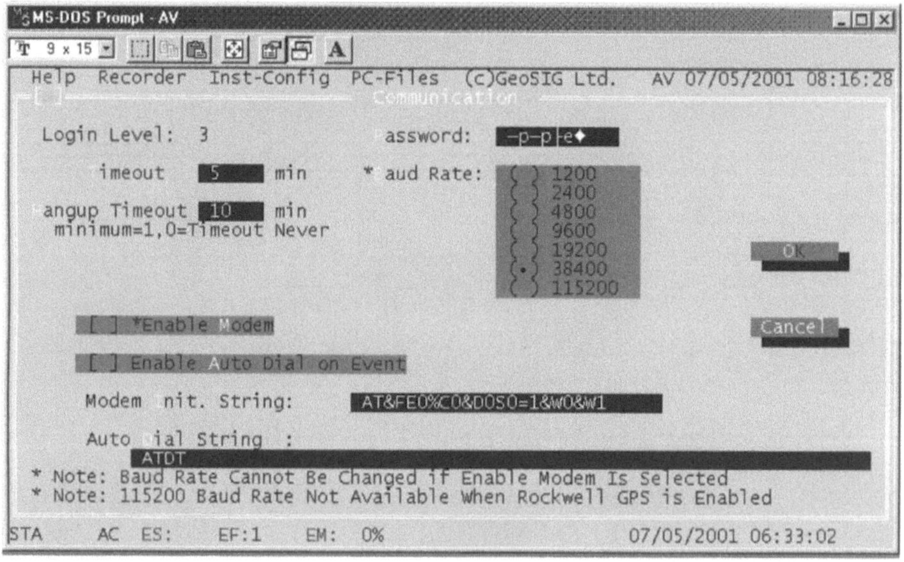

Figure 8.21 Login screen on a GeoSIG data logger. On top is seen the main choices like setting up the recorder (Recorder), general configuration (Inst-Config) or working with the files downloaded to the PC (PC-files). The main screen shown deals with communication, which can set communications speed and choose if an event is automatically downloaded after recording. The permanent bottom line shows recorder status flags of which STA are station name, EF, number of files in memory and EM is % of memory used.

8.11.2 TCP/IP COMMUNICATION

In high-speed local area networks, Ethernet most commonly connects computers. The low-level protocol is not what the user sees directly, he sees some high-level communication protocol, which works on top of the Ethernet protocol. The most widely used protocol is TCP/IP for file transfer and remote login (see 8.4.1). This is also the protocol used by the Internet although the connection between the different Internet nodes might be by other means than Ethernet. The TCP/IP protocol might be used over a serial line, an ISDN or ADSL line (Figure 8.22) and the new GPRS wireless phone links.

The simplest seismic stations are usually not able to communicate with the TCP/IP protocol but as the computer power of remote stations is constantly increasing, more and more have TCP/IP and Ethernet built in. A remote seismic station, which can be reached by TCP/IP either through Internet or dial up ISDN or phone lines, presents the most general purpose and flexible system available.

Figure 8.22 shows the most common way of establishing TCP/IP connections to a central data collection system. Dashed lines between routers indicate that connection is made to one station at a time. Large central routers are also available, which can communicate to many ISDN nodes at the same time.

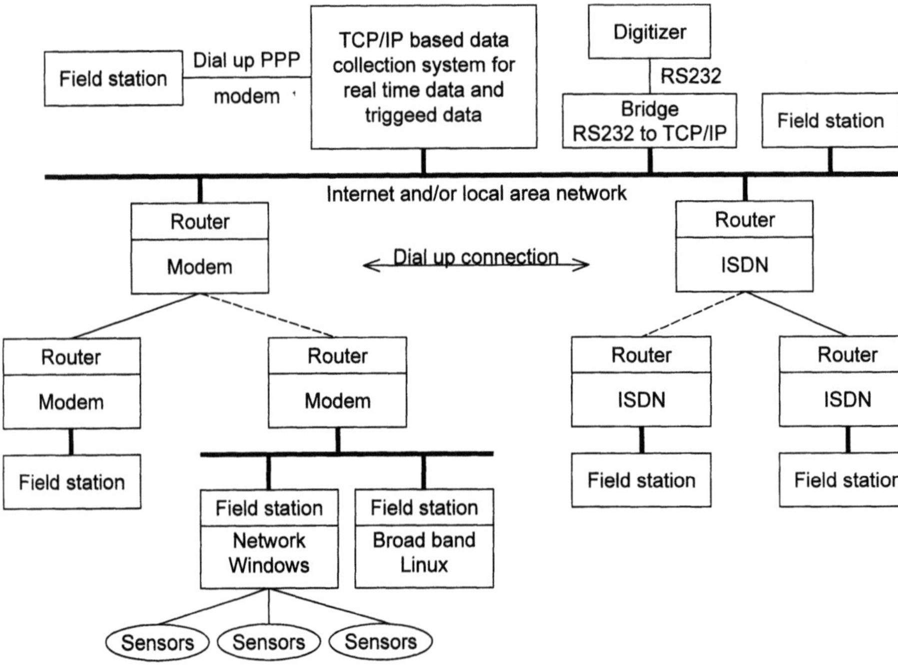

Figure 8.22 Different ways of getting a TCP/IP connection to a central data collection system. The thick solid lines indicate permanent Ethernet connections.

Permanent connection. The simplest is the permanent connection (top right), where the field station might be on the LAN or far away connected to internet.

Permanent real time. The digitizer is connected to a black box (bridge, top right), which sends out all the date entering on the serial line to an internet node, which is also the data collection system. Several data collection systems can receive the data from one digitizer in this way. The receiving software must install a socket to receive the data. Some digitizers have output directly to the internet via built in Ethernet and TCP/IP software (see Chapter 4).

Dial up PPP. In this case the two computers first have to establish contact via the modem and serial lines and then start up PPP. This is easy if the two computers have the same operating system, but might be hard to automate for different operating systems. It is therefore not a desirable solution.

Dialup via intelligent routers, ISDN or normal phone. In principle this works in a similar way as the dial up PPP. The main difference is that the communication is very easy to set up and completely independent of the operating systems of the field stations and the central computer. Once the routers have been configured, the field stations nodes become just another internet node and routers take care of setting up all connections and taking them down again when there is no more communication. At the field station site, there can even be several local nodes (bottom left). This is the most desirable dial up TCP/IP configuration.

Getting seismic data from a GSN station using Internet via a local computer is simple. The user uses the Telnet to login to the station. Once logged in, he can check available seismic data and then use the FTP file transfer protocol to copy the data to the local computer. The process is easy to automate. Figure 8.23 shows how communication is done with the GSN network.

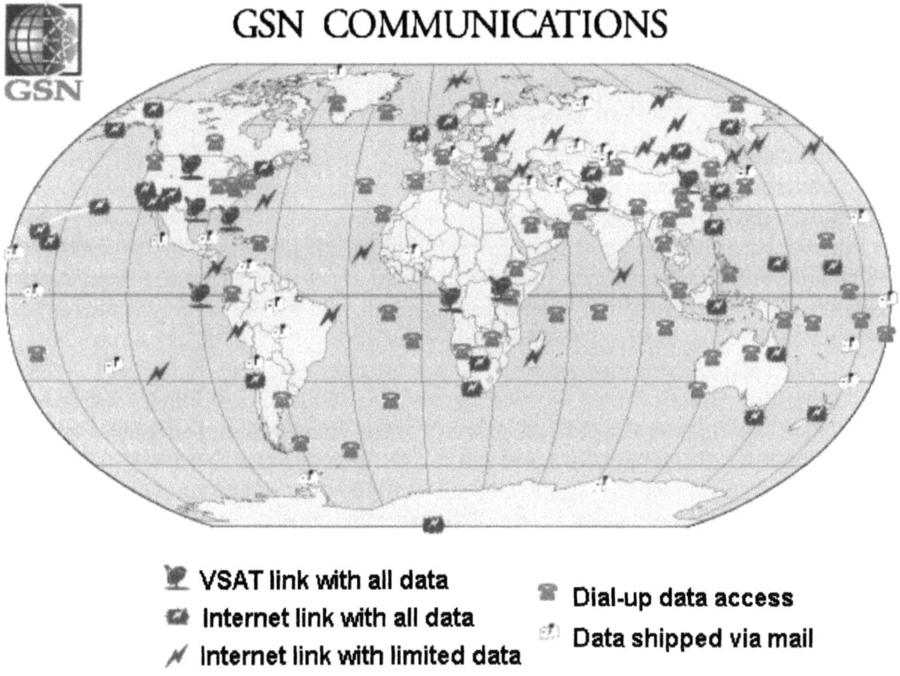

Figure 8.23 Communication to GSN stations in 2001. VSAT is a satellite connection. VSAT and internet have real time connections and potentially all data can be downloaded, while the dial up stations only transmit a limited amount of data. Figure from www.iris.washington.edu.

Many computers do not have direct access to the Internet but are able to send e-mail. Some seismic stations and centers, particularly in Europe, use a shared protocol for providing seismic waveform data semi-automatically by e-mail. This system is called AutoDRM (Automatic Data Request Manager, Kradolfer, (1996)). The user sends an e-mail request for particular data and the remote system automatically ships back the data by e-mail.

8.11.3 COMPRESSION OF DIGITAL DATA

Because of the high data rates from digital seismic stations and the throughput limitations of available data transmission links, data is often compressed before transmission. The compression can generally be expected to halve the quantity of seismic data. After transmission, data must be uncompressed unless it is stored directly without processing. There are several compression routines in use, some of which are in the public domain and others are used only in a particular type of equipment. Generally the public domain compression routines are used for data storage, while the company made compression algorithms are used with specific equipment. If communication is by a telephone line with modem, the compression can take place in the modem with standard agreed protocols and no compression software is needed at the seismic station.

With many compression algorithms, the degree of data compression depends on the amplitude of the seismic signal. Therefore the efficiency of the compression falls sharply during strong (and therefore also long lasting) earthquakes. One should be sure that the links throughput associated with the local temporal data memory of a seismic station suffices also in case of large, long lasting events.

8.11.4 ERROR CORRECTION METHODS USED WITH SEISMIC SIGNALS

All digital communications experience errors. In transmission of seismic signals this is particularly fatal since just one bit of error might result in a spike in the data with a value a million times the seismic signal. Obviously this wreck havoc in trigger systems and one byte missing in an event file might corrupt the whole event file.

One of the principles of error correction is that the data is sent in blocks, e.g. 1 s long, and along with the block of data there is some kind of checksum. If the checksum does not tally with the received data, a request is sent to retransmit that particular block of data. A checksum can simply be the sum of all the sample values in one block or more sophisticated algorithms can be used. Obviously, this type of error correction requires duplex transmission lines and local data memory at the remote station. If only one-way transmission is available, the errors cannot be corrected using the checksum method, but they can be detected and appropriate action taken at the receiving end. However, loss of data is inevitable, unless some type of redundant information is sent to allow the use of some forward correction routine. Error correction can be utilized on different hardware and software levels and can be of different types.

Proprietary error correction is used in some systems on the market. This means that the system operates dedicated links where all the responsibility for transmission and error correction is with the system. An example is digital radio links to remote stations with a manufacturer's protocol for error correction. The protocol is built into the commercial product.

Standardized hardware error correction is another type of error correction. The hardware unit where the data enters the digital link and where it comes out has its own error correction built in. From the user's standpoint, it is assumed that no errors occur between the input and output of this hardware. The most common example of such hardware error correction is a standard telephone modem link, which uses industry standard error correction internally.

Computer networks use their own error correction methods. When computers are linked with common computer network protocols like TCP/IP or Kermit, error correction is built in from computer to computer. This is obviously the best solution; however, it requires that the seismic remote stations can operate quite sophisticated software. Most simple digital remote stations cannot benefit from this type of error correction today, but the trend is to use more powerful processors in the remote stations, so this form of error correction might be more dominant in the future. Nevertheless, several seismic data loggers already exist in the market capable of communication with higher level protocols.

Satellite data transmission links usually use forward error correction (FEC) methods. FEC works on one way links and does not require any retransmission of data blocks in case of errors. It is similar to checksum error detection. It does not only detect corrupted bits, but by comparing transmitted checksum and the one of the received data, also corrects them. Its drawback is an increased data channel bandwidth due to a significant data overhead dedicated to error correction.

If a given seismic system with a built-in error correction uses different communication means, one should carefully consider the interplay between the system's own error correction and different error correction methods, which may be built-in a particular communication equipment.

8.11.5 SEISMIC DATA TRANSMISSION AND TIMING

All digital data acquisition and transmission systems exert a certain time delay. The delay depends on the digitizer, the digital protocol used for transmission and the computer receiving the data. For this reason, most digital field stations time stamp the data at the remote station and subsequent delays in the transmission have no effect on timing accuracy. However, there are also digital network designs where the timing takes place centrally. This can be done if the digital data arrives at the central site with a predictable or measurable time delay. The central computer must then time stamp the data when it arrives in real time and later correct it for the known transmission and digitizer delay. The advantage of this system is that only one clock is needed for timing of the network. A further advantage is a simpler remote station consisting only of a sensor and a digitizer. The disadvantage is that timing accuracy is not as good as with time stamping at the remote sites because time delays are known only with a limited precision and they may also vary in time. Also if the central clock or its synchronization with RF time signals fails, the whole network fails. The current development goes in the direction of time stamping at the sites because GPS clock prices are now a small fraction of total digitizer costs.

8.11.6 COMMUNICATION AND GRAPHICS

Proprietary software will often have possibility of graphics display over the communication link whether RS232 or TCP/IP. Graphics display is important to be able to check signals and to use graphics based programs for recorder settings. For non proprietary software, there are some general ways of obtaining remote graphics.

Unix systems. The most common graphics environment is X so if both the field station has an X-sever and there is a TCP/IP connection, then all graphics working locally will also work remotely.

Windows systems. Until the version XP, there was no built in software to make remote graphics easy, mainly because Windows is a single user system. There is third party software available, like PCANYWHERE©, which allows to export the local screen to a remote PC by internet or modem, so all graphics seen locally can also be seen remotely. Windows XP also has built-in software to do this. There is a free software, VNC, which can do the same across several platforms (*http://www.uk.research.att.com/vnc/*).

Web server. Some field stations are able to install a web server, which allows any web browser to use the web server programmed display. This is a very general solution, which is not yet much used, but might be more common in the future. It is used by at least one commercial portable recorder (Nanometrics). It is currently common in central recording systems and data base servers.

8.12 Examples of networks

This section will give additional examples of seismic networks in operation and the intention is to give examples of different types as described in the preceding section, both with respect to technical solutions and purpose of the network.

The International Monitoring System (IMS). In recent years, a new global network, the International Monitoring System (IMS) has been set up aimed at monitoring the Comprehensive Nuclear-Test-Ban Treaty (CTBT; see *http://pws.ctbto.org*). The IMS consists of 50 designated primary stations, mostly arrays (see Chapter 9), with real-time data transmission to International Data Centers (IDC's), the IMS center in Vienna in particular, and 120 auxiliary stations. The latter provide data on request to the IDC's. Many of the auxiliary stations are members of the Federation of Digital Seismic Networks (FDSN). The IMS network (Figure 8.24) is currently the largest and most modern physical real-time network in the world. However, when requesting data from auxiliary stations, it works like a virtual network where the real-time network makes the detections and preliminary locations and then requests additional information from remaining stations for improving these preliminary findings.

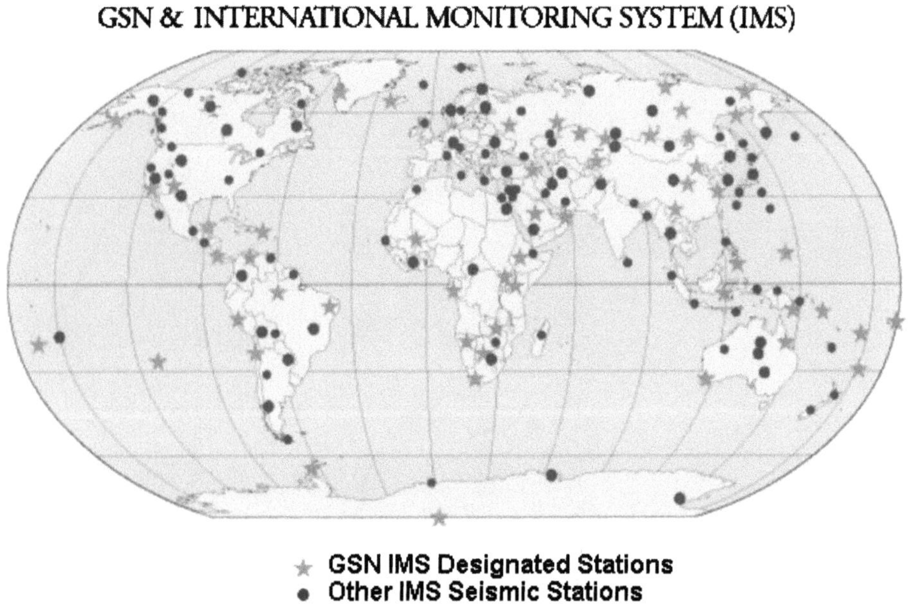

Figure 8.24 Stations in the International Monitoring System, from www.iris.washington.edu.

Southern California Seismic network (SCSN). California probably has the world highest density of seismologists and seismic stations and there are two large networks of more than 350 stations each. The SCSN (Figure 8.25) is one of the largest and most automated networks in the world consisting of a mixture of triggered and continuous systems using a large variety of equipment and communication means. The central recording is taking place at California Institute of Technology, Pasadena (CALTECH). This network and the Northern California network have been early pioneers in setting up local networks (Lee and Stewart, 1981). The network cannot be characterized as either physical or virtual since it is a complex mixture of both. It is interesting to note that, despite the high technological level, there are still some simple robust analog stations in the network. This network is definitely NOT a turnkey network.

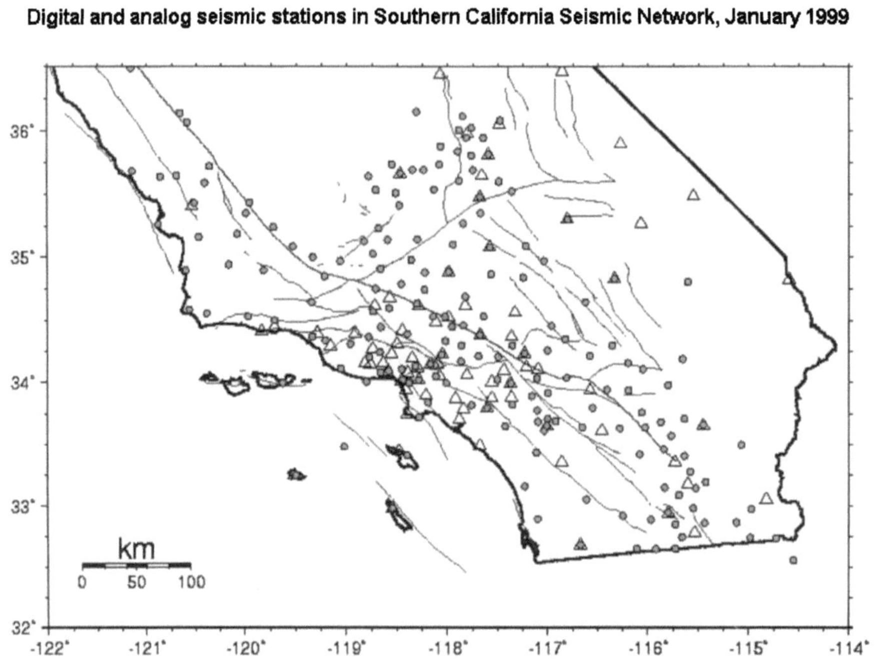

Figure 8.25 The Southern California Seismic Network. The system has several types of stations. Open triangles are digital broadband stations, the round filled symbols are analog stations and the rest different digital stations. The lines shown are faults. Figure from the SCSN home page.

Norwegian National Seismic Network. This network is a typical virtual network operated by the SEISNET data collection system. It consists of 28 stations of which 6 are connected to 2 analog sub-networks with analog transmission (Figure 8.26). Field stations are IRIS GSN or SEISLOG types with Nanometrics digitizer, Earth data digitizer, SEISAD18 digitizer or multi channel boards for the analog networks.

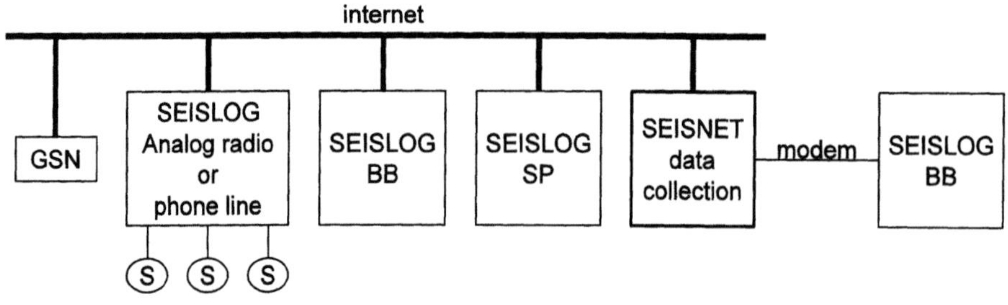

Figure 8.26 Types of stations in the Norwegian National Seismic Network (NNSN). Nearly all stations are connected by internet (fixed or ISDN dial up). Abbreviations are: S: Seismic stations in a local network and GSN: Quanterra type GSN station.

The network covers a large area (Figure 8.27) and communication is by internet (fixed or ISDN dial up) or a simple ASCII modem connection. For most stations, only triggered data is used, while for 4 BB stations, continuous data are collected. Each station has its own trigger and, due to the large area, it is rare that an event is recorded on all stations.

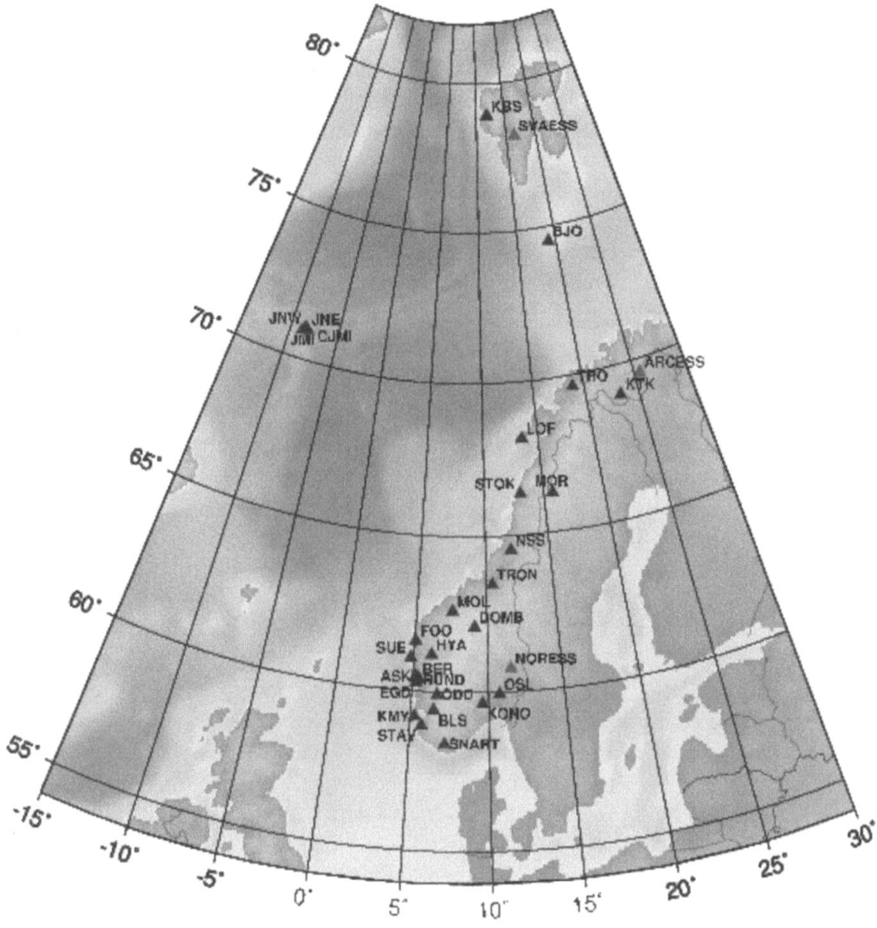

Figure 8.27 Norwegian National Seismic Network and the 3 NORSAR arrays (NORESS, ARCESS and SVAESS).

8.13 Running a seismic network

Setting up a seismic network is the easy part. Making it run for many years is much harder, since routine tasks are hard to keep up and have to be very well organized to continue with a constant quality. NMSOP (Bormann (Ed.), 2002) has detailed recommendations. The operation can be divided into two tasks: Physical continuous operations and data processing. Usually two groups of people do these tasks. It is important that both work closely together since the people analyzing often will spot problems with the stations right away, like that magnitudes suddenly become too small at a particular station. On the other hand, instrument experience is needed in order to set up proper calibration information for the network

8.13.1 PHYSICAL OPERATION

Keeping stations going is an obvious priority and can be organized in many ways so it is hard to make any general recommendations, see NMSOP (Bormann (Ed.), 2002). If stations are nearby, it is often easy to fix problems, but when stations are far away, maybe a day of travel, it is no longer easy or cheap to fix problems, even if it is just a question of pushing a button. So, lets give a few hints on how to prevent excessive travels.

All computer based boxes have a bad habit of stopping without a good reason and most have some kind of reset function, when everything fails, or power off-on when that does not work (common experience with PC's!). Modems are particularly prone to a hang-up, but recorders (PC based and others) and GPS will also hang. The problem is how to activate this reset remotely. Some equipment might have a sophisticated watchdog which will restart a system if e.g. data acquisition fails, but most have none. Some simple solutions:

Automatic check of system operation. Operators are not always aware of a system stop, particular during non working hours. If the field stations are automatically interrogated regularly, the data collection system can try to restart the remote system, if possible, or send a mail to the operator in case of failure. So a first requirement is to actually be aware of a system problem.

Connect timer to critical equipment. This a brute force solution. The timer turns on and off the power for maybe a minute once every 24 hours and thereby restarts the system. For modems and GPS, this is no problem, but some data acquisition systems with disk based recording might not like it, although most will suffer no harm.

Remote telephone operated switch. If a telephone line is available that is not used for communication, it is possible to use a common remote control box, which by a simple telephone call can turn on and off several switches (Figure 8.28). These boxes are usually used to e.g. turn on heat in a remote dwelling, so it is simple cheap equipment. In this way, a reset is made only when needed and only for equipment which has stopped working.

External watchdog timer: This unit (Figure 8.28) will turn on and off the power if it does not receive a certain signal from the recorder within a given time. In the example above, the recorder sends out a pulse every second. If it is absent for more than e.g. a minute, it will activate the switch.

Data acquisition system timer. Some recorders will have the possibility of making a restart at regular intervals. That is better than a power off since regular shutdown can be done, but of course does not help if there is a complete stop. This function might also be used to reset modems and GPS.

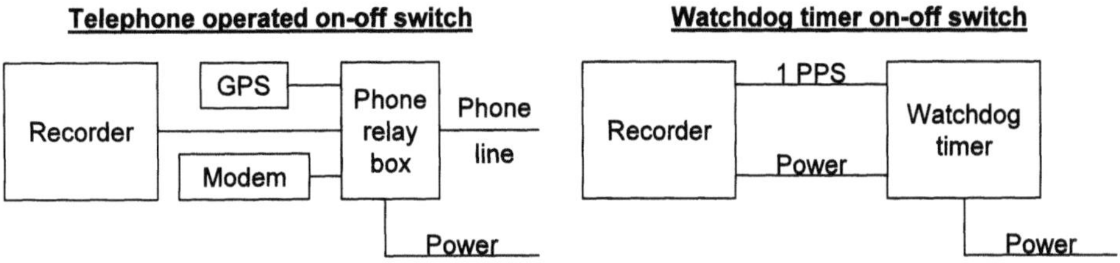

Figure 8.28 Methods for restarting a remote seismic station (see text). To the left is shown a telephone operated switch, which can turn on and off any of the 3 units. To the right a watchdog timer, which continuously monitors output from the recorder (sends a pulse per second (PPS)). If it stops, the power is turned on and off.

8.13.2 DATA PROCESSING

The single most important task, not directly related to the physical network, is to process and store the data and to keep on processing it. Far too often the user is getting a beautiful network where gigabytes of data end up in some directory, but only has some primitive software to handle it. The processing part MUST be planned when the network is planned in order to ensure that adequate hardware and software is available. This might mean substantial extra costs for computer equipment.

Processing. By processing we mean organizing the incoming data and doing basic routine processing like location and magnitude determination. The first thing, which can help here, is that the data collection system by itself organizes the data in some data base and does preliminary analysis. Do not expect too much from automatic phase picking, it should normally be checked by an analyst. Large sophisticated systems can of course do quite well, but even a network like the Southern California Network (Figure 8.25), also has to rely on manual analysis (Figure 8.29). If the data is already correctly stored in a data base, this makes processing much easier. So, decisions about processing system and data format must be made at the same time as the network is planned; unfortunately this is far too often not the case.

Archiving. When one first sets up a seismic network, one needs to think thoroughly about organizing the data that is recorded in light of the fact that, eventually, the network will have many, many years of accumulated records. Often, this crucial aspect of seismic system organization is overlooked or left to on the spot decision by whoever is starting archiving records at the beginning of a new network operation. This may work fine for a while, but eventually everybody will run into serious problems if the archiving system chosen is inappropriate. It is necessary to carefully think through the archiving organization from the beginning and keep the long-term future in mind. This is particularly the case for continuously recording systems.

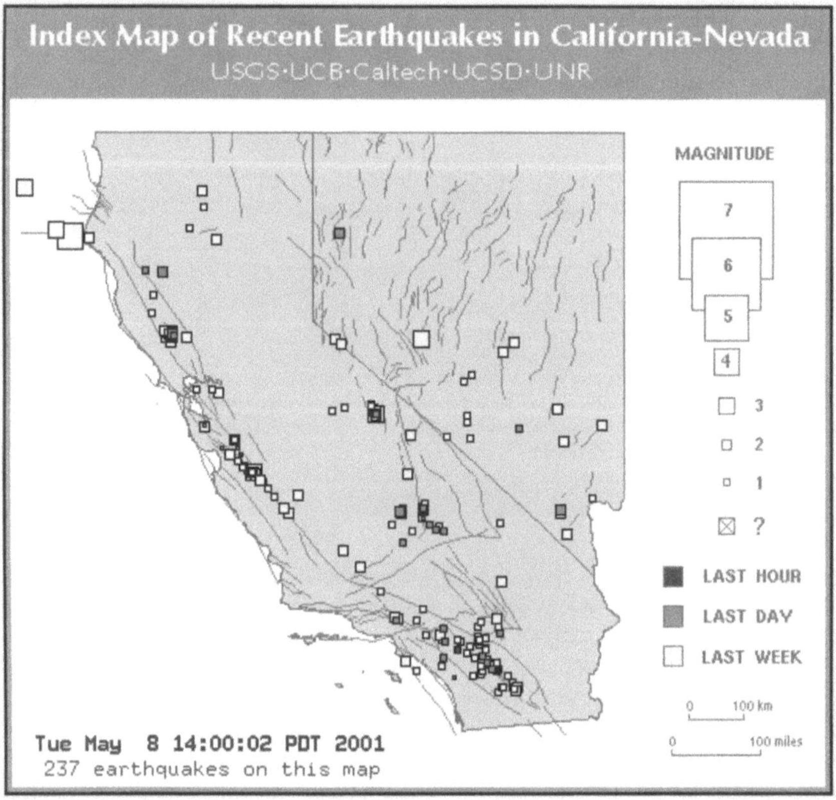

Figure 8.29 Automatic and manual locations from seismic networks in California and Nevada. Figure from Southern California Seismic Network home page, www.trinet.org

In a small, weak-motion network in a low seismicity region that generates only a small number of records each year, or in a small or medium size strong motion network, one can probably get by with a directory tree organization for the data archive. Nevertheless, filename coding of events must be thoroughly thought out to avoid confusion and/or file name duplications and file names should reflect complete date and time of event and source of data. Due to long file names, on UNIX and Windows 95 or higher operating systems, this doesn't present any problems. Larger networks in moderate to high seismicity regions require a better organized database for archiving purposes. One should carefully consider different options used by other seismological observatories and in the market before one begins to log records. It is very painful to change the data coding or archiving method after several years of network operation, once thousands upon thousands of records are already stored.

Very powerful professional databases may not be the most suitable choice for seismology, primarily due to their high initial and maintaining costs, and secondly, due to too many utilities which will never be used in seismology (but which has to be paid for, of course). Special data bases which have been developed in the seismological community for the needs of seismology, thoroughly tested in several existing applications, and accepted by many, seem to be the best choice at the moment, however, very few are available commercially.

Always keep the raw, unprocessed seismic data (raw event files, or sequences of continuous data) in the archive along with the full documentation about the recording conditions (data acquisition parameters and accompanying information). Processing and seismic analysis methods will change and evolve as time passes. Future generations will appreciate having unprocessed seismic data available to further their research and knowledge.

8.14 How to make your own network

This section will try to give some simple guidelines to those who want to set up their own network and not buy a turn-key system.

Most seismic networks today are made of proprietary hardware and software developed and manufactured by a few small companies specialized in seismology. Only in the recent times have technical solutions using little proprietary hardware and software become available, and sensors, digitizers and data loggers are the main units developed and manufactured by seismological equipment manufacturers. All the rest are commercially available standardized products used in other fields and manufactured by much bigger companies. Data transmission is done by commercially available and standardized software and by off-the-shelf hardware components, although at least one company partly makes its own, see Table 8.2.

Using off-the-shelf hardware and software significantly reduces the cost of network maintenance, increases reliability, and guaranties flexibility. The user is much less dependent on an individual manufacturer of the seismic system. Long term maintenance and upgrading of the system has a much brighter perspective. Proprietary electronic equipment is very hard to maintain in the long run today, since average 'life time' of modern electronic devices and components is only three or four years. In such situations, using off-the-shelf standardized products instead of proprietary, guaranties the longest lifetime, because the new products, which come on the market, are usually downwards compatible. This is often not the case with new designs of seismic equipment manufacturers.

Using off the shelf products of course requires that the network owner takes an active part in setting up the network, which anyway is the best way of ensuring that the network can be maintained in the future. In amateur car repair books, they always say “If you lack the skills or tools, go to an authorized dealer”. In seismology, the authorized dealer might be on the other side of the Earth, so the network operator should rather acquire the skills and the tools, although repairing an ADC is not something that is expected of an operator.

The general advice is: Select each element (sensor, digitizer, recorder, communication) of the network independently, if possible, make sure they can work together, and preferably make sure that each element has more than one manufacturer (not always possible).

Communication. This is the main element to consider and should not depend on the seismic equipment manufacturer. The main point is to establish the best possible communication network within the available possibilities, considering the seismic equipment that might be used. The communication protocol should be industry standard.

TCP/IP permanent network. The ideal solution. This can be used with digitizers only in the field and with recorders. Not all hardware supports TCP/IP, but that can be fixed with a conversion box. The physical media can be e.g. phone line, radio link or satellite link. For some real time networks, it might be overkill if the protocol is mounted on a fixed serial line, which could have been used directly.

TCP/IP dialup network. Almost as good, but real time data transmission is not available.

Permanent serial line connection. Good for real time networks with remote digitizers and central recording.

Dial-up serial line connection. A very simple solution, if a phone line is available. A mobile phone will do just as well, but costs more. This requires a remote intelligent recorder. If your network has to cover critical information for emergency management, relying the data transmission on dial-up phones is not a good idea: These lines tend to fail or become saturated during or immediately after earthquakes.

Radio links. This is the most difficult option unless you have special experience. In some countries it is hard to get a permission and for digital links, it must be clear how repeaters and error correction is done. Spread Spectrum transmission could be an easy way out for a virtual network.

Satellite connection. Some systems in some countries can be quite straight forward, since it is just a question of ordering the link (usually VSAT) and you have a permanent computer connection to the field station. The main considerations are installation and operation costs, both of which are going down. A satellite telephone might cost-wise be a future option.

The simplest and cheapest network consists of a network of recorders connected with modem to a central data collection facility, or alternatively to free internet nodes. In this way there is no communication infrastructure to make, but access to a phone line (fixed or mobile) is needed. The network does not have real time capability, but, although it is a great selling point, that is rarely needed. Having an earthquake location within half an hour is in most cases good enough.

Recorder and/or digitizer. The digitizer is the most universal element and will often be able to work with different types of public domain data acquisition systems, so for a real time system with only the digitizer in the field, there is usually not much problem in finding the hard and software. With a commercial recorder in the field, the central data collection system is mostly a one company only system and therefore undesirable, but there might not be much choice. So if the network consists of two types of recorders, there might have to be two independent data collection systems. If the field recorder is based on a public domain system, this aspect is easier.

CHAPTER 9

Seismic Arrays

The advantages of using multiple seismic sensors deployed in a small area were known in seismic prospecting long before these techniques were applied to earthquakes. The present development of seismic arrays is partly due to the recommendation by a group of experts in 1958 of improving the quality of seismic stations worldwide as a mean of detecting possible violations of the agreement of nuclear test suspension. At present, the Comprehensive Test Ban Treaty Organization (CTBTO, Vienna) provides the standards and the coordination for this purpose (see also 8.12). The availability of smaller sensors, better data acquisition systems and powerful computers has extended the use of arrays from global seismicity monitoring to regional and local seismicity studies and even portable arrays are now used. The large global arrays like LASA in Montana or NORSAR in Norway have been described in detail elsewhere (Green et al., 1965; Capon, 1970; Bungum et al., 1971; Bungum & Husebye, 1974). In this chapter, we will mostly deal with the small or portable arrays for local and regional studies.

The seismic arrays or antennas are a powerful tool, not only for improving the signal-to-noise ratio and detecting distant events otherwise masked by the background noise (e.g Frankel, 1994), but also for a number of studies on sources and wave propagation near the seismic source (Gupta et al., 1990; Iida et al, 1990; Niazi & Bozorgnia, 1991), site effects and near receiver wave propagation (Dainty & Toksoz, 1990; Al-Shukri el al., 1995; Aoi et al., 1997; Barker et al., 1997; Bodin et al., 1997), etc. Another application field is represented by the array surveys done in recent years in volcanic areas, where conventional networks are difficult to use. These studies have yielded insight into the nature of seismo-volcanic sources, crack models, volcano structure, etc (e.g., Chouet et al., 1998; Del Pezzo et al., 1997). The analysis techniques associated with arrays allow the location of seismic phases with onsets not well defined or the tracking of sources of almost-continuous signals such as a volcanic tremor (e.g. Almendros et al, 1997). This is not possible with conventional network methods.

The waveform recorded at a given station depends on the source, path, and local site effects. Records on two nearby stations (A and B) with similar site effects, for which the paths are very similar, will show nearly identical signals The slight path difference will produce a relative delay and perhaps a small waveform difference that may be corrected for with the suitable transfer function. Thus the waveform at B may be obtained from the waveform at A with a linear transformation and a delay. Usually this may only be true for each phase of the seismogram separately, since different seismic phases arrive with different apparent velocity (9.1).

If several stations are close enough, the waveforms recorded on them from a distant event (distant in terms of the array size) will be very similar, except for the time delay and the added local noise. The detected signals are correlated, while the noise – both local ground noise and perhaps instrumental noise – is not. This allows the enhancement of the signal-to-noise ratio (SNR), for instance, by summing the records with the suitable delays, so that the wanted signal will be in phase on all channels. Even if there is a non-local correlated noise present, such as microseisms (ocean generated noise, see 3.4), the relative delays will not be the same for signal and noise, since the azimuth and apparent velocity of both will be different. Therefore, the in-phase sum of the signals recorded on all channels will be enhanced above the noise. This principle has been adopted by radio-astronomy, radar technique, acoustics, sonar and seismology for improving the SNR. Arrays will also provide estimates of azimuth (the direction of the ray arrival and apparent velocity of the incident ray (see Figure 9.2 for definition)).

We consider a seismic array *a set of seismic sensors, with common response function, recording time synchronized and deployed in a homogeneous area according to a certain configuration, with the aim of obtaining a coherent spatial sampling of the seismic wavefield in time.* From the analysis point of view, what defines an array is the collective waveform processing. In a network, the phases are picked and the arrival times are determined on each station separately; in an array, all channels are processed together, as different spatial samples of the same time-varying wavefield.

The ray spatial coherency is assumed in this process. This means that each wavefront keeps its shape while it passes through the array. For each spectral component, the phase relation between two points along the ray path depends only on the distance between the points. This makes the size of any practical array a function of the area homogeneity and the wavelengths we are interested in, since earth heterogeneities produce wave scattering and diffraction, which tends to destroy the coherency.

The waveform similarity is frequency-dependent, or better, wavelength-dependent. The signals are expected to be similar only for distances between stations shorter than a few wavelengths. The array size and the spacing and geometry of the stations deployment are conditioned by the wavelength range of interest.

A local network (diameter < 100 km) may be used as an array for teleseisms, since the waveform shape will be very similar at all stations (Figure 9.1), provided that the instruments characteristics are similar or well known so instrument correction can be done, and that the stations have a common time base. Conversely, for local or regional events, the dominant frequencies are higher than for distant events and thus the wavelengths are shorter, so the earth heterogeneity has a major influence in the waveforms; in addition, the azimuth to the source is different for each station. These reasons make coherency to be almost completely lost within the network.

Regarding the array configuration, Asten and Henstridge (1984) suggest four conditions to be met:

-The array diameter D has to be at least as large as the largest wavelength to be recorded.

-Looking at the array from every direction, there must be stations as near as half of the shortest wavelength, to avoid spatial aliasing.

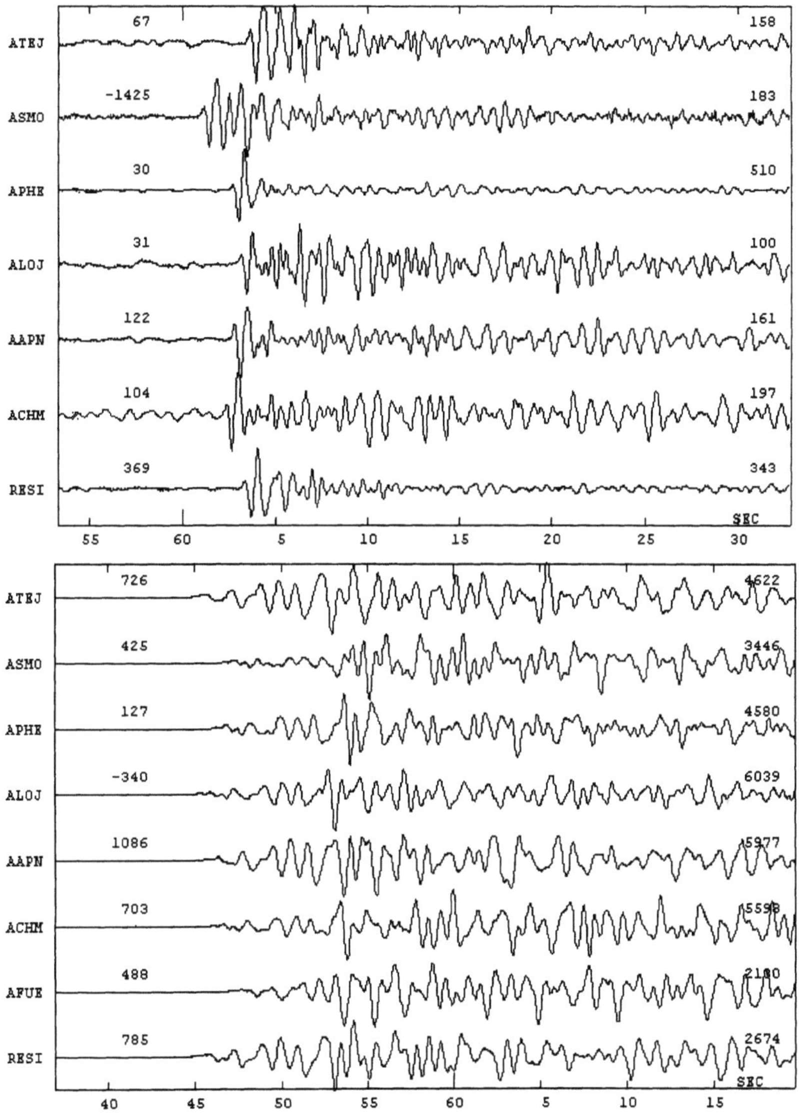

Figure 9.1 Upper: Record of a China nuclear test at the local network RSA in southern Spain. The waveforms are not identical, due the local geological conditions under each station, but their similarity allows the use of array techniques to estimate azimuth and apparent velocity, and therefore distance from travel-time tables. Lower: A teleseism (Coast of Peru) recorded (82 degrees away) by the same network shows emergent onsets, but again array techniques yield the precise determination of relative delays and estimation of azimuth and distance.

-The number of stations must be higher than the number of plane waves that simultaneously arrive at the array. Simple analysis methods may identify a single plane wavefront arriving at the array and estimate its propagation direction and apparent velocity. But quite often, especially for near sources, several wavefronts with different direction and velocity or incidence angle may arrive simultaneously, either due to the presence of noise or to rays multi-path. The ability of the array to resolve these different rays is limited by the number of stations.

- The array installation has to be *logistically* possible.

9.1 Basic array parameters

Array techniques always require digital processing. The resolution of the array is limited by its size and shape. The discrete time and space sampling of the wavefield also imposes some restrictions on the frequency bandwidth and wavelengths that can be resolved according to the Nyquist sampling theorem. Furthermore, the practical topology of connections between stations, by means of cables or radio, limits to some extent the array size and shape.

We will first look at some basic parameters for arrays. Figure 9.3 shows nine stations that are deployed in a rectangular array (this configuration is not practical by reasons explained below). Station s7 will record the wave arrival first, then s4, s8, s1, s5, s9 and so on. If the structure under the array stations is homogeneous, the waveforms recorded on all stations will be almost identical, except for a time delay and the added local noise. The wavefront is moving horizontally across the array with an apparent velocity v. We call this *apparent velocity* since it is what the velocity appears to be in the horizontal plane. The apparent velocity (do not confuse with the horizontal component of velocity!) is related (see Figure 9.2) to the true velocity c in the medium as

$$v = c/sin(i) \tag{9.1}$$

where i is the angle of incidence as measured from the vertical direction. If e.g. the ray comes up vertically, the wavefront arrives at the same time at all stations and the apparent velocity appears to be infinite. This is also what we get from (9.1) since the angle of incidence is 0 degrees. A vector **s** with the ray direction and with magnitude $1/c$ is called *slowness*. Similar to apparent velocity, we define the *apparent or horizontal slowness* (the horizontal component of slowness) as

$$p = 1/v = sin(i)/c \tag{9.2}$$

We can also define a horizontal vector **p** (Figure 9.2) as the horizontal projection of **s**, therefore with magnitude p. If we assume that the arrival time to station S7 is $t=0$ and the coordinate system origin is also at S7, we can calculate the arrival time to the other stations knowing v and the azimuth α. This is most simply done using the horizontal slowness vector **p**. If the vector from S7 to S8 is called **r**, then the travel time difference

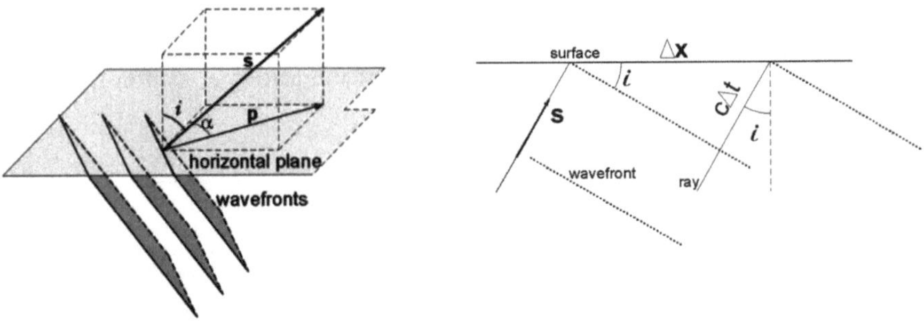

Figure 9.2 Slowness definition. *Left*: Vector **s** is normal to the plane wavefronts and its angle with the vertical is the incidence angle i. The horizontal projection of **s** is ***p***, the horizontal slowness, whose angle with the North direction is the ray azimuth *α*. *Right*: A projection on the incidence vertical plane. As the wavefront advances a distance *cΔ*t, the intersection with the horizontal plane advances *Δx*, with apparent velocity *v*. Relation 9.1 between *v* and c can be seen from the geometry in the figure.

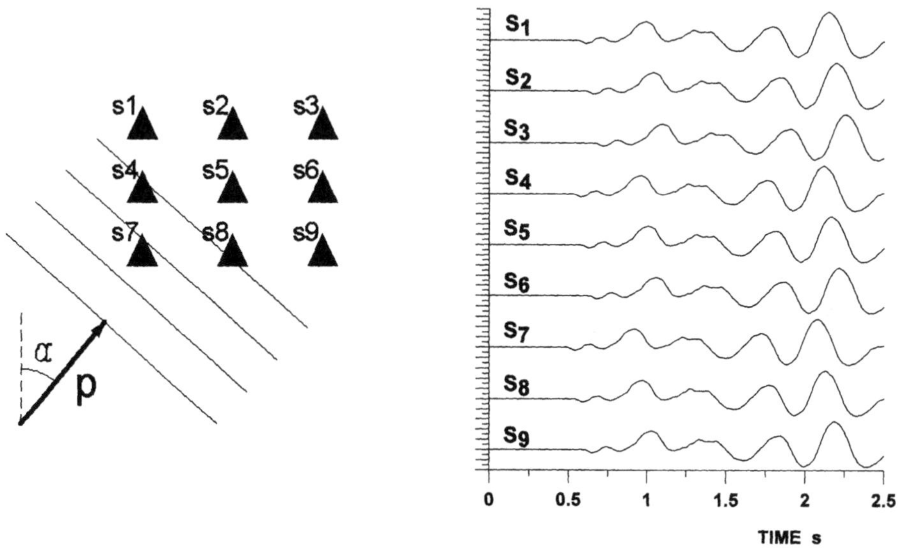

Figure 9.3 Left: A plane wavefront arrives with apparent slowness ***p*** to a rectangular shaped array (the intersection of the wavefront with the surface, where the array is installed, is shown). The angle α with respect to North defines the ray arrival azimuth. Right: The waveforms recorded are ideally identical, except for the relative delays. In practice, they present differences due to the added noise and to the local ground heterogeneity.

τ, between S7 and S8 is

$$\tau = \mathbf{p}\cdot\mathbf{r} = r\cos(\theta)/v \tag{9.3}$$

where θ is the angle between the vectors **p** and **r.** (9.3) contains two unknowns, θ and v, so two equations and 3 stations are needed to determine azimuth and apparent velocity. This means that an array must have at least 3 stations.

(9.3) can be generalized to any pair of stations using an arbitrary origin for the coordinate system. This gives the time delay at any station with position vector $\mathbf{r}_i$ and apparent slowness $\mathbf{p}$. Let $\tau_{ij} = t_j - t_i$ be the delay of the arrival at station j relative to station i. Then

$$\tau_{ij} = \mathbf{p} \cdot (\mathbf{r}_j - \mathbf{r}_i) \tag{9.4}$$

That is, if the horizontal coordinates of each station j are (x_j, y_j) with respect to the origin (normally one of the stations or the array center),

$$\begin{aligned} &\tau_{ij} = p_x(x_j - x_i) + p_y(y_j - y_i) \\ &\textit{with} \quad p_x = p \cdot \sin(\alpha) \\ &\textit{and} \quad p_y = p \cdot \cos(\alpha) \end{aligned} \tag{9.5}$$

If the array stations are not situated on a plane, their height has to be included in the equations, e.g. by using the 3D slowness vector $\mathbf{s}$ instead of the 2D slowness vector $\mathbf{p}$.

If the hypocenter is close to the array, a more realistic approach is to consider a circular wavefront (Almendros et al., 1999), instead of a plane wavefront (at the surface, we observe the intersection of the three-dimensional wavefront with the ground plane). Some of the analysis methods are easily adapted to this geometry. The center of the circle gives an estimate of the epicenter.

The Nyquist sampling theorem holds also for spatial sampling. Therefore, for a given wavelength λ to be sampled by the array without ambiguity, the station spacing d in the ray direction has to be less than $\lambda/2$. Moreover, resolution for the wavenumber k corresponding to this wavelength, $k=2\pi/\lambda$, requires that the array size D to is least λ. These two conditions may be written as

$$2d < \lambda < D \tag{9.6}$$

For instance, a linear array of 21 stations separated by 100 m (d=100 m, D=2000 m) will cover wavelengths from 200 m to 2000 m in the linear direction. For an apparent velocity of 2 km/s, this means the frequency band 1-10 Hz.

The fact that signals are *sampled* in time at intervals Δt imposes some constrains on the array geometry. If the wavefront travels across the whole array in less time than is required to take two consecutive samples, we will not be able to obtain any information of relative arrival times. If the apparent velocity across the array plane is v, and the array aperture in the ray direction is D, one condition to be met is

$$\Delta t << \frac{D}{v} \tag{9.7}$$

For instance, if the array size is 1 km and the expected maximum apparent velocity is 20 km/s, the sampling interval should be much less than 0.05 s.

Apart from the number of stations required, it is useless to install an arbitrarily large array, since the coherence of the wavefield (hence the waveform similarity) is often lost within a spacing of two or three wavelengths.

A number of array configurations have been used, depending on the objective. If the seismic sources to be investigated can be at any azimuth, the array configuration should have a shape like a circle or several concentric circles. Figure 9.4 shows a regional small aperture array of this kind, ARCES, in Norway. It is composed of 25 stations with vertical SP seismometers. Four stations also have horizontal components, and there is a BB instrument at the central site. An identical array is NORES, also in Norway, installed as one of the 42 sites of a large aperture array, NORSAR. This later is arranged as seven groups of six sites each, within an area of diameter about 60 km (Chapter 9 (Schweitzer et al.) in NMSOP, Bormann (Ed.), 2002).

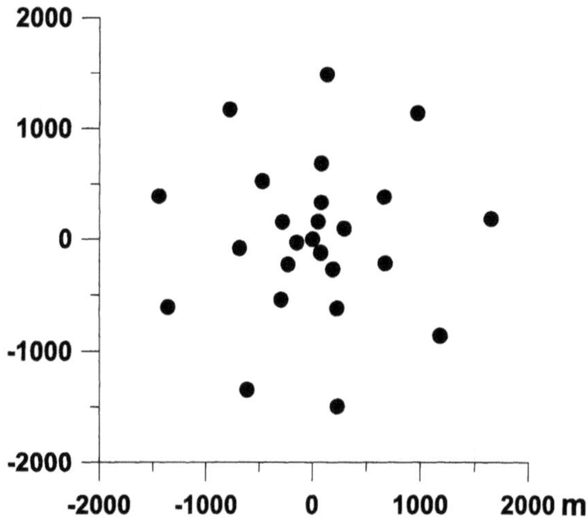

Figure 9.4 ARCES is a small aperture array with 25 stations covering four concentric circles. The diameter of the external ring is about 3 km. The radial spacing of the stations is not uniform in order to deal with different wavelength.

Sometimes, the seismic sources under study are limited to a small region and the array configuration is designed accordingly to optimize the array performance for the waves coming from this location. This may be the case for several forms of semicircular shaped arrays (Ferrazzini et al., 1991). Both for these and the circular arrays mentioned above, the stations of each concentric circle or semicircle are often situated following two criteria:

-The sites of stations on consecutive circles are chosen to avoid radial alignment, to improve the azimuth coverage.

-The radial spacing is not uniform, but shorter for the inner circles, to help avoid the spatial aliasing.

Recent sampling techniques in electronics make use of random sampling for periodic signals. This overrides to some extent the Nyquist limit. Also the array configuration may contain some spatial randomness, within a convenient general geometric shape, and this improves the sampling of spatially periodic wavefields.

9.2 Array transfer function

A discussion on the array data processing methods is beyond the scope of this book, but it is useful to introduce at least the array transfer function to understand the influence of the array geometry on its resolution. The three main objectives of the arrays are to:

- Improve the signal-to-noise ratio for detection of weak seismic signals.
- Determine apparent velocity and azimuth of approach.
- Locate sources of seismic signals, even without clear onsets.

A plane wave arriving at an array with horizontal slowness **p** will be recorded at the reference station of the array as a signal $w_0(t)$. If we call the positions of all the stations relative to the reference station $\mathbf{r}_i$, and the time delay relative to the same τ_i, the records at each station, in the noise-free case and assuming perfect coherence, will be

$$w_i(t) = w_0(t - \tau_i) = w_0(t - \mathbf{r}_i \cdot \mathbf{p}) \tag{9.8}$$

We may add all the signals in-phase to build the *beam*. If **p** is known, the beam w_b is defined as

$$w_b(t) = \frac{1}{N}\sum_{i=0}^{N-1} w_i(t + \mathbf{p} \cdot \mathbf{r}_i) \tag{9.9}$$

where N is the number of stations in the array. In the noise-free case, this *beam forming* has no effect on the signal, but if uncorrelated noise is present in the individual station signals, always the real case, the beam improves the signal-to-noise ratio.

The actual slowness is often unknown and has to be determined from the array records. It is therefore interesting to characterize the array ability for discriminating the true slowness among all the possible ones. Let us imagine a signal that consist of a unit impulse approaching with true slowness $\mathbf{p}_t$. Without loss of generality, we can take the time origin to be when the signal arrives at the reference station:

$$w_0(t) = \delta(t) \tag{9.10}$$

where δ is the Dirac function representing the unit impulse.

From (9.8), the signals at the other stations will be

$$w_i(t) = \delta(t - \mathbf{p}_t \cdot \mathbf{r}_i) \tag{9.11}$$

Now, if we form the beam (let us call it h in this case) using an arbitrary slowness **p**

$$h_b(t) = \frac{1}{N}\sum_{i=0}^{N-1} w_i(t + \mathbf{p}\cdot\mathbf{r}_i) = \frac{1}{N}\sum_{i=0}^{N-1} \delta[t + (\mathbf{p} - \mathbf{p}_t)\cdot\mathbf{r}_i] \tag{9.12}$$

This represents the impulse response of the array, as a function of the slowness relative to the true slowness. If we take the Fourier transform H_b of this signal, it is

$$H_b(\omega, \mathbf{p}) = \frac{1}{N}\sum_{j=0}^{N-1} e^{-i\omega\mathbf{r}_j\cdot(\mathbf{p}-\mathbf{p}_t)} \tag{9.13}$$

Taking into account that the wavenumber is $k=\omega p$, we may substitute the product $\omega\mathbf{p}$ by the apparent (surface) wavenumber **k**,

$$H_b(\mathbf{k}) = \frac{1}{N}\sum_{j=0}^{N-1} e^{i\mathbf{r}_j\cdot(\mathbf{k}_t - \mathbf{k})} \tag{9.14}$$

where $\boldsymbol{k}$ and $\boldsymbol{k}_t$ represent the horizontal wavenumber as a variable and the true horizontal wavenumber of the ray, respectively.

The square modulus of the function H is often called the *array transfer function* or the *array radiation pattern*. It may be shown (e.g. Chapter 9, Schweitzer et al. in NMSOP, Bormann (Ed.), 2002) that the power of any beam formed for any incoming signal with transform $W(\omega)$ will be

$$E(\mathbf{k}_t - \mathbf{k}) = \frac{1}{2\pi}\int_{-\infty}^{\infty} |W(\omega)|^2 |H(\mathbf{k}_t - \mathbf{k})|^2 d\omega \tag{9.15}$$

This means that if the signal power of the beam has to be maximum for the true **k**, and ideally null for all the wrong ones, the function $|H|^2$ should have a peak at the origin of the plane $\mathbf{k}_t$-$\mathbf{k}$ (the maximum value is unity) and no other significant peaks over the plane. These other peaks represent some degree of spatial aliasing, which may yield ambiguous estimation of $\mathbf{k}_t$ or $\mathbf{p}_t$. The sharper the peak at the center, the more precise determination is possible of the true slowness. Figure 9.5 shows some examples of transfer functions for several array configurations. The axis values may be expressed in slowness value for a given frequency, using the relation $\mathbf{k}=\omega\mathbf{p}$.

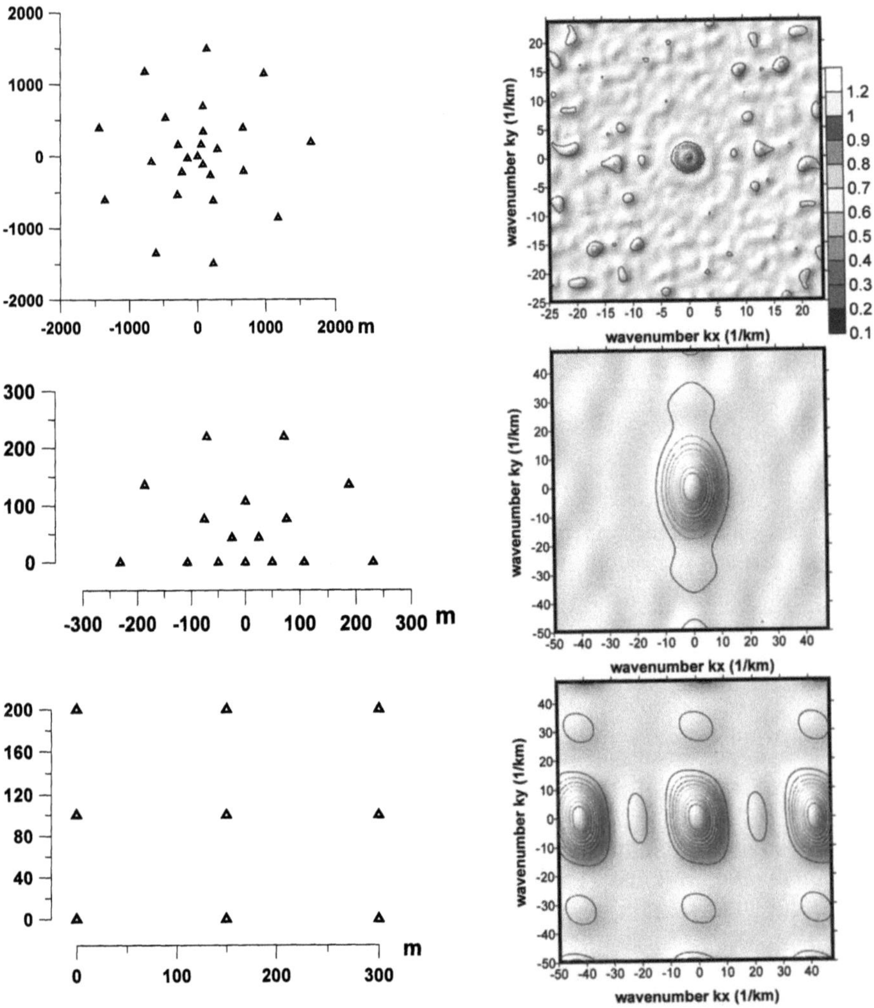

Figure 9.5 Different examples of small aperture array shapes and their transfer functions. Top, the array ARCES, mentioned above, with 25 stations and a diameter of 3 km, has a sharp almost ideal response. Next, there is a semicircular array, a rectangular one with a response showing spatial aliasing (ghost peaks).

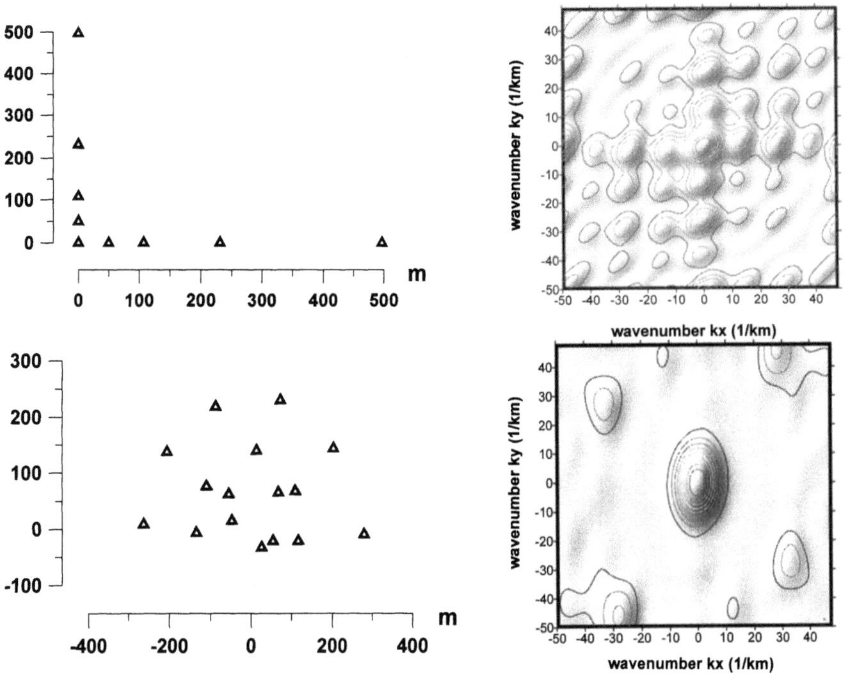

Figure 9.5 (cont.) Different examples of small aperture array shapes and their transfer functions. A bi-axial one, with a sharp response but some degree of multiple aliasing, and a random array derived from the semicircular one with random spatial deviations, which shows a good sharp response, improving the resolution along the *ky*-axis.

9.3 Instruments characteristics

In principle, any kind of seismic instrument suitable for field use may be used in an array. Actually, both permanent and temporary arrays are deployed with broad-band instruments, short-period seismometers and accelerographs. Nevertheless, the time delays involved in the analysis of seismic arrays may be smaller than the characteristic times associated with the instruments (i.e. the free period or the filters time constants). This makes it important that all the instruments used are identical or at least well calibrated (Scherbaum, 2001, personal communication). Furthermore, if the instruments are identical, the analysis may be performed directly on the raw traces, since they are affected in the same way by the instruments. Otherwise, the instruments responses have to be deconvolved or at least the traces reduced to a common response. In practice, most arrays are arranged with a unique type of instrument or two at most.

The group delay $D(\omega)$ of an instrument or dynamic system (including filters) is defined as the derivative of the phase response with respect to the angular frequency

$$D(\omega) = \frac{d\Phi(\omega)}{d\omega} \tag{9.16}$$

The filters often use high-order poles and the phase response, and the group delay, is affected well below the corner frequency.

This function depends on frequency and represents the time delay introduced by the system response for a harmonic signal of this frequency. The system in this context is the sensor, amplifier, telemetry system, digitizer, and digital filters. Some FIR filters implemented on high resolution digitizers are designed with linear phase response and thus have a constant group delay for every frequency; so every signal is delayed a fixed time, which may be automatically corrected at time stamp. Particular attention must be paid to the sensor and the antialias filters if they are of analogue type.

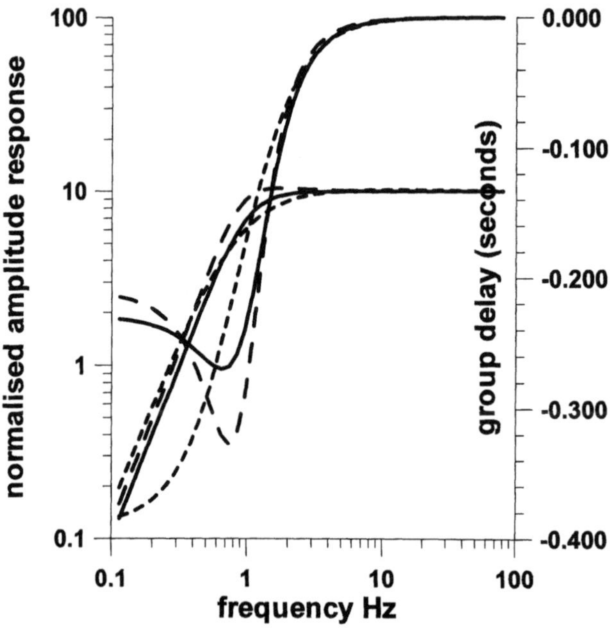

Figure 9.6 The normalized amplitude responses for ground velocity and the group delays are represented for two sensors of nominal frequency 1 Hz. Solid lines: Sensor with free period 1 s and damping 0.7. Dashed lines: Sensor with free period 1.1 s and damping 0.6. Dots lines: 4.5 Hz sensor equalized to mimic the amplitude response of the 1 Hz sensor. The amplitude responses are very similar, but the group delays may be quite different, particularly in a band about the corner frequency. This illustrates the errors that may be introduced by components with non-identical characteristics in an array.

To illustrate this, let us consider the group delay introduced by two passive sensors both of nominal 1 Hz natural frequency, but with small real deviations, together with a 4.5 Hz sensor equalized to have a similar amplitude response to the 1 Hz sensor. The group delays are represented in Figure 9.6, as well as their amplitude responses for ground velocity. Actually, the group delay is the same for displacement, velocity and acceleration responses, since the phase responses differ only by a constant (see Chapter

2). As it can be seen in the figure, in spite of the fact that all amplitude responses are very close, the group delays may differ as much as 0.1 s, which may be of the same order as the relative delays between array stations due to the wave propagation.

Portable arrays often use small 4.5 Hz sensors, due to their low weight and price. They can be used directly or with equalizers to extend the response to 1s (see Chapter 2). As the dominant frequencies for local studies may be around the 1-5 Hz range, a careful selection of the sensors must be made to ensure uniform free period and damping. Alternatively, a fine-tuned individual equalization of the sensors must be done if possible.

9.4 Field arrangement

The similarity of the waveforms, which is essential for array processing, requires a choice of deployment area to be geologically homogeneous, stable and of smooth topography. If an array is intended mainly for teleseismic recording, such a zone will not be difficult to find, since the exact site may be chosen within a wide possible area. However, if the objective is a local study of an active area, the presence of tectonic or volcanic activity is often associated with a topography and geology which is not very homogeneous. In this case, the search should be directed to a nearly flat area, perhaps not horizontal and reasonably far from strong known hetereogenities. For example, if the array is installed in a valley, it should be far from the borders of the surrounding mountains that may cause wave reflections. As far as possible, the structure under the array must be known, or otherwise a survey may yield at least the velocity structure of the shallower layers. This knowledge makes it possible to estimate the incidence angle from the apparent slowness, see (9.1).

Digital elements have an inherently stable and predictable response. So, in view of the above discussion, the signal from the sensors should be digitized directly, without any more analog parts (i.e. analog telemetry or filters). For small aperture arrays, the distance between stations usually makes it possible to use direct cable connections between the sensors and the recording system. The choice of a single multichannel digitizer may be a cost-efficient solution, provided that a preamplifier operates close to each sensor. Alternatively, if the array operates in an electromagnetic noise free environment, there is little danger of noise pickup by the cables and the preamplifier may be at the digitizer end of the cable. This eliminates the need for a power line to the sensors, if they are of passive type. In the case of using a single ADC with channel multiplexing, the channels skew (time delay caused by sampling being sequential, not simultaneous, see Chapter 4) might have to be considered in the data analysis.

Relative time synchronization of array elements is far more stringent than with seismic networks. Typically, the relative timing accuracy needs to be at least one order of magnitude better than for single stations. Therefore, a central recording system with a unique time base is preferable. If the sensor signal is digitized close to the sensor and digitally telemetered to the recording site, there must be a common synchronization for all digitizers. The use of a GPS receiver for each site is expensive, and requires extra power, but it may be the only choice in some circumstances. Actually, many recent temporary arrays have been composed of completely independent seismic stations with digital recording and GPS. In some cases, there has also been real time telemetry to a central site. It should be pointed out that such arrays of several tens of stations, each with a heavy battery, solar panels, recorder, etc, hardly might be considered a truly portable array. There are some practical alternatives (Figure 9.7), though:

- To sample at each sensor site at a high frequency (at least 1000 sps), using a free running clock. The data is sent in real time to a central recording unit, where the samples are precisely time-stamped and possibly decimated after a suitable digital filtering. The delay on the data transmission has to be known at least with one sampling interval accuracy.

- The sampling time is accurately synchronized at all sensors sites, by means of, for example, a pulse transmission from the central unit using a PLL (phase-locked loop) controlled oscillator in the digitizer, instead of a free-running oscillator. The central unit may e.g. send 1 pulse per second and the individual digitizers will take their samples synchronously.

- The use of several small sub-arrays, each being composed of a multichannel digitizer with GPS synchronism. The stations in the sub-array are equipped with a preamplifier in order to transmit the analog signal through a low-impedance differential channel, minimizing the noise pick-up. The preamplified sensors get the power from the digitization module, also through the cable. Actually, a double twisted pair cable of telephone type may transport the signal and the power. This combination is quite versatile for setting up a portable array, as a trade-off between lightweight equipment, cost and performance and has been used efficiently for volcanic surveys (e.g., Del Pezzo et al., 1997).

Portable arrays have to be lightweight and with the possibility to be packed in a small volume. Often the equipment is shipped by air. One of the most bulky and heavy parts is often the cables. Considering that typically several km of cable is used, a possible solution is to order cables locally from the nearest supplier to the deployment site and make the connectors locally.

The present state of technology makes it possible to use wireless digital data transmission at the short distances required for a small aperture array, or even at distances of some tens of kilometers with low power consumption (see also Chapter 8). For instance, a LAN card with Spread Spectrum radio transmission is available at a reasonable cost. Each station may then be a node of a local network. This kind of solution requires, nevertheless, a small computer at each node to implement the LAN protocol or an intelligent digitizer with a LAN output, and an independent power supply, which may be a small battery. The major problem to be solved is the power consumption of computers, and consequently heavy batteries. However, as described in Chapter 5, several recorders now have Ethernet communication capability and newer recorders are often based on standard industrial computers, like the Windows-compatible PC104 single card, so the cable connection will probably disappear from some of the future portable arrays.

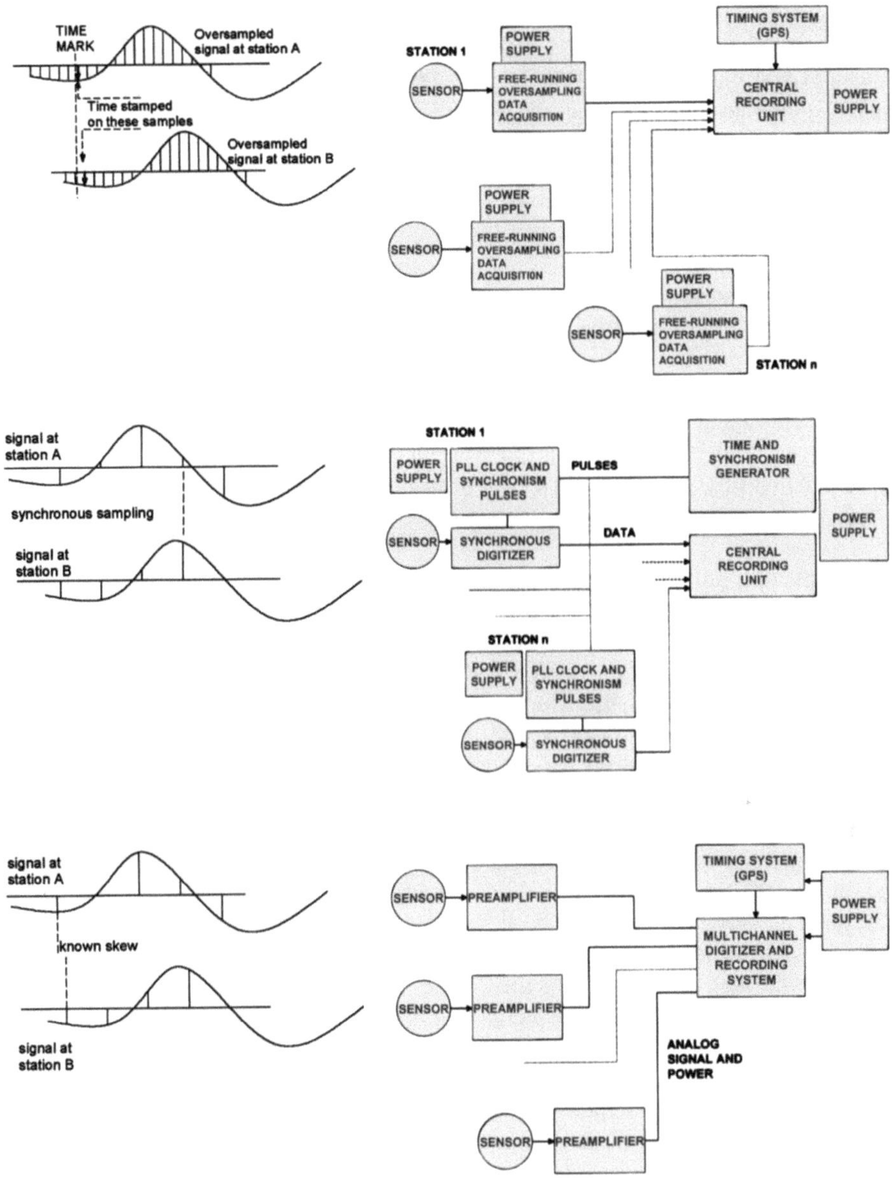

Figure 9.7 Three possible array arrangements. Upper: Each station has an oversampling digitizer with a free-running clock. At the central recording, time is stamped at the next sample following the time mark on each channel. Middle: The digitizers at each station run with a clock locked to the central timing system and all of them sample simultaneously. Lower: Each station has only a preamplifier with the sensor, to provide a low-impedance analog path to the central recording unit, which has a multichannel digitizer. To avoid too long cables, the array is divided in several such subsystems. Each of them has a unique power supply, and each station takes its power from the cable. This is possible also for the other two arrangements, but if the cables are long and the power consumption of each station is not negligible, it is not practical to send the power from the central station, and individual power units must be used for each station.

From a technological point of view, large aperture arrays can be considered as local networks, since they might have the same range of distances (typically tens of kilometers) and the same solutions for data telemetry can be used. Conversely, local networks might be used as large aperture arrays, if the requirements of timing and homogeneity are met, under some circumstances (e.g. for longer wave lengths). Large aperture arrays might be composed of several smaller sub-arrays. In this case, each sub-array center collects the data of its stations and sends them to the central data center. Unlike small aperture arrays, large aperture arrays are seldom portable.

The same recommendations for station installations in the field (Chapter 8) apply for installing the stations of an array.

An additional requirement for arrays is the precise relative location of the stations. Typically, coordinates should be measured with an accuracy of 0.2 m for a small aperture array. This is usually achieved with topography instruments or, if the stations do not have line-of-sight, differential GPS measurement might do the job.

9.5 Example of a portable array

Several examples may be found in the bibliography; one of the first prototypes was called PANDA (Chiu et al., 1991) and used 2 Hz geophones. In 1994, three collaborating groups from the Universities of Salerno and Naples (Italy), Museo Nacional de Ciencias Naturales (CSIC, Spain) and Instituto Andaluz de Geofísica (University of Granada, Spain) developed an array system composed of independent modules of eight channels each, able to operate jointly as flexible portable arrays. This system has operated since then on field experiments at Teide volcano (Canary Islands, Spain), Stromboly, Etna, and Vessuvious (Italy), and Deception Island (Antarctica) (Del Pezzo et al., 1997, Almendros et al. 1997, Alguacil et al, 1999). The block diagram of the system is shown in Figure 9.8. It consists of several sub-arrays, each of them composed of a central 8-channel digitizer with 16 bits resolution, including a fixed frequency (48 Hz) eight-pole antialias filter per channel. The synchronization is provided by a GPS receiver, connected to a PC via the serial port. This PC also controls the data acquisition and records the data on the hard disk. The sampling frequency is 200 Hz. As a single AD converter is used, the sampling is sequential and the time skew between channels is the sampling interval (1s/200) divided by the channels number (usually 8). The sampling clock is synchronized with the GPS 1 pps signal within 10 microseconds.

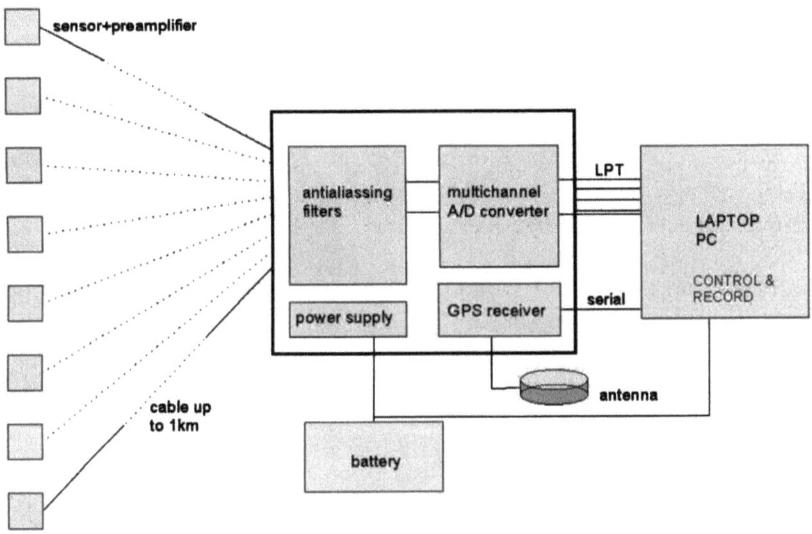

Figure 9.8 Block diagram of an example of portable array. The array is composed of several modules, one of which is represented. A central box contains an 8-channels antialias filter, a 16 bit-digitizer, a GPS receiver and a power supply for all the components. The digitizer is controlled by a PC through the parallel port and is synchronized to the GPS time via the serial port. The seismometers have a preamplifier with a low-impedance differential output, so that long connection cables may be used. The recording is made on the PC hard disk.

The type of seismometer used for most experiments has been the 4.5 Hz Mark L-15 or L-28, with a preamplifier-equalizer close to it, so as to mimic a 1 Hz sensor response. This provides a low-impedance differential output, which may drive a long cable of up to 1 km without significant loss or interference. The cables used are double twisted pairs, one for signal and one to give power to the preamplifier.

Some pictures of this array operating at Deception Island (Antarctica) are shown in Figure 9.9. In this case the array is composed of two 8-chanels modules, with a total of 13 vertical sensors and a triaxial set of sensors, all of the 4.5 Hz type with extended response to 1 Hz. The central units are protected with a tent and the cables are extended on the soil, protected with stones to limit the wind effects. The sensors are buried about 0.5 m for the same reason. In some other cases, the central unit has been protected from excessive cold weather in a simple icebox. The cable connections are made on site with screw terminals, since the cable lengths depend on the field arrangement.

Figure 9.9 Some views of the portable array described above, operating in the Antarctic Deception Island. a) The central unit is composed of a box with antialias filters, data acquisition card and GPS receiver and the laptop PC, which controls the conversion and saves the data on disk. In this case two such sub-array central units are installed in a tent for weather protection. b) An external view of the tent and the sensor cables. c) The cables are simply extended on the surface. In some cases it may be necessary to bury them for protection against the local fauna. d) One of the small sensors used is buried about 0.5 m deep. The geophone is enclosed in a box containing the preamplifier-equalizer, which modifies its response to simulate a 1 Hz sensor.

CHAPTER 10

Calibration and testing

The previous chapters have described how seismic instruments work. It is now time to describe how to calibrate and test the equipment. In the first instance, we often need to do some basic tests to evaluate if the equipment is working properly and, if it is malfunctioning, to try to figure out what is wrong. In the second instance, we need to know the system parameters in order to account for the system response (Chapter 6). In most cases, we can rely on manufacturers' parameters, and in some cases we have to, but instrument parameters might change with time, be wrong due to defects in the equipment or be unknown, so it is useful to be able to determine the parameters yourself.

This chapter will describe some simple tests. Do not expect to become an expert, however it is surprising how much can be done by simple means. The sensors are the most complex part of the seismic station, while the remaining parts have few parameters, so this chapter mainly deals with sensors.

10.1 Test equipment and recording of test signals

The basic test equipment is a signal generator, a recorder and a multimeter. The signal generator should go down to low frequencies, at least to 0.1 Hz for working with SP sensors and down to 0.001 Hz for BB instruments. A lower limit of 0.1 Hz is quite standard, while 0.001 Hz is a bit harder to get. Modern signal generators have digital read out of the frequency. If not, it is advisable to use a frequency counter or period-meter to get accurate frequency measures since the analog calibration might not be very accurate. The best option is a digital low-frequency signal generator, which may cost less than 2000$.

For the recorder, there are several possibilities:

Analog oscilloscope. A standard instrument in most electronic labs. It is sometimes hard to use for slowly changing signals used in seismology and it is not very precise for signals which are not steady.

Digital oscilloscope. A very versatile instrument where the signal can be recorded by freezing it on the screen. It can then be amplified and precise measurements of time and amplitude can be made with cursors. Some of them have a computer interface and the data may be transferred for analysis. The disadvantage with oscilloscopes in general is that they are not sensitive enough to e.g. record the signal of background noise from a standard passive seismometer.

WHEN USING OSCILLOSCOPES, MAKE SURE TO USE IT IN DC MODE, SINCE OTHERWISE THERE IS A HIGH PASS FILTER, WHICH SEVERELY MODIFY THE SIGNALS AT LOW FREQUENCIES.

Analog chart recorder. A simple way of getting a paper record, able to show low frequencies; usually has the same problem as oscilloscopes with respect to sensitivity.

Seismic recorder. If available, this is a good choice since the user will normally be familiar with the operation. There might also be a real time display available and sensitivity is high enough for measuring the sensor output directly. Note that the anti-alias filter might affect the measurements at high frequencies. This can be avoided by using a high sampling rate.

It is important that the recording device has a high input impedance (> 1 Mohm) so as not to affect the measurement (particularly seismometer free period and damping tests), and that it has a DC response in order to not introduce a low frequency cutoff and associated phase shifts. Instruments for measurements usually fulfill these criteria, however be aware that a seismic recorder might have a built in damping resistor or, even without damping resistor, it might have a relatively low input impedance in order to achieve low input noise.

10.2 Sensors

For sensors, the basic parameters to check or determine are listed below. Not all sensors have all parameters.

- Free period
- Damping, open circuit and damping used
- Generator constant
- Generator coil resistance
- Calibration coil (if available) motor constant
- Calibration coil resistance
- Polarity

10.2.1 SENSOR FREE PERIOD

The free period is best measured without any external damping, since the damping changes the apparent period (2.39). The free period can be determined in several ways depending on the type of sensor.

LP or BB sensor: Simply give the sensor an impulse or a step to make it swing, a small push or tilt will do. For traditional LP sensors, it is possible to see the mass swing, while for BB sensors a measuring device must be connected to see the swinging, like e.g. a voltmeter. Some BB sensors have control units with a built in voltmeter. Usually these instruments will make several complete swings before the amplitude becomes too small to see. Measure the time of several swings and calculate the average.

SP sensor: Some SP sensors have low mechanical damping and will swing freely for a long time. In this case, it is possible simply to count the number of swings within e.g. 20-50 s and then calculate the average free period. Many SP sensors and geophones have too high open circuit damping to observe more than a few swings of the signal (Figure 10.1) and the signal must be recorded to measure the period.

This free transient motion of the mass produces a transient output signal referred to as a calibration pulse. The analysis of this is the base of several calibration techniques. Usually, the initial shift of the mass is produced by a step-in-force, by injecting a current in the calibration coil or even in the signal coil of an electromagnetic seismometer. From the sensor point of view, this is equivalent to a step-in-acceleration of the ground, see later.

The release test is performed in two steps: a) The mass is forced away from its equilibrium position by an external force, opposite to the spring restoring force, and it is held at the new equilibrium position; b) The external force is suppressed and so the mass moves "freely" from its position under the action of the spring and the damping forces.

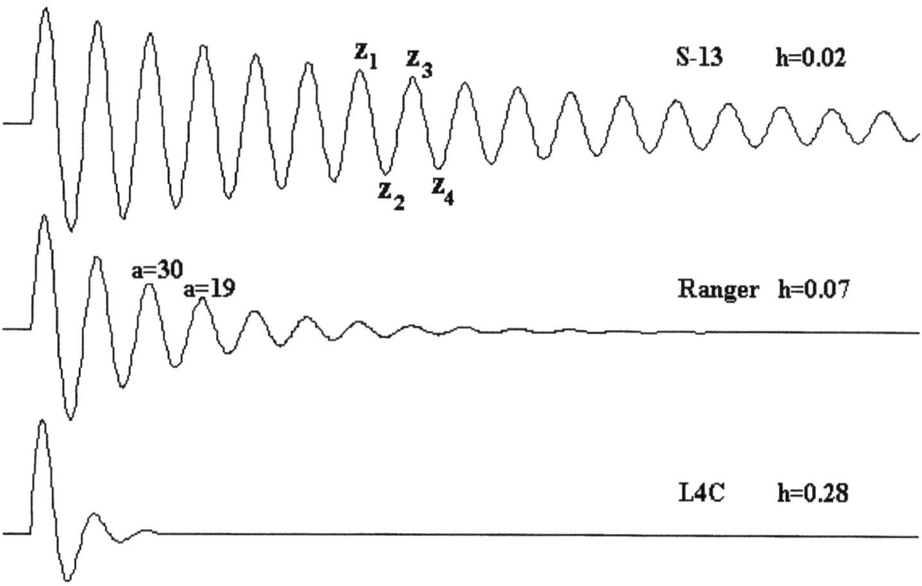

Figure 10.1 Free swinging of three typical 1 Hz seismometers with different open circuit damping h. Traces have been generated synthetically. The decaying extremes (peak amplitudes) are labeled z_1, z_2 etc (top trace). On the middle trace, the amplitudes of two maximums following each other are given in an arbitrary scale.

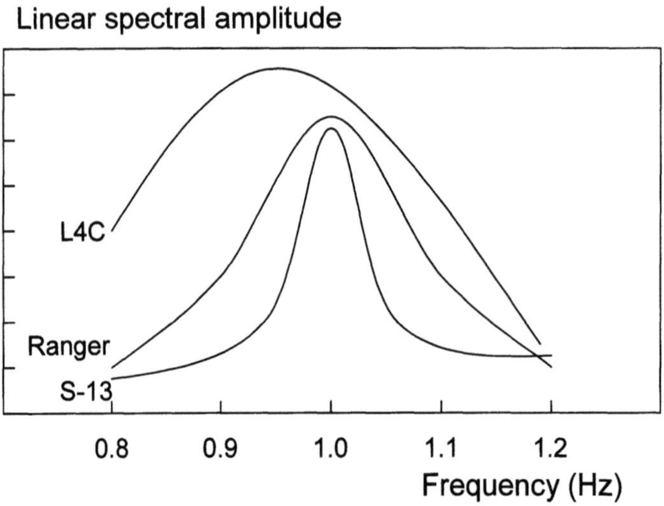

Figure 10.2 Spectral analysis of the signals in Figure 10.1. A 100 s time window has been used to get sufficient frequency resolution. The three spectra are plotted with linear amplitude; however, the amplitude scaling for each one is different.

Make sure the recorder has a high input resistance in order not to affect the free swinging (not the case for BB sensors). Figure 10.1 shows the free swinging of three common 1 Hz seismometers (see Table 2.3) with quite different open circuit damping. For the S-13, it is easy to just count the number of swings, while for the other two sensors, a recording must be made. From a recorded time trace, the period can be read directly. From a digital recording, it is also possible to make spectral analysis to get the free period, see Figure 10.2. From Figure 10.2, we can clearly resolve the natural frequency. It is important to choose a long enough time window to get sufficient frequency resolution since the frequency steps are *1/T* where *T* is the time window (Chapter 6). Note also the lowering of the measured apparent frequency for the highest damped sensor (L4C).

Another method of measuring the free period is using harmonic drive and Lissajous figures, see later section.

The spectra shown in Figure 10.2 can also be calculated theoretically. A mass-release test is equivalent to a ground step in acceleration. The Fourier transform of a step of amplitude a_0 is $a_0/i\omega$. Thus, the Fourier transform of the voltage output $v(t)$ will be the product of the Fourier transform of the input $a_0/i\omega$ and the sensor voltage response for acceleration, T_a^v (see (2.37))

$$V(\omega) = \frac{a_0}{i\omega} \cdot \frac{i\omega \cdot G}{\omega_0^2 - \omega^2 + 2h\omega_0\omega \cdot i} = \frac{a_0 G}{\omega_0^2 - \omega^2 + 2h\omega_0\omega \cdot i} \tag{10.1}$$

The amplitude of $V(\omega)$ has been plotted in Figure 10.3 for the three sensors used in Figure10.2 and the same damping values as indicated in Figure 10.1. The spectral amplitudes are represented for a narrow band around the natural frequency for the same sensors as in Figure 10.2, with their output open (no external damping resistance). It is clear that the lower the damping, the higher and sharper is the spectral peak. Note that for the L4C, with an open circuit damping of 0.28, the peak is clearly shifted down relative to the theoretical undamped natural frequency.

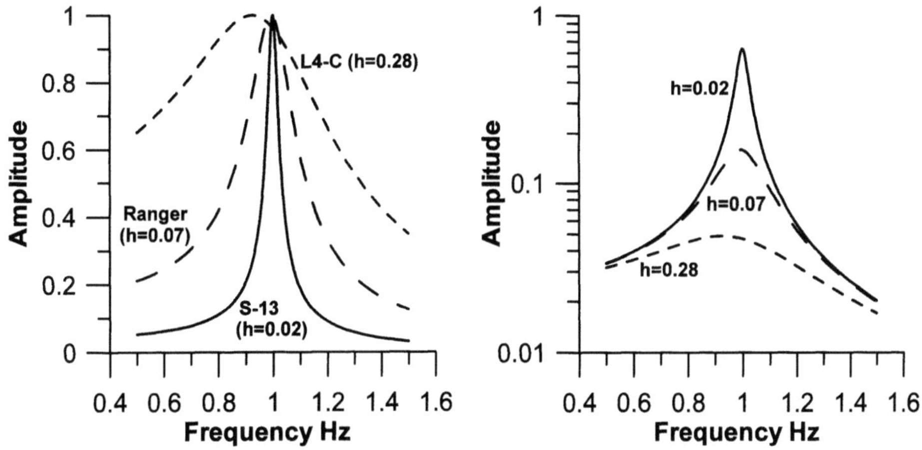

Figure 10.3. The spectral amplitudes of the calibration pulses with the sensor output open, plotted in a narrow band centered on the sensor natural frequency. Left: The amplitude scale is linear and the amplitude of each trace has been normalized with respect to its maximum. Right: The amplitude scale is logarithmic and the factor a_0G is taken as unity.

10.2.2 SEISMOMETER DAMPING

The next thing to determine is the open circuit damping and we can again use the signals from the free swinging system as shown in Figure 10.1. Recall that the relation between two extremes (absolute values) has been given as (2.42)

$$\ln(z_1 / z_n) = \frac{\pi h}{\sqrt{1-h^2}} n \tag{10.2}$$

where z_1 is extreme 1 and z_n is extreme n+1 (so n=1 for two following extremes, Figure 10.1) and h is the damping. This equation can be rewritten as

$$h = \frac{\ln(z_1 / z_n)}{\sqrt{n^2 \pi^2 + \ln(z_1 / z_n)}} \tag{10.3}$$

If the damping is small, (10.3) can be replaced by

$$h = \frac{\ln(z_1 / z_n)}{n\pi} \tag{10.4}$$

If we measure z_1 and z_2, h can now be determined directly. In Figure 10.1, two following maxima (n=2) have approximate amplitudes 30 and 19 respectively. This gives a damping of

$$h = \frac{\ln(30/19)}{\sqrt{2^2\pi^2 + \ln(30/19)}} = 0.0723 \tag{10.5}$$

Using (10.3) gives $h=0.0727$, so in this case we can use (10.4). This method of determining damping of course also works if a damping resistor has been connected, in which case the sum of the electrical and open circuit damping is being measured. However, if the seismometer has a damping of more than about 0.8, there is practically no over-swing and this method cannot be used. We will label the total damping h and the open circuit damping h_m (*m* for mechanical).

For BB sensors, there is no damping resistor and the damping is preset from the factory. Usually, the BB sensor has an option of making it operate without damping (similar to open circuit damping) for testing.

Determining critical damping resistance (*CDR*) for standard velocity sensors.

We recall from Chapter 2, that *CDR* is defined as the sum of the resistance of the external resistance R and the coil resistance R_g, which will give a damping of h=1.

$$CDR = R_g + R \tag{10.6}$$

We cannot simply adjust R and use (10.3) until $h=1.0$ since there will be no over-swing. However, from (2.44) we have that the ratio between two electrical damping constants is inversely related to the ratio of the damping resistances:

$$R_{T2} = R_{T1}\frac{h_{e1}}{h_{e2}} \tag{10.7}$$

R_{T1} and R_{T2} are two different total damping resistances and h_{e1} and h_{e2} the corresponding damping coefficients caused by the resistances. In addition comes the open circuit damping, so for h_1 we get

$$h_1 = h_{e1} + h_m, \tag{10.8}$$

so writing (10.7) in terms of the total damping gives

$$R_{T2} = R_{T1}\frac{h_{e1}}{h_{e2}} = R_{T1}\frac{h_1 - h_m}{h_2 - h_m} \tag{10.9}$$

When $h_2 = 1$, $R_{T2} = CDR$ and we get

$$CDR = R_{T1}\frac{h_1 - h_m}{1 - h_m} \quad \text{or} \tag{10.10}$$

$$R_{T1} = CDR\frac{1 - h_m}{h_1 - h_m} \tag{10.11}$$

when CDR is known. If h_m is small, (10.11) is reduced to $CDR = R_{T1}h_1$. So, the procedure to determine CDR is

1: Determine h_m as described above (10.2).
2: Connect a damping resistor R_{T1} of "suitable size", meaning that the over-swing is substantially less than when measuring h_m but large enough to get an accurate determination of h_1.
3. Use (10.10) to determine CDR.

Figure 10.4 shows an example using the Ranger seismometer from above.

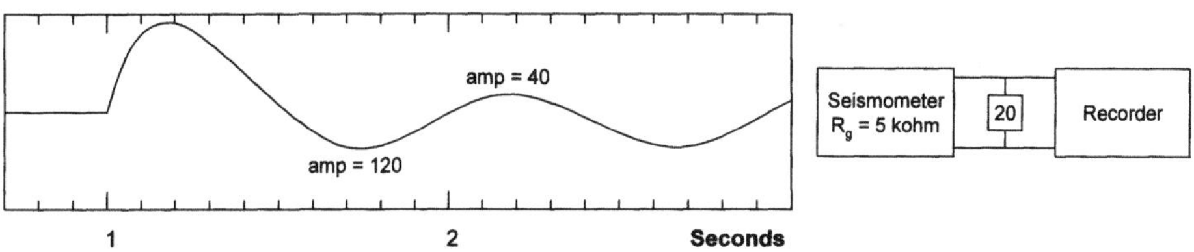

Figure 10.4 The Ranger seismometer with an external damping resistor of 20 kohm. The approximate maximum amplitudes of two following extremes are marked.

We can now calculate CDR. From (10.3) we get the damping, using amplitudes 120 and 40, to be 0.33. Using (10.10) CDR is then

$$CDR = (20+5)\frac{0.33-0.07}{1-0.07}k\Omega = 7.0\ k\Omega \tag{10.12}$$

while the nominal value given by the manufacturer is 6.5 kohm. The total resistance to get a damping 0.7 would then be 7.0/0.7 kohm=10.0 kohm and the external damping resistance is then (10.0-5.0) kohm = 5.0 kohm.

Generator constant for a standard velocity sensor. The generator constant G can be calculated from (2.43) when the mass m is known:

$$G = \sqrt{2m\omega_0 h_e R_T} \tag{10.13}$$

In the above example with the Ranger, the mass is known to be 1.45 kg so with the above measurements, G would be

$$G = \sqrt{2 \cdot 1.45 \cdot 2\pi \cdot 1 \cdot (0.33 - 0.07) \cdot 25000} = 344\ V / ms^{-1} \tag{10.14}$$

Generally, several values of R_T and h_e should be determined to get the most accurate G. From (10.13) it is seen that plotting h_e vs $1/R_T$ will give a straight line with slope $G^2 / 2m\omega_0$. If the mass is not known, and it is not easy to take the seismometer apart and measure it, then other methods for determining G are more suitable, like using a shaking table or comparing to another known sensor, see section on absolute calibration later.

Generator and calibration coil resistances. When measuring the generator coil (or signal coil) resistance, the mass must be clamped in order to get a stable value since the current injected into the signal coil will cause a motion of the mass. The resistance of the generator coil gives a hint on the generator constant, since higher resistance often indicate a higher generator constant (Table 2.3). Nevertheless, it should be pointed out that coil resistance changes with temperature even if G does not. The temperature coefficient of copper resistivity is about 0.004 $°C^{-1}$, so its resistance may change by as much as a 10 % typically for a temperature change of 25° C. Active sensors do not have a generator coil, or their terminals are not externally available, and only sometimes a calibration coil, see below.

10.2.3 DETERMINING CALIBRATION COIL MOTOR CONSTANT

In order to determine the motor constant of the calibration coil, we must have access to the mass of the seismometer in order to put a known force on the mass. This is most often not possible with modern sealed seismometers. A typical seismometer where this still is possible is the S-13. On the other hand, if the manufacturer gives a value, this is unlikely to change unless the coil is damaged by e.g. using a too large calibration current. The procedure for determining the motor constant with a weight lift is (Figure 10.5) as follows:

1. Displace the seismometer mass with a small weight with mass m_w.
2. Lift the weight rapidly and measure the pulse height z_w
3. Apply a DC step current i to the calibration coil to get a similar size pulse amplitude, z_i
4. Calculate the motor constant K for the vertical seismometer as

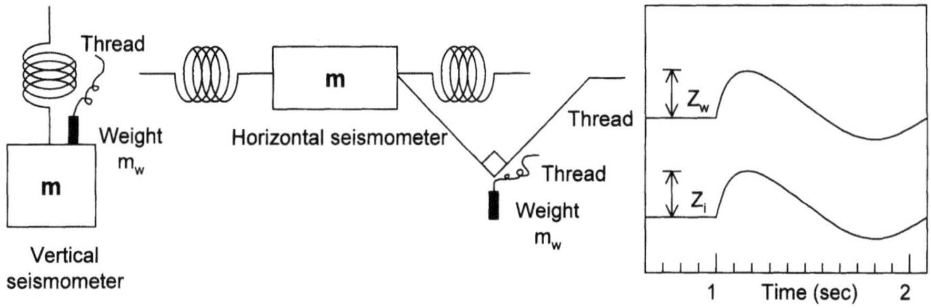

Figure 10.5 Determining calibration coil motor constant using weight lift. The weight m_w, (vertical seismometer) is placed on the mass m, and removed rapidly with the connected thread, giving rise to the calibration pulse with amplitude z_w. For the horizontal seismometer, the weight is pulling at the mass at an angle of 45 degrees, which means that the horizontal force is half the vertical force. The pulse made with the current step has amplitude z_i (right). The seismometer in the example above is a 1 Hz seismometer. Ideally, the mass is pulling or pushing centrally on the seismometer mass, which often is possible.

$$K = \frac{gm_w z_i}{iz_w} \tag{10.15}$$

Since the force generated by the current is the same as the force generated by the weight lift scaled by the relative amplitude; g is the gravitational acceleration. For the setup above with the horizontal seismometer, the force on the seismometer in the horizontal direction is only half, so

$$K = \frac{gm_w z_i}{2iz_w} \tag{10.16}$$

If e.g. z_i=z_w, m_w=0.05 kg and i=0.01 A, then K=9.8•0.05/0.01 = 0.5 N/A, which is a typical high value (Table 2.3).

10.2.4 POLARITY AND SENSOR ORIENTATION

The polarity of the sensor output is simply defined as positive when the ground moves up, North or East. For a complete recording system, the polarity can be flipped at several places and each must be checked independently. The first unit to check is the sensor. This might seem as a simple test, but it is hard to make a signal with a clean sharp onset. Knocking on the sensor rarely gives a clear first polarity signal. One of the simplest mechanical tests is to tilt the sensor (Figure 10.6). A small tilt assures that the sensor moves only one way.

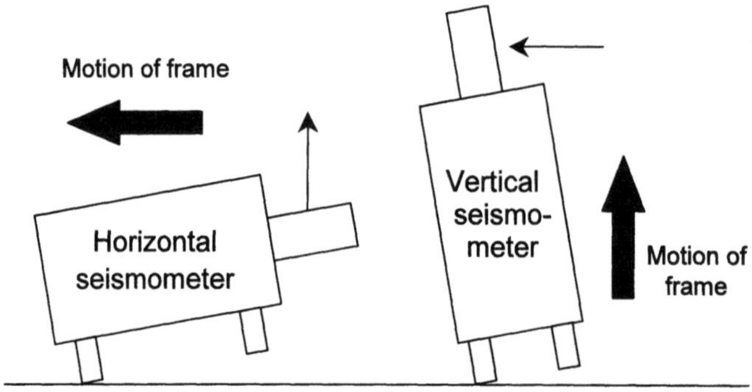

Figure 10.6 Polarity test. By tilting the sensor in the direction of the thin arrow, the seismometer mass moves only in the direction of the thick arrow. For the horizontal component, the tilt should be made in a time shorter than the natural period; otherwise the relative motion might be in the opposite direction, due to the weight of the mass, which will naturally go down.

This test can be made with both horizontal and vertical sensors. The output will be quite high and can be measured with an insensitive recorder, even in some cases just with a voltmeter.

Once a seismometer has been installed, it might not be so easy to make tilt tests without disturbing the installation. For a network, a simple test is to use the polarity of a teleseismic recording. If the network is only a few hundred km across, it can be assumed that the polarity is the same for all stations when the event is far away. The onset might not be very sharp, but by filtering or instrument correction to ensure similar looking waveforms, it is easy to compare polarities (Figure 10.7).

In the example shown in Figure 10.7, where the traces have also been instrument corrected, it is in addition also possible to get an idea if the correct instrument corrections have been used by comparing the amplitudes at the different stations. The first P-wave onset amplitude can easily vary by a factor of 2-5 over a distance of 100 km. In this case, there is only about a factor 2 difference between the stations with the maximum and minimum amplitude.

The recorder polarity can simply be checked by connecting a constant voltage to the input and observing the polarity on the output (do not saturate the system). If the recorder is not DC coupled, connecting a constant DC to the input would result in a pulse on the output, which however should have the same initial polarity as the input DC.

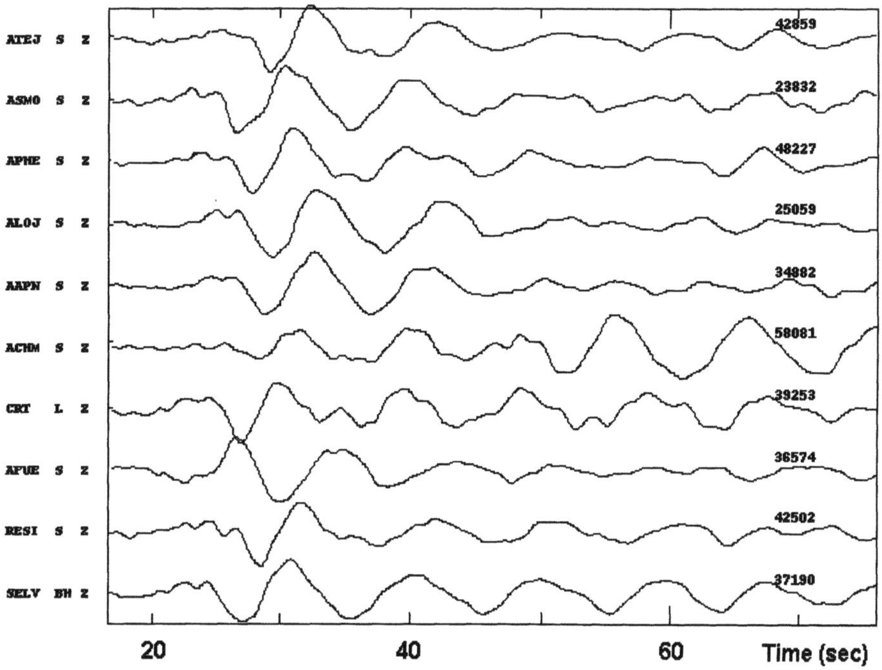

Figure 10.7 A distant earthquake recorded on a microearthquake network (200 km diameter) with eight SP stations (S Z), one LP (L Z) and one BB (BH Z). The traces have been filtered in the frequency band 0.1 to 1 Hz and converted to displacement. The numbers to the right are maximum displacement in nm. Note that station AFUE clearly has inverted polarity.

The sensor orientation can be checked by calculating the direction of approach (azimuth) to a know event source. The azimuth is calculated from the very first P-onset and compared to the azimuth given by the event location. Most popular earthquake analysis programs provides this facility. If e.g. the Z-component has wrong polarity, the azimuth will have an error of 180° while a polarity error in one of the horizontal components will result in an azimuth error of 90°. This assumes that the response of the 3 components is equal. If there is a known difference in response, it must be corrected for. On the other hand, if there is an error in the response (like the 3 components are not equal as usually assumed), this will also show up in the azimuth test.

10.2.5 FREQUENCY RESPONSE USING HARMONIC DRIVE METHODS

So far we have only dealt with static constants, free period and damping, from which we can calculate the complete frequency response (Chapter 6). However, we can also measure the complete response by injecting a known harmonic current into the calibration coil, or alternatively the generator coil, at different frequencies and measure the output. Since a known harmonic current will correspond to an equivalent harmonic ground acceleration, the frequency response can be determined. For a sensor with a calibration coil, an example of a setup is shown in Figure 10.8.

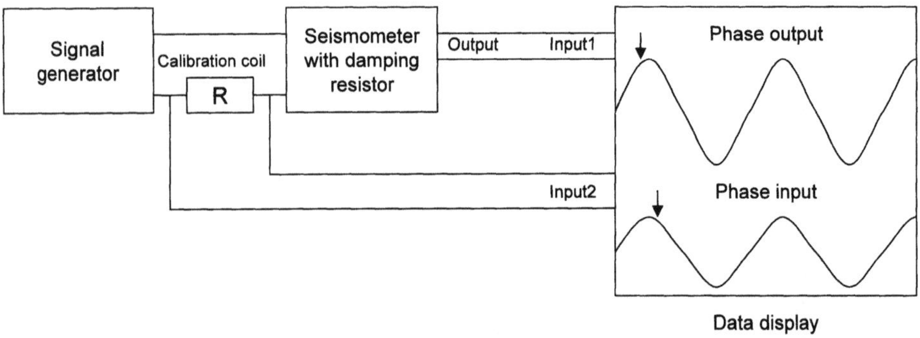

Figure 10.8 Calibration for frequency response with a seismometer with a calibration coil. The resistor *R* is used to have simple way of measuring the current through the calibration coil by measuring the voltage over *R*. The recorder shows the input and output signals which have different amplitude and phase. The output has a positive phase shift relative to the input.

Note that for active sensors of the FBA type, the force transducer coil will have the same function as the calibration coil.

A signal generator is connected to the calibration coil of a sensor through a resistor *R*. By measuring the voltage *V* over the resistor, we can determine the current $I=V/R$ through the calibration coil. A small voltage will be induced in the calibration coil as the mass moves, but since this coil usually has a small electrodynamic or motor constant K, this voltage is typically 10^{-4} to 10^{-6} times the voltage applied and can be neglected. The recorder can be any of the ones mentioned above, which are able to record both signals at the same time. From the output signals, we can directly measure the phase shift (Figure 10.8). The amplitude response is calculated as follows: If the amplitude of the current in the calibration coil is I_c, then the equivalent ground acceleration is KI_c/m and the equivalent ground displacement amplitude is $KI_c/(m\omega^2)$. If the amplitude of the seismometer voltage output is V_s, then the displacement amplitude response (V/m) for this frequency is:

$$A_d^v = \frac{V_s m\omega^2}{KI_c} \tag{10.17}$$

The procedure is then:

1: Set up a circuit as in Figure 10.8, set the frequency a little above the natural frequency to be in the middle of the frequency range to be used and adjust *R* and generator output to a high enough output, so that the natural background noise is not seen (the signal looks clean). For instance, with $K=0.2$ *N/A*, $m=1$ *kg*, $f=1$ *Hz* and $I_c=10$ *mA*, the equivalent ground displacement amplitude is 0.05 mm, which is well above background noise, and well below sensor clipping level.

2: Use a sweep of frequencies (e.g. 0.1 to 50 Hz for a 1 Hz seismometer) and measure the phase shift, I_c and V_s. Unless the recorder and seismometer have a high dynamic range, the input current probably will have to be adjusted for frequencies far from the natural frequency to keep the output level well above the noise but under the clipping limit.

3: Calculate the frequency response function using (10.17), this is directly FAP (frequency, amplitude and phase), which can be used by most analysis programs. Note that, even if the recorder response is not known, the calculation will be correct since we are using the ratio of output and input signals, which are both recorded with the same recorder.

If only the amplitude response is to be measured, the recorder can be a voltmeter with a peak-hold function. This means that, in DC mode, the voltmeter records the largest amplitude, which in our case is the amplitude to be measured. Make sure there is no DC component since that is added to the peak value. An AC voltmeter may not be suitable, because most of them have a low-frequency limit (~20 Hz) under which the measures are inaccurate.

If the calibration coil and signal coil are close to each other, signals can be induced into the signal coil from the calibration coil, particularly at high frequencies. To check this, just clamp the seismometer mass and measure the output from the signal coil when a signal is introduced into the calibration coil. This induction between calibration and signal coils is quite common and does not appear in the manufacturers' specifications! Nevertheless, some clamp devices are not rigid and the mass can move even when clamped: Test if there is signal while moving the seismometer clamped; this should not be confused with the mutual induction cited above. Steck and Prothero (1989) suggest a procedure to correct the measures from this effect of undesirable coupling.

In case the sensor does not have a calibration coil, it is still possible to make a harmonic drive calibration by using a bridge circuit, (Wilmore, 1979). Figure 10.9 shows the principle.

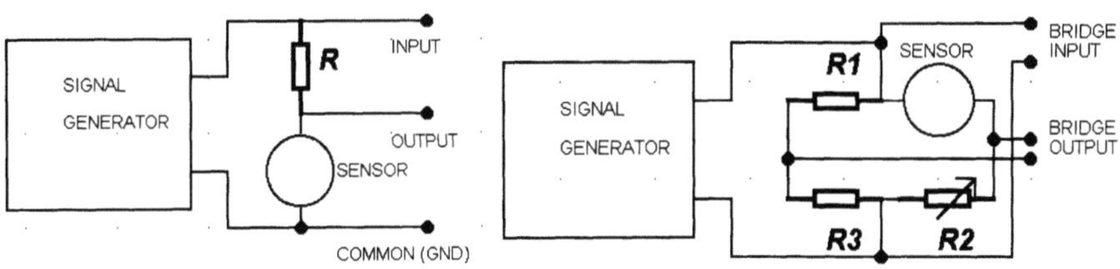

Figure 10.9 Bridge circuit for setting up harmonic drive when there is no calibration coil. The bridge circuit to the right will balance the signals so the output signal only contains the signal generated by the signal coil. Note that both the input and the output of the bridge are differential and thus must be connected to true differential inputs in the digitizer. In the half bridge (left), the output signal contains both the input and the generator coil signals and digital processing is required to remove the input signal. In this case, both the bridge input and output are referred to a common terminal (ground).

In the bridge (Figure 10.9, right), the 3 resistors are chosen to balance the bridge ($R_1/R_3=R_g/R_2$) by having one of them adjustable. R_1 should be selected in relation to R_g (the signal coil), so that the differential output is almost the same as the sensor signal. R_2 and R_3 are higher to avoid loading the sensor. The adjustment is done with the mass locked. When the bridge is balanced, the output signal will only be the signal generated in the signal coil since, for a clamped mass, input and output signals are equal. In practice, it is often necessary to add a capacitor to balance the coil self-inductance, which may vary depending on the mass position (i.e. mass clamped in one end). This makes the balance operation quite tricky and hence other calibration techniques, such as the use of transients, may be more advisable.

In the half bridge (Figure 10.9, left), the output signal contains both the signal coil generated signal superimposed on the input signal. However, if the input signal is also recorded digitally, the input signal can be subtracted from the output signal to recover the sensor output signal. Software to do to this is described by Wielandt in NMSOP, Bormann (Ed.), 2002.

FBA accelerometers usually do not have a calibration coil, since the force transducer coil can serve the same purpose. By inputting a sine wave calibration current directly in the force transducer coil, a force proportional with the current is generated, just like for the seismometer. So we can do a frequency response calibration of the accelerometer just like for the seismometer except that (10.17) is simplified since we now want the acceleration amplitude response A_a^a by inputting a known acceleration

$$A_a^a = \frac{V_s m}{K I_c} \tag{10.18}$$

Broadband sensors have in some cases a separate calibration coil (like the STS2) or the force transducer coil can be used through a calibration input. So, broadband sensors can in principle be calibrated just like any other velocity sensor with a calibration coil.

10.2.6 LISSAJOUS FIGURES

A Lissajous figure is generated by putting the input to the X-deflection of an oscilloscope and the output to the Y-deflection (Figure 10.10). This can also be done with an X-Y-plotter.

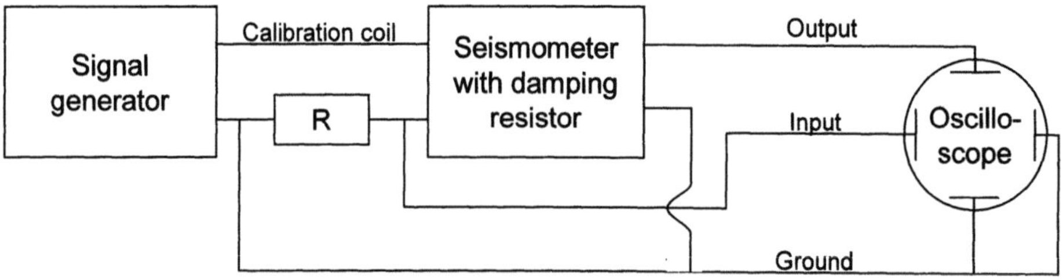

Figure 10.10 Setup for making Lissajous figures.

It is well known that this arrangement will generate an elliptical figure, Figure 10.11. Lissajous figures can be used for several purposes apart from measuring phase shift. We saw above that for an electromagnetic sensor, we do not measure the natural frequency directly with a free-swinging test when damping is not small. However, at the natural frequency, the phase shift between ground acceleration and voltage output is zero (see (2.17) and Figure 2.4). So by varying the frequency until the Lissajous figure becomes a straight line, we can determine the natural frequency. Knowing both the damped apparent frequency and the natural frequency, the damping can be determined with (2.39).

Non-linearity of the system will deform the ellipsoidal shape of the figure, so this is a simple test for the linearity.

Now that digital recording is available, Lissajous figures are less used. For more detail on how to use Lissajous figures, see Wielandt, Chapter 5 in NMSOP, Bormann (Ed.), 2002 and Mitronovas and Wielandt, (1975).

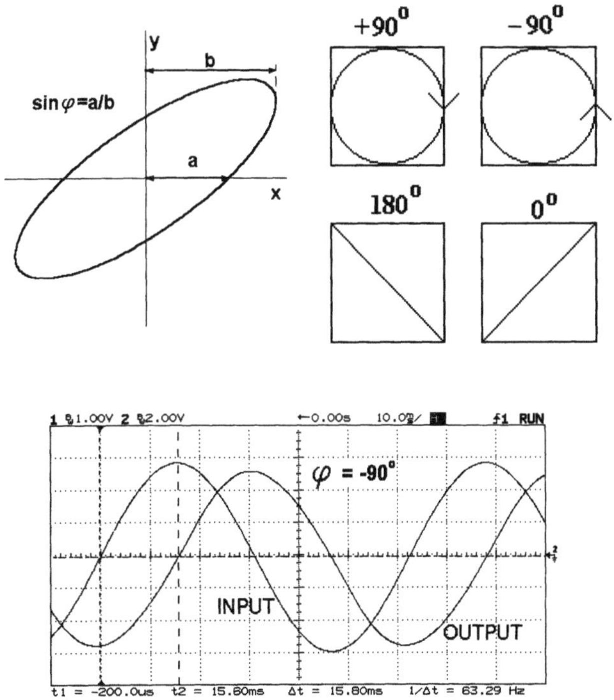

Figure 10.11 Lissajous figures. Upper: To the right is shown the cases of phase shift 90, -90, 0 and 180 degrees. To the left the figure shows how to measure the phase shift φ. The phase shift is positive if the motion is clockwise and negative if the motion is anti clockwise (left). Lower: Two signals with a phase shift of –90 degrees are shown as they appear on the oscilloscope screen. These figures assume that only the phase is different for input and output signals. If also the amplitude is different, the length of the x and y-axis will scale with the input and output amplitudes respectively.

10.3 Methods for absolute calibration of a sensor

So far, we have dealt with determining the frequency response shaping parameters. All the methods require the mass, and for some cases the calibration coil motor constant, to be known. However, for several reasons (for instance, modern sealed sensors), sometimes it may be preferred to perform a completely empirical calibration with no assumptions at all about a physical model of the sensor. Essentially the problem is simple; all we have to know is the ground motion and then measure the output of the sensor. Therefore, it all boils down to somehow using a known input.

10.3.1 SHAKING TABLE

Theoretically, the ideal solution to a controlled input source is to have a shaking table on which to put the instrument. The shaking table can move vertically and horizontally in a controlled fashion so the ground input is exactly known. All we have to do is measure the output and divide by the known input to get the response function. It is not as simple as it sounds. Making a precise shaking table, particularly for horizontal motion, without introducing tilt is complicated and expensive, so shaking tables are rarely available for the general user and only used in special laboratories.

10.3.2 USING THE GROUND AS A SHAKING TABLE

The ground is always moving (Chapter 3) so if we know that motion, we have a shaking table moving both horizontally and vertically. Our problem is then to determine this motion so we have an absolute measure. This is done by using a well calibrated sensor. The experiment then consists of setting up the two sensors close to each other and assuming that they are subject to the same ground motion. If the known and unknown sensors have recorded signals of the ground displacement z_1 and z_2 respectively, and corresponding spectra are $Z_1(\omega)$ and $Z_2(\omega)$, we can calculate the unknown response function T_2 as

$$T_2(\omega) = T_1(\omega)\frac{Z_2(\omega)}{Z_1(\omega)} \tag{10.19}$$

where T_1 is the response function of the known seismometer. Note that T_1 and T_2 can be for either displacement, velocity or acceleration, but both are of the same type. It is also possible to use an accelerometer to calibrate a seismometer or vice versa, but then factors of $i\omega$ must be used. Thus by measuring the signals of the known and unknown sensor, it is simple to determine the response function of the unknown sensor. Whereas the generator constant can be determined quite accurately by this method, the shape of the response function requires some more data processing, see below. In order for the method to work, the sensors self noise must be well below the signal levels generated by the ground motion. That will clearly limit the usefulness of the method at lower frequencies, where e.g. for a geophone, the output signal level will be low. So, contrary to normal sensor installation, these tests should be made in a high ground noise environment like the top floor of a building, although some care is required to avoid air-coupled acoustic noise, which may affect both sensors in different ways (Pavlis and

Vernon, 1994). In any case, the frequency response is usually determined by other methods, and the most important parameter to determine is the generator constant. If e.g., we have two velocity sensors, we can get the unknown generator constant G_2 from (10.19) if we use frequencies above the natural frequency of the two sensors and thereby assume that G is proportional to T:

$$G_2 = G_1 \left| \frac{\langle Z_2(\omega) \rangle}{\langle Z_1(\omega) \rangle} \right| \tag{10.20}$$

where G_1 is the generator constant of the known sensor and the spectra are averaged. In a simpler way, we can measure G_2 directly from the analog signals, see Figure 10.12. A more sophisticated way is described under cross spectrum below.

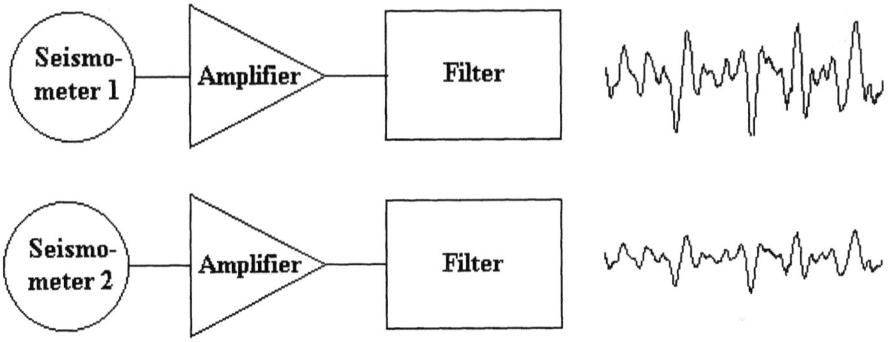

Figure 10.12 Comparison of signals recorded from two different seismometers with the same recording equipment. The filters and amplifiers are identical and filter out signals below the seismometer natural frequency. The ratio of the output amplitudes (right) indicates the ratio of the sensor generator constants.

By filtering the signals with identical frequency bands, well above the seismometer natural frequencies, we will get signals that resemble each other and their amplitude ratio gives the ratio of the generator constants. Note that if the filter is within a factor of 2 – 4 of the seismometer natural frequency, and the two seismometers do not have identical damping and natural frequency, the two output signals will have different phase (see Chapter 2). In the experiment shown in Figure 10.12, it is of course also possible to use a digital recorder, and filter the signals digitally.

Cross spectrum. If the spectra $Z_1(\omega)$ and $Z_2(\omega)$ of the two sensors outputs are contaminated with instrumental noise (a non-causal output), a simple ratio Z_2/Z_1 may yield a biased or unreliable estimate of the response T_2. By using only the correlated part of the signals, i.e. the part due to the common input of both sensors, the ground noise in this case, a more reliable estimate will be obtained. This may be done by using a known relation between input and output of linear and causal systems: The input-output cross-spectrum is the product of the transfer function and the input power spectrum (see, for instance, Lee, 1967, p. 341). In this case, we can imagine a linear system whose input is the output of seismometer 1 (Z_1) and its output is the output of seismometer 2 (Z_2). This system would have a transfer function $T_2/T_1=P_{21}(\omega)/P_{11}(\omega)$, or Z_2/Z_1 if these two spectra were noise-free and linearly caused by the same input. Therefore, a more robust estimation of the unknown response T_2 is obtained using

$$T_2(\omega) = T_1(\omega)\frac{P_{21}(\omega)}{P_{11}(\omega)} \tag{10.21}$$

where $P_{11}(\omega)$ represents the power spectrum of the output of sensor 1 and $P_{21}(\omega)$ is the cross-spectrum between the outputs of sensors 1 and 2. The advantage of this equation is that, ideally, it cancels out the contributions of uncorrelated components (i.e. self noise) of the input and output signals. An alternative procedure was proposed by Pavlis and Vernon (1994). They calculate the spectral ratios Z_2/Z_1 in a series of overlapping windows, and use a robust estimation of the mean to minimize the influence of outliers, due mainly in their method to air-coupled acoustic noise in both sensors.

10.3.3 CALIBRATION BY STEPWISE MOTION

The main idea behind this method is to (1) move the sensor a known distance (like 1 mm) in its direction of sensitivity, (2) record the signal and (3) correct the signal for the known displacement frequency response. Theoretically, we should now have calculated a displacement, which can be compared to the actual displacement. The generator constant may then be corrected by a factor so as to obtain the true motion. The method is described in detail by Wielandt in Chapter 5 in NMSOP (Bormann, (Ed.) 2002), which claims that the method works well for broadband sensors and even 10 Hz geophones. Although the method sounds simple, it is not trivial to correct for the instrument response down to DC, particularly for SP instruments. The effect of tilt can severely contaminate the signal, is unknown and cannot be removed for the pulse itself. The pulse displacement should therefore take place over a short time interval like 1 s. NMSOP gives guidelines (and a program), on how to process the recorded signal from a BB sensor:

1. De-convolve the trace with the known velocity response function of the sensor.
2. De-trend the trace piecewise so it is close to zero in the motion free intervals. Interpolated trends are removed from the interval of motion.
3. Integrate the trace to get displacement and compare with the actual motion.

So now, the only problem is how to move the sensor, horizontally or vertically, a controlled distance. NMSOP is suggesting to use either a milling machine, a lathe or a mechanical balance. The milling machine and the lathe can move very precise distances both horizontally and vertically, but are not the standard tools found in seismological labs, while a mechanical balance is a common instrument (Figure10.13).

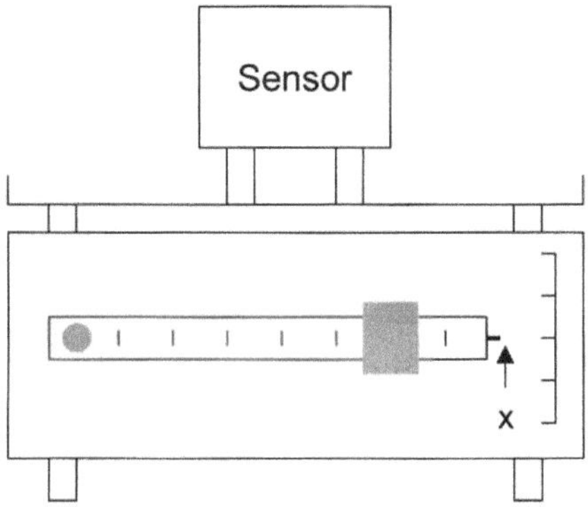

Figure 10.13 Moving a sensor a controlled distance vertically with a mechanical balance. The ratio between the motion at x (where we can measure it) and on the balance table can be determined by placing a mass *m1* at *x* that will balance a mass *m2* at on the balance table. The ratio of the two motions is then *m1/m2*. Redrawn from Figure 5.30 by E. Wielandt in NMSOP, Vol. 1, Bormann, (Ed.), 2002; copyright granted by IASPEI.

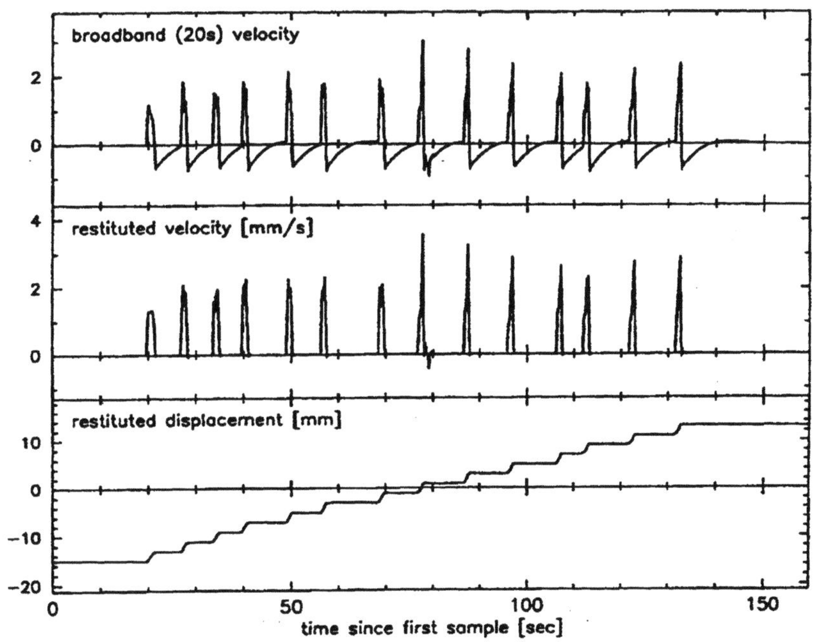

Figure 10.14 Absolute calibration of a 20 STS1-BB using a milling machine. The sensor was moved 14 steps of 2 mm each. The top trace shows the original recording. The middle trace shows the pure velocity record (de-trending and corrected for response) while the bottom trace shows the displacement obtained by integrating the middle trace. Reproduced from Figure 5.29 by E. Wielandt in NMSOP, Vol. 1, Bormann, (Ed.), 2002; copyright granted by IASPEI.

Placing the sensor on the mechanical balance table, the seismometer can be moved a controlled distance as measured on the balance arm vertical scale. In principle, it is enough to make one displacement, but usually several are made to take average measurements. Figure 10.14 shows an example using a 20 s seismometer. As it can be seen, the original displacement could be recovered very well using the known absolute response function for the seismometer. NMSOP estimates that by doing several such calibrations, the accuracy of the determination of the generator constant is better then 1%.

10.3.4 DETERMINING SENSITIVITY OF ACCELEROMETER BY TILT

An accelerometer has an output that is proportional to the static acceleration. By tilting the accelerometer, the effective force on the three components can be determined for different tilt angles and the sensitivity determined, see Figure 10.15.

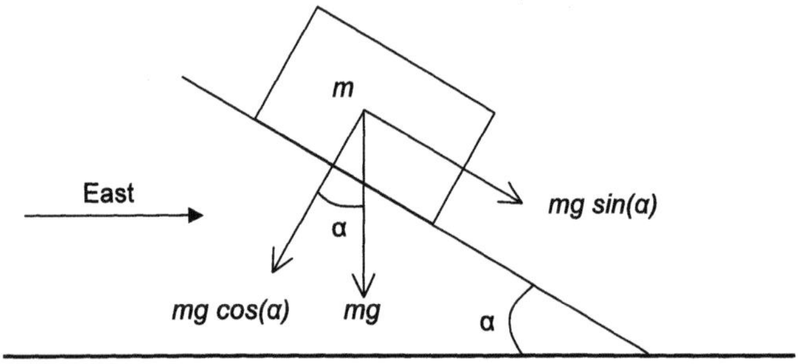

Figure 10.15 Tilting an accelerometer to determine generator constant. It is assumed that the sensor horizontal direction is towards East.

The accelerometer is tilted an angle α. The force in the horizontal direction is now *mgsin(α)* while in the vertical direction it is *mgcos(α)*. If the voltage output in the vertical direction is V_z and in the horizontal direction V_h, we can calculate the generator constants (V/ms^{-2}), G_z and G_h, in the two directions. To get V/g, divide by g. Considering that in the horizontal position the output is supposed to be zero and that the vertical force has been *decreased* from *mg* to *mgcos(α)*, we get

$$G_z = \frac{-V_z}{g(1-\cos(\alpha))} \quad \text{and} \quad G_h = \frac{V_h}{g\sin(\alpha)} \tag{10.22}$$

V_z will be negative since the vertical force is decreased and V_h will be positive since the force is in the East direction. This method is very simple and can determine the generator constant accurately, but will not give a dynamic calibration. The horizontal components can be tested for symmetry by inclining in the opposite direction.

10.4 Use of the calibration pulse

It is common practice to generate a calibration pulse by applying a step current into the calibration coil as a daily check (Figure 10.16). As it was shown in the section on the determination of the generator constant *G*, also the signal coil of passive sensors may be used for this purpose (10.2.5), however here we will use a simpler method since only a short pulse is needed. The pulse is recorded and one has a permanent record of the calibration status of the system. This pulse can be used in several ways to obtain sensor parameters as we saw in 10.2.

A constant current i_0 injected into the calibration coil at time t_0 will exert a force on the mass $F_0=Ki_0$, and for the sensor this will be equivalent to a ground step-in-acceleration from 0 to $a_0=Ki_0/m$. As a result, the mass starts swinging, if the damping is sub-critical around its new equilibrium position $z_0=F_0/k$, where k is the spring stiffness.

Once the mass is at rest again, the calibration current is suppressed by opening the circuit of the calibration coil. This will result in a mass release test, identical but opposite to the previous transient. As the initial position of the mass is

$$z_0 = \frac{F_0}{k} = \frac{Ki_0}{k} = \frac{ma_0}{k} \tag{10.23}$$

the voltage output for a velocity transducer will be (2.40)

$$V(t) = \frac{G_E K i_0}{\omega_1 m} e^{-h\omega_0 t} \sin(\omega_1 t) \tag{10.24}$$

where $\omega_1 = \omega_0 \sqrt{1-h^2}$. The sign of the voltage depends on the polarity chosen for the coil output.

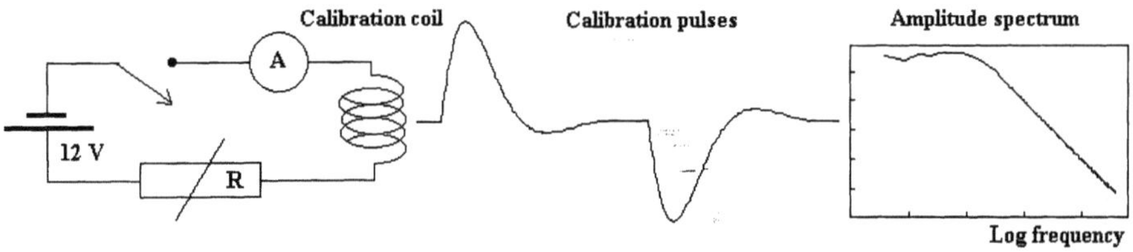

Figure 10.16 Set-up for routinely making a calibration pulse for recording. The current is measured by the amp-meter (A) and regulated by the variable resistor R. Alternatively a regulated current source may yield a more stable calibration current. The calibration pulse is generated assuming a damping of 0.7 and using a velocity sensor for recording. The pulse is generated by closing the switch long enough to record the complete pulse. When the switch is opened again, a new pulse is generated with opposite sign. The amplitude spectrum of one of the pulses is seen to the right (log-log).

If the damping coefficient h is well under critical, f_0 and h may be obtained from two consecutive peaks as explained in 10.2.2. Then, from the first peak amplitude V_1 and time t_1,

$$V_1 = V(t_1) = \frac{G_E K i_0}{\omega_1 m} e^{-h\omega_0 t_1} \tag{10.25}$$

G_E may be obtained. This formulation has the same disadvantage as (10.13) that the mass must be known.

Some passive sensors either lack calibration coil or the one fitted presents a spurious coupling with the signal coil that disturbs the output pulse. In these cases, calibration pulses may still be generated using the signal coil itself. A constant current i_0 is made to flow through the signal coil (if the load resistor is kept connected, this may be difficult and a constant voltage V_0 may be applied instead). After a time sufficient to allow the mass motion to cease, the current source is disconnected (shorted) or the voltage source is opened, see Figure 10.17.

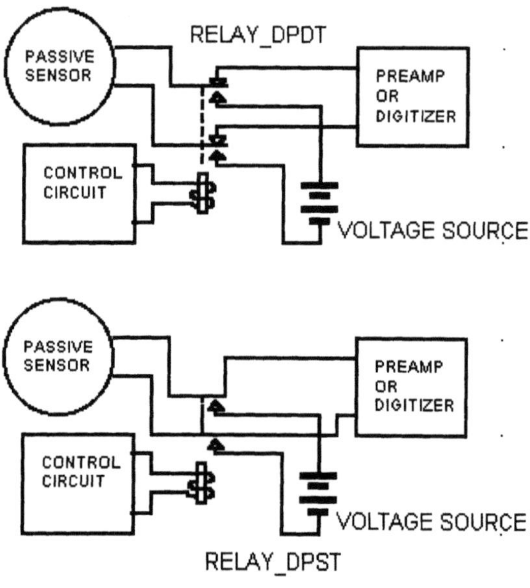

Figure 10.17 Two possible arrangements to introduce a calibration pulse in the signal coil of a passive sensor. Top: The coil is disconnected from the recording system while the current is flowing in the coil. Bottom: The calibration current is switched on while the recording system is kept connected. In this later case, it is advisable to use a voltage source, since otherwise the exact part of current that actually flows in the coil may not be known. This second circuit should not be used with extended-response sensors, because the input voltage step will produce a long transient output superimposed to the mass free motion. The voltage or current source has to be stable and known.

The resulting excitation, when the mass is released, is equivalent to a ground step in acceleration of amplitude of $a_0 = -\frac{Gi_0}{m}$ or $a_0 = -\frac{GV_0}{R_g m}$. The minus sign means that a positive voltage V_0 applied to the coil causes a mass displacement in such a sense that the voltage generated when released is negative, see Figure 10.18. Using this value of a_0 and $G_E = G\frac{R_e}{R_e + R_g}$, where R_g is the generator coil resistance and R_e the external load resistance, the voltage output for this "mass release" will be obtained from the equation for $V(t)$ above as

$$V(t) = -\frac{G^2 V_0}{\omega_1 R_g m}\frac{R_e}{R_e + R_g}e^{-h\omega_0 t}\sin(\omega_1 t) \tag{10.26}$$

An example of this pulse is plotted in Figure 10.18.

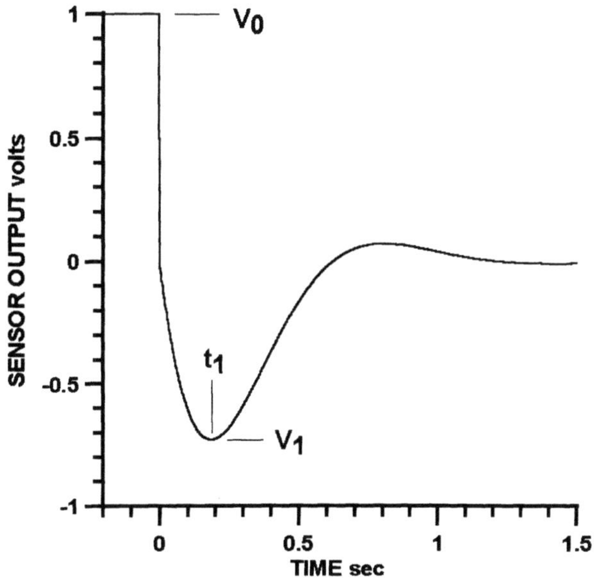

Figure 10.18 A calibration pulse obtained by first connecting a voltage of 1 V to the signal coil of a 1 Hz sensor (in this case a Mark L4-C). At time *t=0,* the voltage is switched off and the mass-release transient is recorded. The values of the parameters used are G=270 V/(m/s), R_g=*5.6* kohm, R_e=*13* kohm, *T=1 s, h=0.6,* V_0=*1 V.* From the measure of the first peak's time and voltage, the value of *G* may be obtained. The accuracy of the *G* value estimation depends on the precise timing of the peak relative to the mass release onset.

If V_1 is the voltage of the first peak and t_1 the time of this peak relative to the pulse onset (the instant at which the mass is released or the current removed), it can be shown (Asten, 1977; MacArthur, 1985) that G may be obtained as

$$G = \left[-\frac{V_1}{V_0} R_g m \omega_1 \frac{R_e + R_g}{R_e} e^{h\omega_0 t_1} \right]^{\frac{1}{2}} \tag{10.27}$$

This formula gives stable estimates of G and, due to the square root involved, the uncertainties are less than in other estimation methods. The constants ω_0 , h and ω_1 must be known or may be obtained from the same transient by the procedures described in this Chapter.

Alternatively, the whole voltage pulse may be fitted to the equation of $V(t)$ by a suitable non-linear least squares procedure and the unknown parameters G, h, ω_0 (ω_1 depends on them) estimated.

Calibration pulses may be repeated several times and added together to improve signal-to-noise ratio. It is essential that the precise relative onset times are well known so they can be aligned synchronously. This is not difficult if the calibration sequence is driven by a crystal clock. Otherwise, if calibration pulses at arbitrary times have to be added, the precise relative onset times may be determined by cross-correlation.

Some systems automatically generate the pulses every 24 hours. By always injecting the same current, and therefore expecting the same pulse, an operator can quickly spot if there is any significant change in the system response.

With digital recording, the pulse can also be used to calculate the response function. From signal theory (Scherbaum, 2001) it is known that the Fourier transform of the impulse response $x_i(t)$ of a linear system, is the frequency response function

$$T(\omega) = \int_{-\infty}^{\infty} x_i(t) e^{-i\omega t} dt \tag{10.28}$$

However, we cannot generate an impulse and we use a step $x_s(t)$ instead. The relation is then

$$T(\omega) = i\omega X_s(\omega) = i\omega \int_{-\infty}^{\infty} x_s(t) e^{-i\omega t} dt \tag{10.29}$$

since the time derivative of a step function is an impulse. By applying a step calibration current, we are trying to generate the response to an acceleration step. From Figure 2.7 we have the acceleration response for a velocity sensor (Figure 10.19), however in order to get what we observe in Figure 10.16, the curve has to be divided by the angular frequency ω and we get the response observed in Figure 10.16 (Figure 10.19).

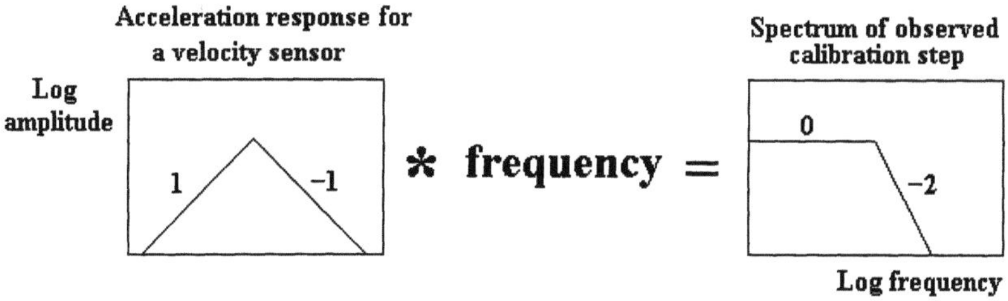

Figure 10.19 Frequency response from a calibration pulse. To the left is shown the acceleration response for velocity sensor and to the right the expected spectrum of the calibration pulse. The numbers show the slope on the log-log graphs.

By making the Fourier spectrum of the calibration pulse, we can then determine the response function of the system. If the Fourier transform of the calibration pulse of a velocity sensor is X_s, then e.g. the acceleration response is

$$T_a^v(\omega) = i\omega \; X_s(\omega) \tag{10.30}$$

and the velocity response is

$$T_v^v(\omega) = (i\omega)^2 X_s(\omega) \tag{10.31}$$

We see that, if the spectrum in Figure 10.19 is divided by ω^2, we would get the expected shape of the velocity response.

In practice, as mentioned above on the determination of G from the calibration pulses, the signal-to-noise ratio is improved by summing several calibration pulses with their onsets well aligned. The background noise, both the instrumental and the ground one, is decreased this way, while the signal due to the calibration pulses is increased.

FBA accelerometers often have an input whereby 12 V in will inject a given current into the force transducer coil, which is equivalent to sending a calibration step into the seismometer calibration coil. The current will correspond to a fixed acceleration like 0.25 g. Apart from the transient response at the moment of closing or opening the switch, the sensor output will be a constant voltage since the FBA accelerometer responds to a static change in acceleration, like in the tilt test. This is a convenient way of testing an accelerometer since all that is needed is a 12 V source and a voltmeter.

10.4.1. USING PSEUDO RANDOM SIGNALS FOR CALIBRATION

The main drawback of using a step input signal for obtaining the response function is that the spectral amplitude decreases with frequency and therefore the spectrum of a calibration pulse becomes noisy as the frequency increases, since noise then dominates the signal. This problem may be partly overcome by summing several pulses, as stated above, but an alternative method may be to use a broad-band calibration signal (Berger et al., 1979) to excite the calibration coil or the calibration input of a feedback sensor.

One such signal is the so-called pseudo-random binary signal (PRBS). This kind of signal is easy to generate by a simple recursion rule (a programmable generator may be used or a simple circuit may be wired), having only two levels of limited amplitude, so that the system under calibration may be driven to a controlled high amplitude well above the noise, and its bandwidth may be as wide as desired (MacWilliams and Sloan, 1976). If the input signal (in this case the PRBS) is x(t) and the system output is $z(t)$, the system response function $T(\omega)$ will be obtained as

$$T(\omega) = \frac{P_{xz}(\omega)}{P_{xx}(\omega)} \tag{10.32}$$

where $P_{xz}(\omega)$ is the cross-spectrum of input and output and $P_{xx}(\omega)$ is the power spectrum of the input. Remember that the cross-spectrum of two signals is the spectrum of their cross-correlation. In this case, if the input peak-to-peak amplitude is V and the PRBS sequence is clocked at Δt time increments, its power spectrum is (Berger et al., 1979)

$$P_{xx} = \frac{V^2 \Delta t}{4}\left[\frac{\sin(\omega\Delta t/2)}{\omega\Delta t/2}\right]^2 \tag{10.33}$$

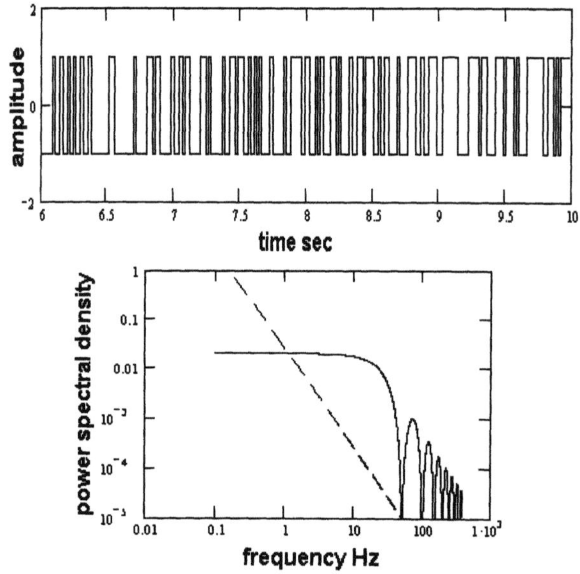

Figure 10.20 A section of a pseudo-random binary signal (PRBS) (upper) and its power spectral density (lower, solid trace), compared with the PSD of a step with the same amplitude (dashed line). The PRBS presents a flat level until a corner frequency that may be as high as desired, depending on the generator clock interval.

10.5 Recorder

The recorder basically consists of the amplifier-filter and the ADC. For purely analog systems, the analog recorder replaces the ADC. For a complete recorder, the two units are usually integrated, so calibration is done with all units together, while for other systems, particularly analog systems, each unit is separate and can be tested individually. However, the ADC must be connected to some recorder to observe the signals. The basic test setup for the amplifier unit is seen in Figure 10.21.

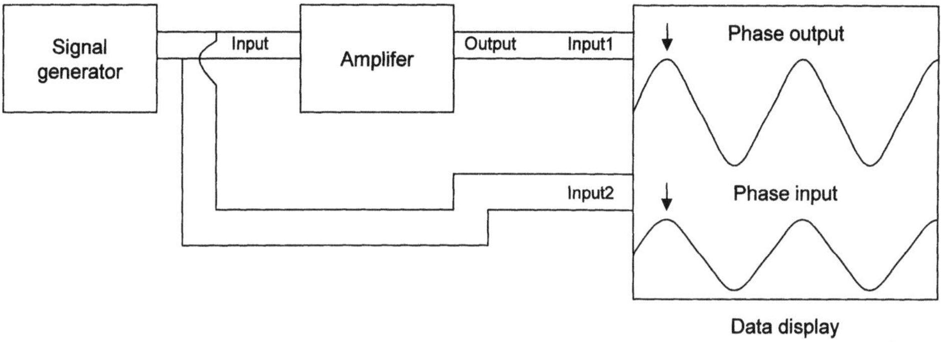

Figure 10.21 Basic setup for testing and calibrating an amplifier. The data display can be a recorder for which the calibration is known. The display shows that the output signal has a positive phase shift (leads the input).

The procedure for the frequency calibration of an amplifier is:

(1) Connect the frequency generator to the input of the unit and adjust the output signal to a level well above the unit self noise. For some units with limited dynamic range, it might not be possible to adjust the signal generator to a low enough value. If e.g. an amplifier has a gain of 80 dB, and a maximum output of 10 V, then the maximum input can only be 1 mV. It is then necessary to use a voltage divider at the input to reduce the signal generator output.

(2) Record or display the input and output signals. If a standard recorder is used, use a sample rate high enough so that there is no phase shift or amplitude change in the frequency band of interest.

(3) Vary the frequency over the frequency band of interest and measure the output amplitude and phase shift relative to the input phase. Usually the signal generator is stable enough so that the input amplitude stays constant. Note that for ADC's with FIR filters, the frequency response is very sharp (Figure 4.11), so very small steps in the frequency variation must be used near the filter cut off.

(4) Calculate the gain of system as the ratio of the output signal to the input signal and make a table of gain and phase shift.

Usually, modern equipment shows very little response changes along time, therefore the initial calibration may be considered valid, unless some change or repairing work has been done.

The procedure for the frequency calibration of a recorder:

Essentially, the procedure is the same as for the amplifier. The main difference is that the output signal is registered by the recorder itself (Figure 10.22). So, if the phase shift has to be measured, the input signal must be recorded on an instrument without phase shift and the two recorders must have identical timing. The signals from the two recorders can then be compared to observe the phase shift. This can be done using the same procedure described above with wide bandwidth signals or using sinusoidal signals and measuring the phase shift for each frequency. Alternatively, a signal generator synchronized with the data acquisition (referred to a common time base, i.e. GPS), may allow phase delay measurements for each frequency.

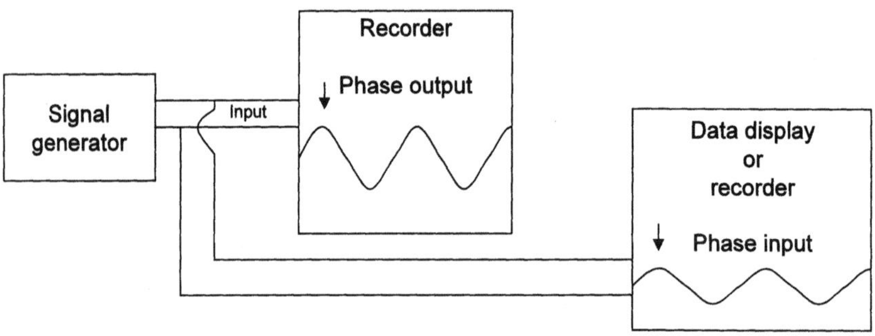

Figure 10.22 Basic setup for testing and calibrating a recorder. The data display can be a recorder for which the calibration is known. Note that the output signal is recorded with the recorder itself since the output is the digitized signal.

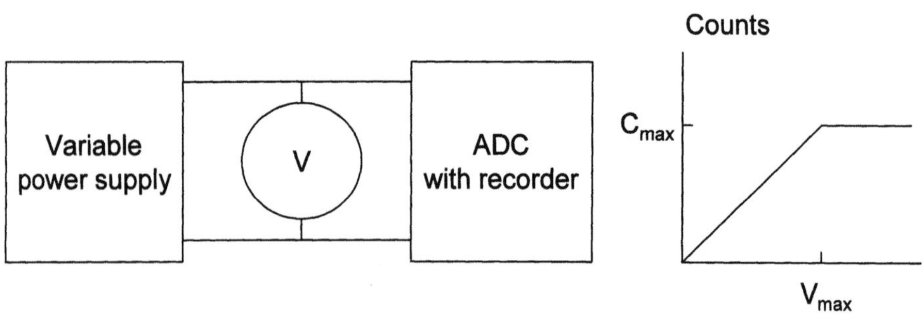

Figure 10.23 Setup for measuring ADC sensitivity and range. The counts output versus the voltage input is plotted to the right.

Digitizer LSB and range. For a digitizer it is often enough to measure only the least significant bit (LSB) and voltage range. Figure 10.23 shows the setup. A variable voltage source is connected to the ADC and the voltage is measured with a voltmeter

(V). The counts output is measured with the recorder. The voltage is slowly increased until the output reaches its maximum counts level C_{max} with a corresponding voltage V_{max}.

The ADC maximum input is the V_{max} and the LSB is

$$LSB = \frac{V_{\max}}{C_{\max}} \tag{10.34}$$

The ADC is not noise free and a simple test to get the ADC noise is to short the input or connect a resistor with the value of the resistance of the input device. Now measure the output count level C_{noise} (as noise is a random signal, it is better characterized by its RMS value). The theoretical dynamic range (dB) of the ADC is $20log(C_{max})$ while the useful dynamic range is approximately $20log(C_{max}/C_{noise})$. Gross errors in the ADC linearity can be observed from the plot (Figure 10.23, right), however for most ADC's any deviation from linearity is likely to be much smaller than what can be measured with this simple test and it is very hard for a user to determine if the ADC lives up the manufactures specifications without special test equipment.

Recorder cross talk. This is the overplay from one channel to another and is always present to some degree, however it should be less than -80 dB or a factor of 1/10 000. The simplest way of testing it is:

(1) Put a large AC signal like half the digitizer maximum voltage into channel 1.

(2) Connect a resistor of a size equivalent to the input device resistance across each of the remaining channels.

(3) Record all channels.

(4) Measure the amplitude ratio of the test signal between channel 1 and the remaining channels.

If the cross talk is so small that it is not clearly seen above the digitizer noise, the cross talk ratio can be determined by making a cross correlation of a long time series between two channels.

For a 3 component station, the cross talk is not such a serious problem since all 3 signals usually have similar amplitudes and are highly correlated. However, if a recorder is used to record different channels in a network, a high amplitude on one station might appear as a precursor to the P-phase on another station a bit further away. This can happen with popular 16-64 channel recorders used with analog networks, if the ADC has not been properly terminated.

10.6 Measuring instrument self noise

In addition to calibrating instruments, its is also important to know the amount of noise generated by the instruments themselves, so that when setting up an instrument, we can be sure that what we are seeing is real seismic signal and not just electronic noise. This topic has already been discussed under sensors and analog to digital converters.

Recorder noise. The simples test is to short the input, or better, connect a resistor with the impedance of the input device. The output gives a good idea of the recorder noise and the sensor must give out a signal, due to the ground natural noise, that is larger than the recorder self noise.

Recorder and sensor. A very simple test of the self noise in the complete recording system is to clamp the seismometer and measure the output of the recording system. This output will give a good idea of the noise generated within the recording system and the sensor. Hopefully, that will be much less then the unclamped signal; otherwise we are just measuring recording system noise and not ground noise. The noise should be very close to the measurement of recorder noise above, since, at least for passive sensors, the main noise source in the complete system originates in the recorder. Clamping an active sensor may not be possible. If it is possible, it might modify the feedback loop such that the noise changes. In that case we can only test the recording system without a sensor as described in the previous section. Be aware of the possibility of electrically induced noise into the sensor or cable. If the clamp test does not reduce the noise to the level measured with an input resistor, there is probably electrically induced noise present.

Seismic sensors. The self-noise in passive sensors is so low (see Chapter 2) that it is hard to measure and therefore has no practical importance. Active sensors can be quite noisy. Since these sensors are of the force-balanced types, they cannot be tested for self-noise by locking the mass. The self-noise must therefore be determined in the presence of seismic noise. The following description largely follows Wielandt, Chapter 5 in NMSOP (Bormann, (Ed.), 2002).

The natural background noise can be used as a calibration signal. It is like putting the sensor on a shake table with a very small amplitude motion, smaller than can be made with a real shake table. The problem, as discussed above, is that we do not know the exact input, although it might be measured by a reference instrument. The marine microseisms are observed almost everywhere and should be visible on any seismometer with a natural frequency of 1 Hz or less if installed at a site with relatively low cultural noise (see Figure 3.1). For broadband sensors at quiet sites, the earth tides yield a stable signal with a period of around 12 hours and amplitudes of 10^{-6} m/s^2. The tides can actually provide a reasonable good absolute calibration signal since the tide can be predicted. In order to see the tides, the signal must be low pass filtered with a corner frequency at around 0.001 Hz. The ability to distinguish earth tides is often used as a criterion of whether a BB station is up to international standards.

In order to get a quantitative measure of the instrument noise, the coherency method can be used (Holcomp, 1989; Holcomp, 1990). Under the assumption that two sensors have identical response to ground motion, the output from the two sensors recording the same ground noise will differ only in the instrumental self noise. One then determines the coherence between the two records and assume that the coherent signal is the seismic signal and the incoherent signal is instrumental noise. This is not quite as simple as it sounds, see Wielandt, Chapter 5 in NMSOP, Bormann (Ed.), 2002, for more details.

Transient sensor noise: Many new broadband seismometers produce spontaneous transient disturbances caused by stresses in the mechanical components. These look like a sudden and permanent change in the spring force and appear on the seismogram like a step pulse has been input to the recording system. The signal is sometimes mistaken for electrical disturbances into the recording system. For some instruments, these disturbances can present a serious problem for the first months or even years of operating a new seismometer. If the disturbances do not decrease, corrosion must be suspected. It is sometimes possible to eliminate transient disturbances by hitting the pier (on which the seismometer is fixed) with a hammer, and this is a procedure recommended for each new installation.

10.6.1 EXAMPLE OF A PRACTICAL NOISE MEASUREMENT

(1) Connect a SP seismometer to a recording system, place the sensor on the instrument pier and measure the background noise from the earth. Make sure the gain is high enough so that the signal amplitude can resolve the noise (amplitude larger than 500 counts). Check that ground noise is recorded by observing the microseism in the 0.2-0.5 Hz band, filtering might be needed if the site is noisy. Record at least 60 s of noise.

(2) Clamp the sensor and record the signal (at least 60 s). This signal represents the recorder system noise. The noise should now be considerably smaller.

(3) Make the equivalent noise power spectral density plot for both the clamped and unclamped signal. The spectrum of the clamped signal will give the limits of the ground noise resolution for the given combination of recorder and sensor. The frequency band for which the amplitude of the noise spectrum of the unclamped signal is above the amplitude for the clamped signal indicates the frequency band where the sensor-recorder combination can resolve ground noise (Figure 10.24).

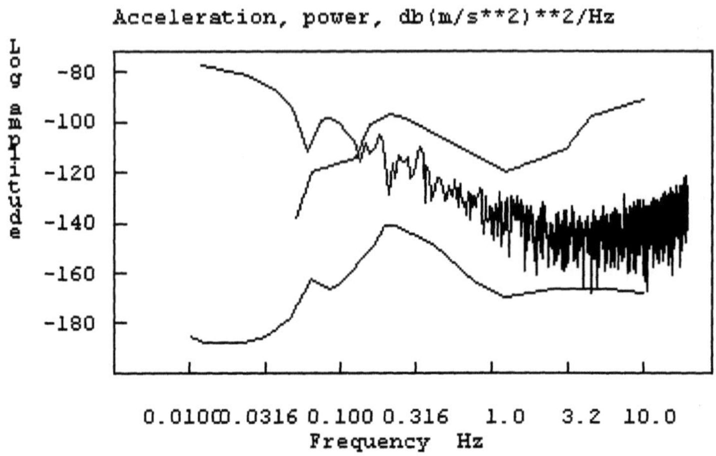

Figure 10.24 The equivalent ground acceleration noise spectrum for a digitizer with its input shorted, using a low-sensitivity virtual sensor of 4.5 Hz and G=30 V/(m/s). The record of the digitizer noise has been reduced to the equivalent ground motion using its response combined with an imaginary low-sensitivity sensor. The spectrum then shows the worst-case sensitivity for ground motion that can be achieved with this digitizer and sensor. Of course, a more sensitive sensor would give a lower equivalent ground noise. The smooth curves on top and bottom are the Peterson New Low-Noise and New High-Noise models, for reference.

APPENDIX

Basic electronics

Seismic instruments, like most scientific instruments today, have a fundamental electronic basis. Whereas the design of these systems is mainly a matter for specialists, the users should understand the basic components, laws and rules involved as well. Many questions about the choice of equipment, its interconnections, the solution of minor operation problems, etc., depend on this knowledge.

A.1 Basic terms

A.1.1 ELECTRIC CHARGE

The electric charge is defined by means of the *Coulomb Law*, which states that the force F between two point charges q and q' is proportional to their product and inversely proportional to the squared distance r

$$F = k\frac{q \cdot q'}{r^2} \qquad \text{(a.1)}$$

The constant k depends on the medium between the charges and for vacuum it is k=9·10^9 Nm^2/C^2. For air, k is slightly less and usually taken at the same value. A coulomb (C) is thus the charge of a particle that, situated at a distance of 1 m from another particle with an identical charge, is repelled with a force of 9·10^9 newton (N) (or equivalent to the weight of 1 million ton !).

A.1.2 ELECTRIC POTENTIAL, CURRENT AND POWER

The potential difference or *voltage* v between two points is the work required for a unit charge to move from one point to the other. A voltage of 1 volt between two points means that 1 joule is required to move a charge of 1 coulomb between them.

A material containing free charges is called a conductor. When a voltage is supplied by an external power source between two points of a conductor, the charges move. The charge that passes through a given section of the conductor per unit time is the current i

$$i = \frac{dq}{dt} \qquad \text{(a.2)}$$

A current of 1 ampere (A) transports a charge of 1 coulomb per second. The total charge is conserved and thus, in a given portion of a circuit, the same current that enter must flow out (the current is assumed to propagate instantaneously, a reasonable approximation in practice for low-frequency circuits). From the definition of the potential difference and the current, it follows that the instantaneous power p used to keep a current i between two points with a potential difference v will be

$$p = v \cdot i \tag{a.3}$$

A power of 1 W is required for a current of 1 A to flow between two points with voltage difference of 1 volt.

A.1.3 KIRCHHOFF LAWS

Two useful rules to analyze circuits are known with this name. The first Kirchhoff law is a consequence of the charge conservation and states that the sum of the currents entering a junction is equal to the sum of the currents leaving the junction. If a positive or negative sign is assigned respectively to them, we can simply say that the algebraic sum of the currents in a junction must be zero.
The second law states that the algebraic sum of the potential differences around a closed circuit is zero.

A.2 Passive components

When energy is supplied to a circuit element, it may be transformed to heat: In this case, the element is a resistor. Alternatively, the element may store the energy as an electric field and this is a capacitor. If the energy is stored in a component as a magnetic field, it is an inductor. Real circuit components usually have one or more of these behaviors, but one of them predominates within a limited frequency range.

A.2.1 RESISTORS, INDUCTORS AND CAPACITORS

An ideal *resistor* follows *Ohm's Law*: The voltage between its terminals is proportional to the current that flows. The proportionality constant is the resistance R (ohms, Ω)

$$v = i \cdot R \tag{a.4}$$

Both v and i may be constant (DC) or time-dependent.

A moving charge (a current) creates a magnetic field around it. If the current in an inductive circuit varies, the magnetic flux through the closed circuit also varies and a voltage is induced proportional to the flux variation rate. The constant of proportionality L between the current change rate and the induced voltage in the inductive element is called the self-inductance coefficient or simply the *inductance*

$$v(t) = L \cdot \frac{di(t)}{dt} \qquad \text{and so,} \qquad i(t) = \frac{1}{L} \int v(t)dt \tag{a.5}$$

The value of L is measured in *henry* (H).

An ideal *capacitor* stores a charge $+q$ and $-q$ in its plates, which is proportional to the voltage applied to its terminals,

$$q = C \cdot v \tag{a.6}$$

where C is the capacity and is measured in *farads* (F). Only if the voltage varies with time, will the charge of each plate change. A rise of voltage will cause a positive charge flowing from the external circuit to the positive plate and a negative charge towards the negative one, which is equivalent to a positive charge flowing out of this plate. That is, a current *i(t) passes through* the capacitor

$$i(t) = \frac{dq}{dt} = C \cdot \frac{dv(t)}{dt} \quad \text{or} \quad \Delta v(t) = \frac{1}{C}\int_0^t i(t)dt \tag{a.7}$$

where $\Delta v(t)$ is the change of capacitor voltage from the time origin to time t.
If we connect a resistor R in parallel with a capacitor initially charged with voltage V_0, the voltage on both will be v(t). The same current flowing from the capacitor will pass through the resistor.

$$-C \cdot \frac{dv(t)}{dt} = v(t)/R \tag{a.8}$$

The minus sign means that a charge flowing from the capacitor will cause a voltage drop on it. This equation may be written

$$\frac{dv}{v} = \frac{-1}{RC}dt \tag{a.9}$$

By integration between time t=0 and arbitrary time t,

$$\int_{V_0}^{v} \frac{dv}{v} = -\frac{1}{RC}\int_0^t dt \tag{a.10}$$

This gives

$$\ln\frac{v}{V_0} = \frac{-t}{RC}, \quad \text{or} \quad v(t) = V_0 e^{-t/RC} \tag{a.11}$$

Therefore, the charged capacitor will discharge exponentially over the resistor and the voltage reaches 37 % of its initial value in time $\tau = RC$. *RC* is called the *circuit time constant*.

A.2.1.1 Non-ideal components

Real components always differ somehow from the ideal model. For instance, a resistor has an associated series self-inductance and a parasitic capacitance in parallel (Figure a.1), which may be important for high frequencies. A capacitor also has some inductance and a small series and large parallel resistance. A coil (inductor) presents some parallel parasitic capacitance and a series resistance, since it is made from conductor wire.

Furthermore, the way in which the components are situated on the circuit board and the printed circuit (PCB) layout itself introduce stray couplings, parasitic capacitances and inductances. Most of these effects may be neglected for low frequency circuits, but they are an important design issue for high frequencies. The layout of circuits dealing with low-level signals, like preamplifiers or high resolution ADC's, have also to be carefully designed so as to avoid such undesirable effects.

For a conductor, the resistance is proportional to its length *l* divided by its transverse section area *a*

$$R = \rho \frac{l}{a} \tag{a.12}$$

The constant ρ is the *resistivity* and depends on the temperature. For a metal conductor, ρ increases with temperature almost linearly from a few Kelvin to more than 1000 degrees and is usually modeled as

$$\rho = \rho_0 (1 + \alpha \Delta T) \tag{a.13}$$

where ρ_0 is the resistivity at a given reference temperature, α is the temperature coefficient (almost constant) and ΔT, the temperature difference with the reference. For example, for copper wire, $\rho_0 = 1.7 \cdot 10^{-8}$ Ω·m at 25° C and $\alpha = 0.0039$ °C^{-1}. A coil made with 500 m of copper wire of 0.2 mm diameter will have a resistance at 25° C of approximately (see (a.12)) $R = 1.7 \cdot 10^{-8} \cdot 500 / [\pi \cdot (0.1 \cdot 10^{-3})^2] = 270$ Ω. At a temperature of 0° C, the resistance will become $R' = R \cdot (1 - 0.0039 \cdot 25) = 244$ Ω (a change of about 10 %).

A.2.1.2 Practical values

Resistors used for small power are usually carbon film type. For low-noise or precision circuits, metal film type is preferable, at a higher cost. The commercial values most used are the so called series E-24 for carbon type (with 5% tolerance of nominal value), E-48 (2%) and E-96 (1%) for metal film type. E-24 series, for example, means that 24 nominal values are available within a decade, with an approximate logarithmic spacing (e.g. 10, 11, 12, 13, 15, 16, 18, 20, 22, 24, 27, 30, 33, 36, 39, 43, 47, 51, 56, 62, 68, 75,

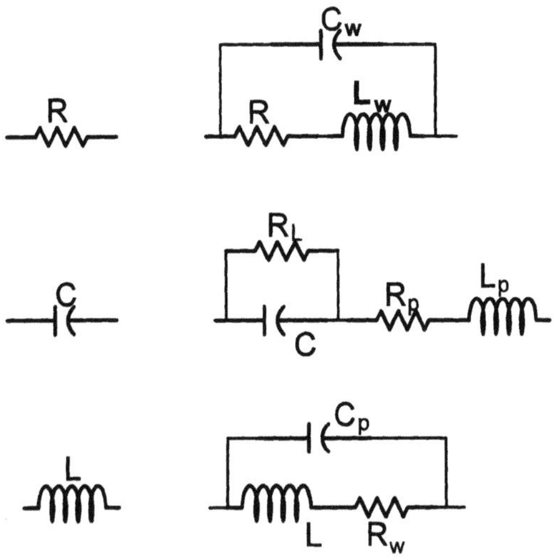

Figure a.1 Symbols of resistor (upper left), capacitor (middle) and inductor (bottom). The right column represents models for real components: A resistor has some parallel capacitance C_w and a serial inductance due to the resistive elements (carbon or metal film, wire winding) and the pins. A capacitor also presents a parallel leakage resistance R_L, a small series resistance R_p from plates and terminals and some inductance due to the same elements. An inductor is made of wire with some resistance R_w and the winding has a parasitic capacitance C_p. All these elements have usually small importance in low frequency circuits if the suitable type of component is chosen, but have to be considered for high frequency designs.

82, 91, 100). Most usual values range from 10 Ω to 1 MΩ. Power dissipation rates most commonly used are ½ W, ¼ W or 0.125 W.

Capacitors are made with a variety of dielectrics (the isolating material between the plates), which present different characteristics. Some examples are shown in Figure a.2. Ceramic types have low inductance and are suitable for high frequency applications, including power supply decoupling (high frequency ripple filtering). Their tolerance and temperature coefficient make them unsuitable for precision circuits. Polyester and polycarbonate types have good stability and linearity. They are used in low or medium frequency circuits (e.g. filters). Their inductance prevents their use for high (i.e. radio) frequencies. Electrolytic types are used also for low frequency and have large capacitance/volume ratio. Liquid electrolyte's have wide tolerances, poor stability and leakage, but high capacity, and are mainly used in power supply circuits. Solid tantalum types are suitable also for filters, because of their stability and low leakage. Both electrolytic types have to work under fixed sign polarization. This restriction may be overcome using two opposed capacitors in series, although the resultant capacity is half the capacity of one. These types of capacitors are rated for maximum voltages of e.g. 10 V, 16 V, 25 V, 40 V…, which cannot be exceeded or the devices may be destroyed. All other capacitor types also have a maximum voltage limit, but usually higher than electrolytics, except for the miniature versions.

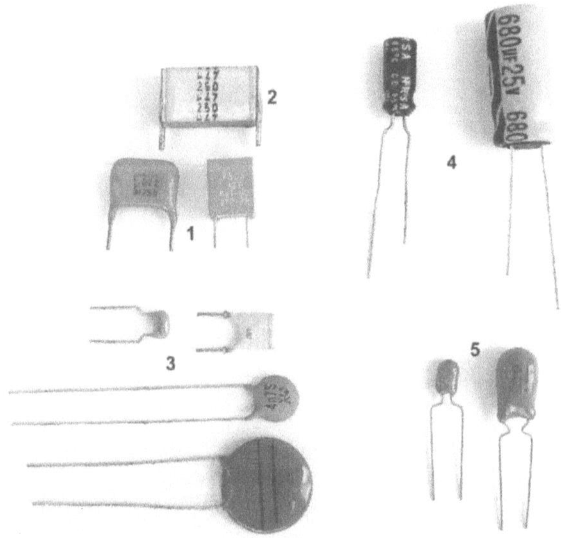

Figure a.2 Capacitors made with several types of dielectrics. 1: Polyester types of medium capacity 10-1000nF. 2: Polycarbonate. 3: Ceramic disk or multilayer (low capacity, high frequency). 4: Electrolytic, high capacity. 5: Solid tantalum. 4 and 5 are polar capacitors and have a limited voltage range.

A.2.2 CONNECTION OF COMPONENTS IN SERIES AND PARALLEL

Two resistors, R_1 and R_2, connected in series (the same current flows through both) have the total resistance $R = R_1 + R_2$. In parallel (they have both terminals in common and hence the same voltage drop), the inverse of their resistances is added, and thus the equivalent resistance is $R = R_1R_2/(R_1 + R_2)$. For two capacitors C_1 and C_2, connecting in parallel gives $C=C_1+C_2$ while in series, $C=C_1C_2/(C_1+C_2)$. For inductors, the rule is the same as for resistors.

A.2.3 A PASSIVE NON-LINEAR COMPONENT: THE DIODE

The *diode* is a semiconductor passive device that generally only permits flow of current in one direction. Ohm's law does not hold for it, since its voltage-current characteristic is not linear. Figure a.3 represents a piecewise linear model of the volt-ampere function of a diode. For small forward voltages, the current is very small (typically nA for a silicon diode). At a given threshold voltage v_t (0.6 V for silicon, 0.2 for germanium with small currents), the forward current increases and the dynamic resistance $R_f=\Delta v/\Delta i$ remains almost constant for a limited current. The static resistance v/i is current dependent and has little practical importance. R_f is typically 5-20 ohms, depending on the current.

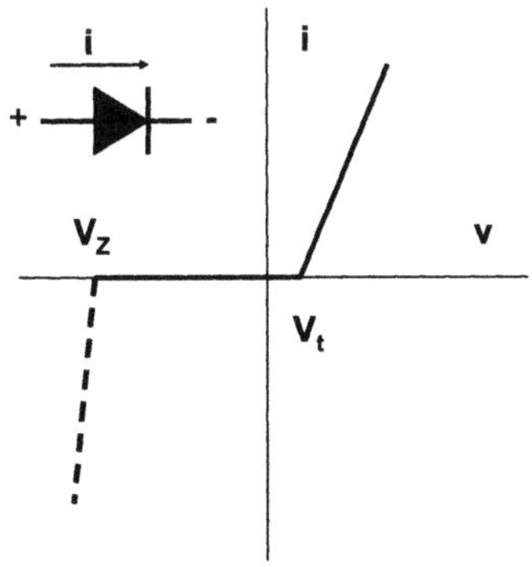

Figure a.3 A piecewise linear model of the diode volt-ampere characteristic. v_t is the threshold voltage, below which the diode is not a conductor. For an inverse voltage v_z the current increases quickly with a small voltage increment (dashed line). There exist Zener diodes designed to operate in this zone as voltage references.

Diodes are often used as rectifiers in power supply circuits. They also perform a variety of functions in radio and logic circuits. They are useful too for transient protection of inductive load drive circuits, such as relay control. A special kind of diode is the LED (light emitting diode). Its threshold voltage is about 1.6-1.7 V (for red, yellow and green types) and the maximum average current is typically 20-60 mA. Solar cells are also diodes, and act as voltage sources when illuminated.

For reverse voltages, the current is also very small until a breakdown, avalanche or Zener voltage V_z is reached, for which the current raises suddenly and the diode may be destroyed, unless it is a Zener diode, designed to operate in this avalanche zone, and the current is limited (for example by a series resistor). This kind of diode is used, e.g., for voltage reference, clipping circuits or voltage regulation.

A.3 DC and AC signals

DC (direct current) refers to power sources or signals of constant sign. AC (alternate current) is applied to power sources or signals whose sign changes periodically with time. By extension, any variable signal that takes positive and negative values along time may be considered as AC.

Power lines use sinusoidal voltages and currents. Any other periodic signal may be expressed as a sum of sinusoidal ones by the Fourier Theorem (see 6.2). Transient signals may also be represented by a Fourier integral of variable frequency sinusoids. A pure sinusoidal voltage *v(t)* is

$$v(t) = v_0 \cos(\omega t + \varphi) \quad \text{(a.14)}$$

The voltage amplitude is v_0. The frequency is $f=\omega/2\pi$ and φ is the initial phase (the angle for $t=0$). For sinusoidal voltages or currents, the amplitude is also the *peak* value. The *peak-to-peak* value is the two extremes difference, that is $v_{pp}= v_0-(-v_0)=2v_0$.

Sometimes, an AC voltage contains a DC value or *offset* v_{os}, which is its mean value. A pure AC signal has no DC component.

$$v_{OS} = \frac{1}{T}\int_{\tau}^{\tau+T} v(t)dt \qquad (a.15)$$

where T is the period. If the signal is not periodic, this expression has to be modified by taking the limit when T tends to infinity.

The same concepts apply to currents, with the correspondent $i(t)$, i_0, i_{pp} and i_{os}.
A current $i(t)$ in a pure resistor R dissipates a heat power $p(t)=i(t)\cdot v(t)=[i(t)]^2\cdot R$.
Its average value P is

$$P = \langle p(t) \rangle = \frac{R}{T}\int_{\tau}^{\tau+T} [i(t)]^2 dt \qquad (a.16)$$

The constant current (DC) value that would produce the same heat power is called the *RMS* (root mean square) or *effective* value I_{rms}

$$P = I_{rms}^2 R \qquad (a.17)$$

thus

$$I_{rms} = \sqrt{\frac{1}{T}\int_{\tau}^{\tau+T} [i(t)]^2 dt} \qquad (a.18)$$

Of course, the same relation applies to rms voltage V_{rms}. For a pure sinusoidal voltage or current, its rms value is related to its peak value and can be calculated by inserting $i(t) = i_0 \cos(\omega t)$ in (a.14):

$$i_0 = I_{rms}\cdot\sqrt{2} \quad \text{and} \quad v_0 = V_{rms}\cdot\sqrt{2} \qquad (a.19)$$

For a triangular wave, the relation is $v_0 = V_{rms}\cdot\sqrt{3}$ and for a square wave it is $v_0 = V_{rms}$.

When a voltmeter measures an AC voltage, its readout is usually a rms value. Cheap instruments simply measure the peak value and use the relation (a.15), which is accurate for sinusoidal waveforms, but not for others. If a voltmeter measures *true rms*, it calculates *rms* as (a.14), valid for any waveform. If an AC voltage is measured by using an oscilloscope, you will get directly the peak value, so do not be confused by the mismatch!

If the circuit is not purely resistive (there is some capacitance or inductance), in general the current will not be in-phase with the source voltage. As power is dissipated only by resistance (the real part of the circuit impedance (see A.3.2)), the average power will be $\langle p \rangle = I_{rms} V_{rms} \cos(\theta)$, where θ is the phase angle between the voltage and the current.

A.3.1 DECIBELS

The *bel* (B) was named in honor of A. G. Bell. It is used to compare two power levels. A power p_1 is one *bel* above another power p_2 if $p_1 = 10 \cdot p_2$. The *bel* number was defined as the decimal logarithm of the ratio of two power levels $B = \log (p_1/p_2)$. Therefore, a positive value represents power gain and a negative value power loss. More used is the *decibel* (dB), ten times smaller, thus the number of decibels is

$$dB = 10 \cdot \log(\frac{p_1}{p_2}) \tag{a.20}$$

The amplitudes ratio (for example voltages v_1 and v_2) may be used instead of the power ratio. As power is proportional to squared amplitude (for a constant load)

$$dB = 10 \cdot \log\left(\frac{v_1}{v_2}\right)^2 = 20 \cdot \log\left(\frac{v_1}{v_2}\right) \tag{a.21}$$

For example, a signal-to-noise ratio of 40 dB means that the signal power is $10^{40/10} = 10^4$ times the noise power, or that the signal amplitude is $10^{40/20} = 100$ times the noise amplitude, which is equivalent. Another example: An amplifier gain of 66 dB means that the input is amplified $10^{66/20} \cong 2000$ times. As a rule of thumb, every increase of 6 dB adds a factor of nearly two and every 20 dB adds a factor ten to the amplitude.

A.3.2 COMPLEX IMPEDANCES AND RESPONSE

A sinusoidal voltage of amplitude v_0 and angular frequency ω may be expressed in complex form

$$V = v_0 \cdot e^{i(\omega t + \varphi)} \tag{a.22}$$

where i is the imaginary unit $i = \sqrt{-1}$ (in electric circuits, it is customary to use j instead of i, to avoid confusion with current, but there is little danger of misinterpretation here) and the phase φ is the angle for $t=0$. Of course (see also 6.1), only the real part of (a.22) represents the actual voltage.

We can group the constant factors of (a.22) as the complex voltage amplitude V_0

$$V_0 = v_0 \cdot e^{i\varphi} \tag{a.23}$$

that is a complex number, whose modulus is the real voltage amplitude and its phase is the initial angle of the voltage.

For current, we can also define a complex amplitude $I_0 = i_0 \cdot e^{i\theta}$, where i_0 is the current amplitude and θ its initial phase. With this notation, and keeping in mind that in a linear circuit the voltage and currents must have the same frequency, but in general different phase, we can write

$$V = V_0 \cdot e^{i\omega t} \quad and \quad I = I_0 \cdot e^{i\omega t} \tag{a.24}$$

Therefore, in a capacitor, (a.7),

$$I = C\frac{dV}{dt} = C \cdot i\omega \cdot V \tag{a.25}$$

Simplifying the equation by removing the time-variable factor $e^{i\omega t}$, the relation between the complex amplitudes is

$$I_0 = V_0 \cdot i\omega C \tag{a.26}$$

This leads to the introduction of the complex impedance of a capacitor $Z_C = \frac{1}{i\omega C}$, and a generalized Ohm's Law

$$V_0 = I_0 \cdot Z \tag{a.27}$$

Similarly, a complex impedance for inductance Z_L is found using (a.5) as

$$Z_L = i\omega L \tag{a.28}$$

and for a resistance, simply $Z=R$. The relation (a.27) holds for any kind of impedance Z. The rules for the equivalent series and parallel components can also be generalized now: Impedances in series are added ($Z_S = \sum_i Z_i$) and the sum of inverses of several impedances in parallel is the inverse of the equivalent ($Z_P^{-1} = \sum_i Z_i^{-1}$). From these, the case for resistors, capacitors and inductors can be easily derived.

Let us show the use of these concepts to analyze a circuit response for a variable frequency input voltage. The simple *RC* circuit of Figure a.4 has an input, represented by the voltage sine source $v_i(t)$, with a complex amplitude V_i. The output is the voltage

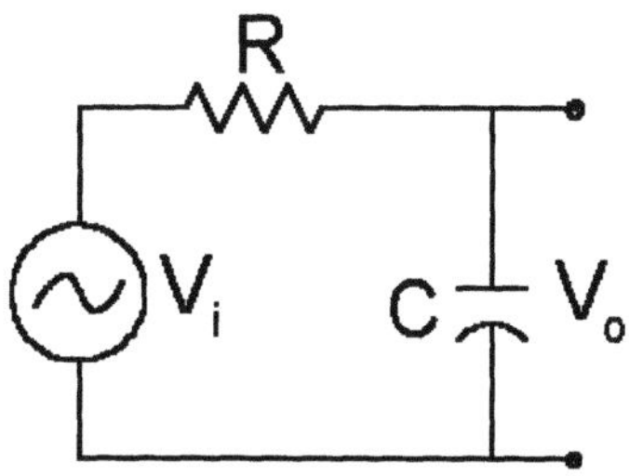

Figure a.4 An RC circuit to demonstrate the use of complex impedances, voltages and currents to find the circuit response as a function of frequency.

between the capacitor terminals, with the same frequency as the input and complex amplitude V_o.

The current I through the resistor and the capacitor will be, using (a.27),

$$I = \frac{V_i}{Z_R + Z_C} \tag{a.29}$$

and the output voltage,

$$V_o = I \cdot Z_C = \frac{V_i \cdot Z_C}{R + Z_C} \tag{a.30}$$

If we define the circuit response $T(\omega)$ as the ratio between the output and input complex amplitudes (see also Chapter 6),

$$T(\omega) \equiv \frac{V_o(\omega)}{V_i(\omega)} = \frac{1}{i\omega RC + 1} \tag{a.31}$$

This relation explicitly shows the dependence on the frequency. This will be the case in general when imaginary impedances are present.

The amplitude response is the relation between the output and input amplitudes and is just the amplitude or absolute value of $T(\omega)$, that is $|T(\omega)|$. The phase response $\Phi(\omega)$ represents the angle that the output is advanced relative to input. It is the (a.31) numerator phase minus the denominator phase, which in the present case is

$$\Phi(\omega) = 0 - \tan^{-1} \frac{\omega RC}{1} \tag{a.32}$$

In this example, the output phase is delayed with respect to the input signal.

A.4 Power supply

Any electronic equipment, with a few exceptions as e.g. simple passive filters, needs a power source to work. Most systems work with one or more DC voltage sources, which may be obtained from AC power lines, batteries (rechargeable or not), solar cells, etc.
A power supply must provide *power*, that is not only a suitable voltage, usually between two limits, but it also has to be capable of delivering a specified current without significant voltage drop. For example, a 12 V, 5 W power supply cannot feed a seismic recorder that needs 12 V, 1 A (12W).

The basic operations that a power supply performs are to change the voltage level, rectify AC, filter-out ripple (see Figure a.5) and regulate the output voltage. DC-DC converters use a DC input voltage (which may have a range) and supply another DC regulated voltage at a different level, which may be lower or higher than the input or even of inverse polarity. Figure a.5 represents a typical circuit of an unregulated power supply and the waveform obtained when it is connected to a load. A transformer changes the input AC voltage level, then the signal is rectified by a diode bridge and filtered by a high-value capacitor to smooth ripple to an acceptable amplitude. Keep in mind that transformers are specified for *rms* voltages: A 12 V transformer for mains will provide 12 V_{rms} or $12\cdot\sqrt{2}=17$ V peak, which, once rectified, may be some 16V DC.

Transformers are heavy parts and many modern power supplies operating from mains AC use active switching techniques for the same purpose: For example, rectifying directly the mains voltage and then obtaining the suitable DC voltages with switching DC-DC converters. The signal is cut up in small time duration DC pulses (kHz frequency), which are then fed to a charging capacitor through an inductor to make an average lower voltage. Sophisticated regulation electronics is used to make a stable voltage.

Voltage regulated power is required for most circuits. A simple monolithic linear voltage regulator, such as the LM7805 (National semiconductors), a three pins integrated circuit with input, output and ground, may provide regulation and short-circuit protection. Linear regulators are simple and have good ripple rejection, but have poor efficiency: for example, a 5 V regulator driving a 1 A load, whose input is a 12 V battery, will drain from the battery more than 1 A (the regulator itself consumes some current), so its efficiency will be less than 5/12 or 41 % while the rest of the power (in this case 7 W) is wasted as heat in the regulator.

DC-DC converters use a variety of techniques and are available for different voltage input ranges, output voltages and powers. Their efficiency may reach a 90% for some units, an advantage for power saving and low heat dissipation. Many of them operate as black boxes that the user just has to connect to his circuit. They generally accept a wide range of input voltages and provide one or several regulated output voltages. Some models have complete insulation between input and output (i.e. the input ground may be different of the output ground). For a good efficiency, all of them use switching techniques, usually at frequencies of tens or hundreds of kilohertz. The output has a high-frequency ripple, which may cause noise in analogue circuits unless filtered out by suitable external circuitry.

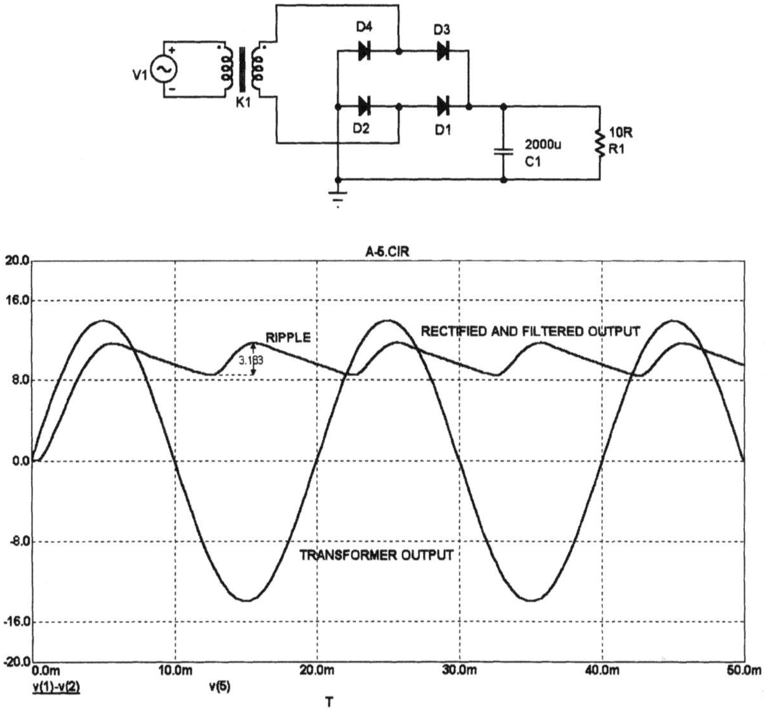

Figure a.5 A simple unregulated power supply. Upper: The circuit consists of a transformer to obtain a low voltage from the mains (in this case 50 Hz), a diode bridge full-wave rectifier and a capacitor filter. The load is represented by a resistor, draining about 1 A. Lower: The waveform obtained at the transformer output and at the circuit output. The ripple increases with the current load. A small delay is observed due to the RC filter formed by the diodes dynamic resistance and the capacitor.

A.5 Common laboratory instruments

The most used instruments at a test laboratory are the multimeter, the oscilloscope and the waveform generator.

The multimeter may measure voltage or current DC and commonly also AC, resistance and sometimes capacity.

The range may be selected by the user or automatically by the most sophisticated instruments. The voltage is measured between two points in parallel with the existing circuit, while for a current measurement, the circuit has to be interrupted and the meter inserted in series (Figure a.6). For resistance measurements, the circuit has to be powered off and the resistance must be disconnected from the circuit at least from one terminal, so the rest of the circuit resistance does not affect the measurement. The multimeter feeds a known current through the resistor and measures the voltage drop, thereby calculating the unknown resistance.

Analog multimeters, nowadays almost completely outdated, used microamp-meters to measure current, voltage and resistance. Digital multimeters, on the other hand, are internally just voltmeters. The current is measured by the voltage drop along a *shunt* resistor in series with the circuit and in parallel with the internal voltmeter.

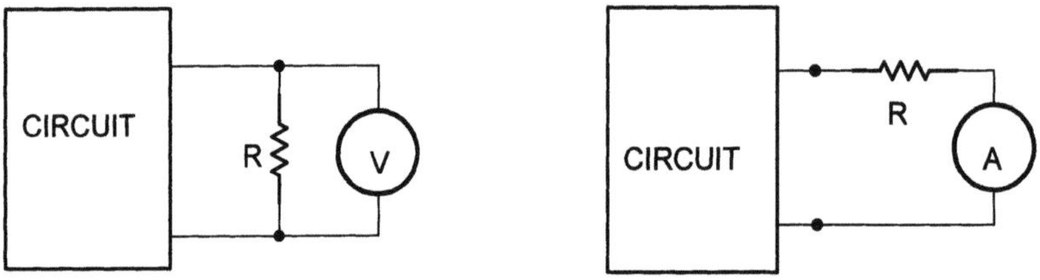

Figure a.6 Voltage and current measurement. For a voltage measurement (left), the meter is connected in parallel with the element through which the voltage drops. Right: Current measurement needs the circuit to be interrupted and the meter connected in series (AC current may be measured with a magnetic probe without circuit interruption).

Different ranges of measurements are selected by using voltage dividers in the instrument (Figure a.7). This may make the input impedance dependent of the range selected, which may itself affect the circuit under measurement. For the voltage divider of Figure a.7, the ratio between output and input voltages is

$$\frac{V_2}{V_1} = \frac{R_2}{R_1 + R_2} \tag{a.33}$$

and the input impedance (in this case pure resistance) R_i *is* $R_i = R_1 + R_2$, unless other non-negligible loads are connected to the output.

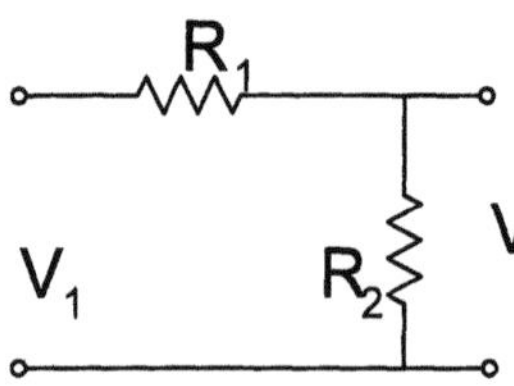

Figure a.7 Voltage divider used in a multimeter. The input voltage is V_1 and the meter measures V_2.

Oscilloscopes are essentially dynamic voltmeters that represent graphically the variable input voltage as a function of time. Most instruments have at least two input channels for voltage and some of them a separate synchronization input. Different y and x scales can be selected. Using a probe, the oscilloscope measures the voltage at a point referred to a common ground point. The probe ground may be connected to any point of the circuit under test, but if two probes are used or the oscilloscope ground is common to the power supply ground of the circuit, care must be taken to avoid a short-circuit. A better and safer practice is to always connect the probe ground to the circuit ground and use both probes to measure in differential mode, if this is required.

As the PC-based data acquisition cards have become reasonably cheap, a virtual oscilloscope is easy to implement in the lab or with a laptop for field use. Figure a.8 represents an example of such an instrument, as seen on the PC screen. All the equivalent controls of a real oscilloscope are present.

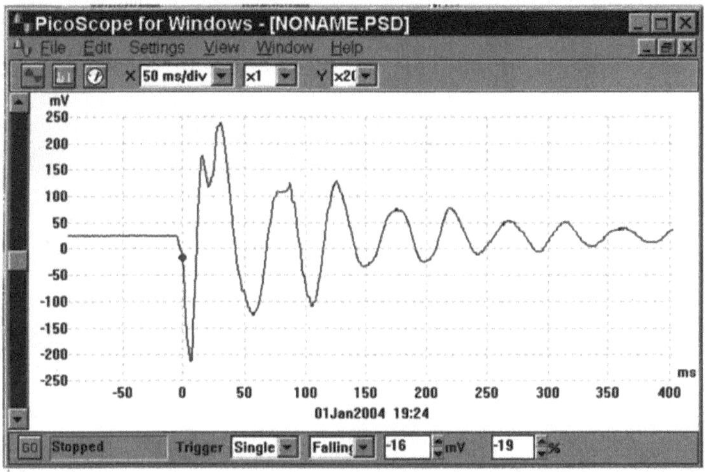

Figure a.8 An example of a virtual oscilloscope made with a laptop using a cheap module of data acquisition (from Pico Technology, www.picotech.com) and the accompanying program. Almost all the controls normally available in a real oscilloscope are present.

At present, many circuits can be tested, after being designed, without the need of building a real prototype, by using simulation programs that include schematic capture, components libraries, analogue and digital simulation algorithms, and even random parameters variations from nominal values to test the reliability of the circuit. An example is represented in figure a.9. An analog to digital converter (ADC) followed by a DAC (digital to analog converter) is simulated with Microcap, a program from Spectrum Software (www.specusa.com).

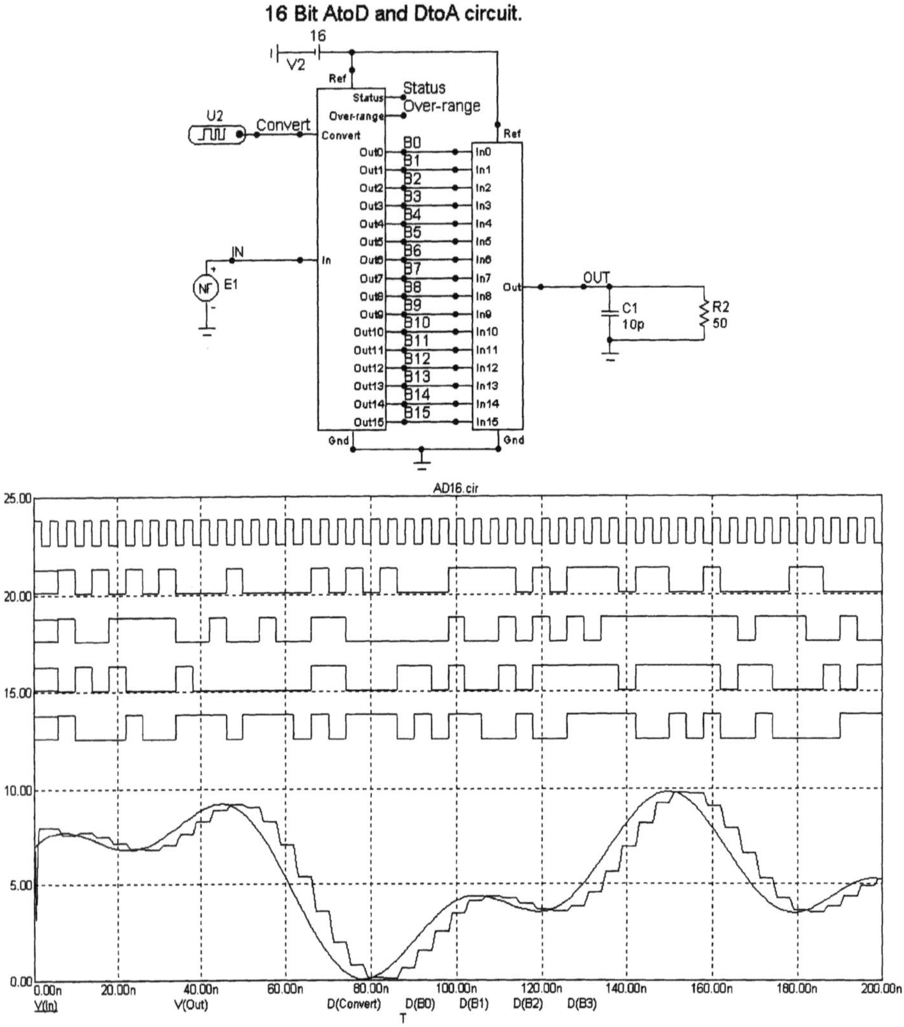

Figure a.9 An example of circuit simulation using a computer-aided design program. Upper: The schematics used for testing as defined by the user. The program has built in the functionality of all the components and can therefore calculate the outputs corresponding to a given user input (lower figure).

A.6 Amplifiers

Some sensor outputs are too low level to be used directly by digitizers or analog recorders, because the digitizing or recording device sensitivity is poor, and an amplifier must be used. As modern digitizers have improved resolution, this problem is less frequent nowadays. Another reason to use an amplifier is when the sensor is not located close to the recording system, and noise from external sources masks the signal. A low-noise preamplifier may adapt the output level of the sensor to the input of a digitizer or provide a low-impedance path to transmit the signal through a long cable. Sometimes the amplifiers are also used to implement some kind of filtering or analog processing of the signal. For instance, the feedback loop of a force balanced accelerometer (FBA) or broadband seismometer (BB) use amplifiers to integrate, add or subtract signals, etc.

The basic building block of an instrumentation amplifier is the *operational amplifier (op amp)*. Its schematic symbol is represented in Figure a.10.a. It has a pair of differential input terminals (v_i+ and v_i-) and an output terminal v_o. This means that the output voltage is the voltage difference between the input terminals multiplied by the gain, provided that both input voltages, with respect to ground, are within the *common mode input range*, generally limited by the supply voltage. The amplifier usually has a flat response from DC to some product specific cut-off frequency. The power supply is usually bipolar, so the output spans from near the negative supply voltage to near the positive supply voltage. The ideal operational amplifier has a gain of infinity (open loop gain), an input impedance of infinity and a null output impedance.

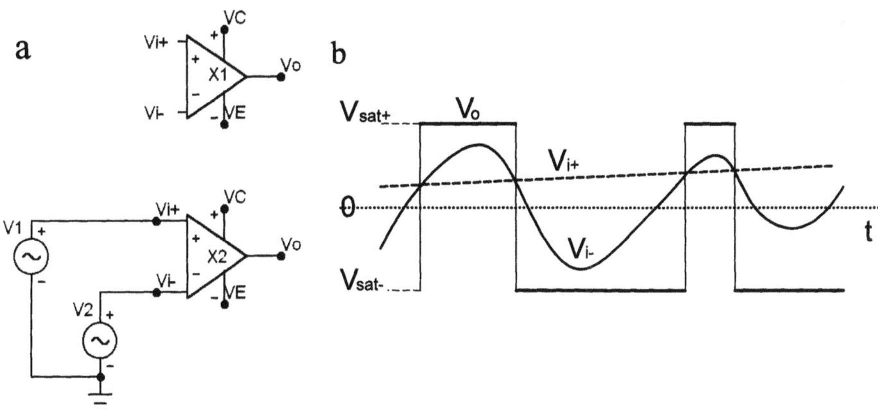

Figure a.10 a) Upper: the symbol of an operational amplifier (*op amp*). The terminals VC and VE represent the inputs for positive and negative power supply. Lower: In an open loop, the op amp acts as a voltage comparator, due to the very high open-loop voltage gain. b) An example of input waveforms of the comparator. The output is saturated at the positive end (V_{sat+}) of the output voltage swing if the voltage at the non-inverting input v_i+ is above the voltage at the inverting input v_i- ; it becomes saturated negative (V_{sat-}) in the opposite case.

Real amplifiers have an open-loop gain A_v of several million, their open-loop input impedance is typically tens of mega-ohm to giga-ohm and the open-loop output impedance is a few tens of ohm. They have also offset errors: for the output to be null, there must be a small voltage difference between the inputs (for the instrumentation models, this input offset is typically less than 1mV which can be further reduced by suitable external resistors).

Open loop configuration is only used when the op amp operates as comparator (Figure a.10.a left). Its differential input stage and high open loop gain drives the output to positive saturation (Figure a.10.b) if the non-inverting input v_i+ voltage is above (even a *fraction of a millivolt)* the inverting input voltage v_i-. The output becomes saturated negative in the opposite case $v_i+ < v_i-$. The output transition occurs within a few microvolt change at the inputs, but in practice, if the input signals contain some noise, this may lead to multiple edges transitions. In this case, a comparator with some hysteresis[1] is more suitable; that is, a comparator with a higher threshold voltage for the input rising edge than for the falling edge. A technique to implement this consists of applying a small positive feedback to the non-inverting input (a circuit known as *Schmitt trigger,* see, e.g. Millman, 1987).

A.6.1 BASIC AMPLIFIER CONFIGURATIONS

When the op amp is used as a linear amplifier, a much smaller gain is used (even less than 1 in some special cases) which is achieved by feeding back some of the output to the negative input. Since the feedback opposes the input signal, a reduced gain is obtained. This negative feedback thus determines the gain, the so called closed-*loop gain.* For the analysis of the circuit, it is now helpful to introduce the concept of *virtual ground.* In normal operation, the output is not saturated. This means that, due to the very high open loop gain, the voltage difference between the input terminals must be almost null. If one of the inputs is at zero level (ground), then the other input has to be null as well. This is called a *virtual ground.* We will show how this idea helps to analyze the closed-loop gains of the four basic configurations of the op amp as amplifier represented in Figures a.11.a-d: Inverting amplifier (a), non-inverting amplifier (b) differential amplifier (c) and a summing amplifier (d). Amplifiers (a) and (b) are also called single ended input amplifiers, since they amplify the voltage at their single input referred to ground. The four examples are single ended output, since their output voltages are at their terminals relative to ground.

All the configurations shown in Figure a.11 use negative feedback and each closed-loop gain is determined by the feedback network. Let us calculate the voltage closed-loop gain for these amplifiers.

The circuit represented in a) is an inverting amplifier since the output has opposite sign to the input. The non-inverting input is tied to ground, so, because of the *virtual ground,* the voltage at v_{i-} must also be zero. Since the input impedance is very high, the current flowing into the input is negligible, so the same current must flow through R_1 and R_2. By applying Ohm's law, the current is

$$i = \frac{0 - v_{O1}}{R_2} = \frac{v_1 - 0}{R_1}, \qquad \text{(a.34)}$$

and the voltage gain with feedback,

[1] Hysteresis is present in some physical systems able to "remember" somehow their past states. For example, many elastic materials have a non-unique stress-strain relationship: their present strain depends on how the stress has increased or decreased and not only on its present value. This means that the elastic work is not fully reversible and on a stress cycle a fraction of the work is dissipated as heat. Another known example of hysteresis is presented by ferromagnetic materials, whose magnetization depends on the history of the external field applied to them.

$$A_{vf} = \frac{v_{O1}}{v_1} = \frac{-R_2}{R_1} \tag{a.35}$$

The input impedance in this circuit is simply R_1, due to the virtual ground. In practice, the non-inverting input is connected to ground by a resistor (no effect on the gain) of equivalent value to the parallel of R_1 and R_2, to minimize offset error due to the input bias currents.

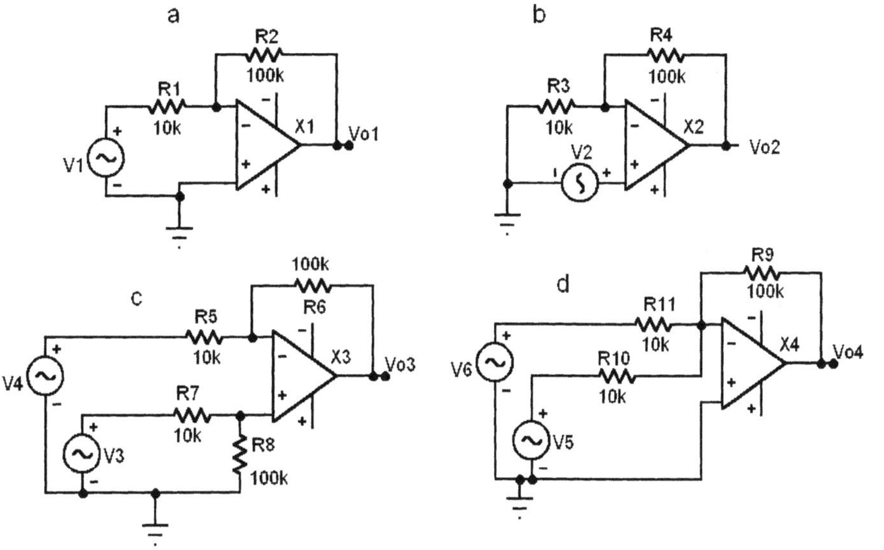

Figure a.11 Several linear applications of the *op amp* as amplifier. a) Inverting amplifier. The voltage gain in this example is –10. b) Non inverting amplifier. The gain is (1+10)=11. c) Differential amplifier. The output voltage is ten times the difference v3-v4. d) A summing amplifier. The output in this example is v_{04}=-10(v_5+v_6).

The circuit in a.11.b is a non-inverting amplifier. Since the voltage at v_{i-} has to be equal to the input at $v_{i+}=v_2$ (due to the virtual ground) and the same current flows on R_4 and R_3,

$$\frac{v_{O2} - v_2}{R_4} = \frac{v_2}{R_3}, \tag{a.36}$$

thus the voltage gain is

$$A_{vf} = \frac{v_{O2}}{v_2} = 1 + \frac{R_4}{R_3} \tag{a.37}$$

The input impedance in this case is very high, since the signal source is only connected to the *amplifier input, which draws a negligible current.* In practice, the signal source is connected to the input by means of a resistor, to minimize DC errors due to small input bias currents. If the signal source is a geophone, it will have a parallel damping resistor to also be considered.

We thus see that it is very simple to construct an amplifier with a desired gain by just adjusting the values of two resistors.

A particular case of a non-inverting amplifier is obtained using the circuit of Figure a.11.b by choosing R3 as infinity, that is, an open circuit. Then the value of R4 becomes irrelevant and in fact may be just a short. In this case, the gain is unity and the circuit is called a *voltage follower*. This is used as an impedance adapter, since the high input impedance does not affect the signal source and the voltage follower has very low output impedance.

By using the *superposition principle*[2], the output of the circuit in Figure a.11.c may be considered as the sum of an inverting amplifier with input v_4 and a non-inverting one whose input is $v_3 \cdot R_8/(R_7+R_8)$ (R_7 and R_8 form a voltage divider). Using (a.35) and (a.37), this output will thus be

$$v_{O3} = -v_4 \frac{R_6}{R_5} + v_3 \frac{R_8}{R_7 + R_8} (1 + \frac{R_6}{R_5}) \tag{a.38}$$

If we choose the resistors values so as $\frac{R_6}{R_5} = \frac{R_8}{R_7}$, it may be shown by using (a.38), that the output then depends *only on the difference* v_3-v_4 and is

$$v_{O3} = \frac{R_6}{R_5}(v_3 - v_4) \tag{a.39}$$

Hence, the *common mode gain* (gain of the same signals relative to the true ground) is null and the differential gain is R_6/R_5. The differential amplifier will thus suppress noise induced into a signal cable if the same amount is induced into both wires. The input differential impedance can be shown to be R_5+R_7. Again, in practice, making $R_5=R_7$ and $R_6=R_8$ helps to avoid DC errors. If a higher input impedance is needed, a circuit with three opamps is commonly used, generally at the cost of a higher output noise (e.g. each input of the differential amplifier is buffered with a high-impedance non-inverting stage).

Some active seismometers have differential output in order to match the differential input of the digitizers and to help rejection of external noise. This output may be easily implemented from a single-ended (referred to ground) output using in parallel two

---

[2] The superposition principle states that a general linear system responds to two inputs as the sum of the individual corresponding outputs. Applied to a circuit, if two independent signal voltage sources are used as inputs, the output is obtained as the sum of the individual outputs that the circuit will give for each source when the other is shorted.

output stages, one non-inverting and another one inverting. Differential inputs and outputs are illustrated in Figure a.12.

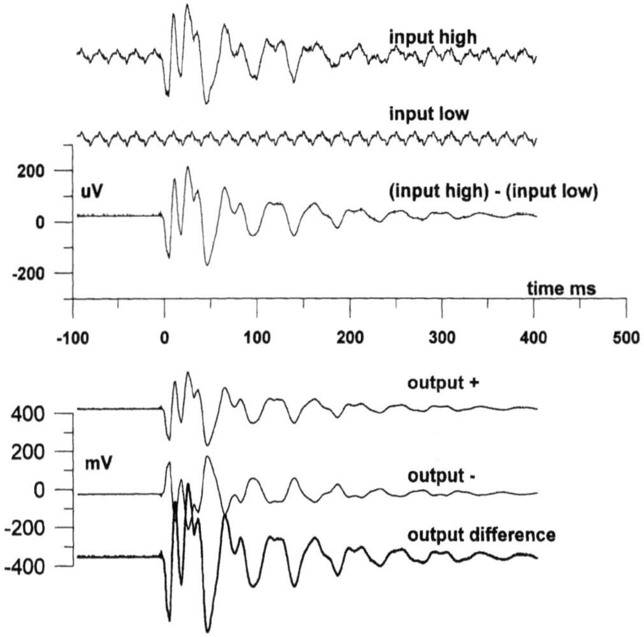

Figure a.12 An example of differential input and differential output. Top: A seismic pulse is contaminated with mains noise, picked-up in both wires (two top traces) from sensor to amplifier. The difference eliminates this noise (also a good cable shielding would do it).
Below: A differential output of the same amplifier with gain 60dB (x1000). Output- is output+ inverted. If only one output (single ended, relative to ground) is used (+ or -), the gain is x1000, but if the differential output is used by connecting to, e.g., a differential digitizer, the gain will be x2000 (bottom signal).

A 6.2 ACTIVE FILTERS

A passive RC filter has been described earlier. Since the RC filter has no amplifier, it attenuates the signal. The opamp may be used as a building block for so called active filters that do not attenuate the signal and even might have a gain. Multistage (multipole) active filters may be implemented with a variety of responses that can be sharper than the passive RC filters.

Filter theory is beyond the scope of this book, and many such circuits can be found in the application notes of opamp manufacturers. Let us show two simple examples: an integrator circuit and a second-order filter. Figure a.13.a shows an integrator (integrates the signal).

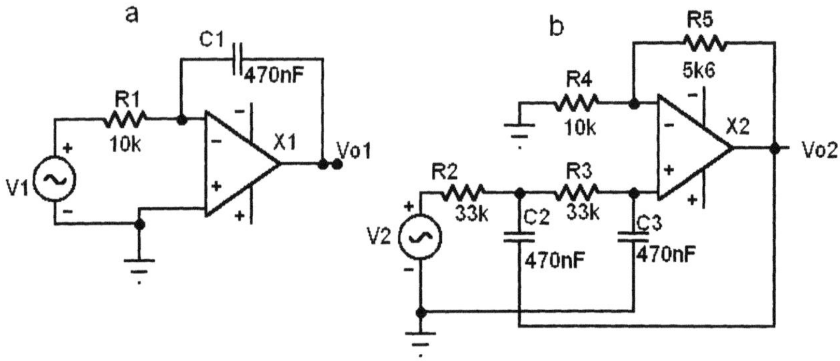

Figure a.13. Two examples of active filters. a: An active integrator. b: A second-order Butterworth low-pass filter.

Integrator. From the concept of virtual ground explained above, the current through R1 has to be just v_1/R_1 and this same current will flow through C_1. The voltage drop over it, v_{o1}, changes with time as

$$\Delta v_{o1} = \frac{1}{C_1}\int_0^t i dt = \frac{-1}{R_1 C_1}\int_0^t v_1 dt \tag{a.40}$$

In this expression, the time origin is taken when the voltage on the capacitor is null. This may be controlled by a temporal short on the capacitor at this instant. Integration in the time domain is equivalent to dividing by iω in the frequency domain, see (2.8), so the frequency response will be

$$H(\omega) = \frac{-1}{R_1 C_1 \cdot i\omega} \tag{a.41}$$

The amplitude frequency response is inversely proportional to the time constant R_1C_1 and to the frequency. The phase is delayed 90 deg (remember that a complex division by i is equivalent to a phase delay of $\pi/2$) and inverted, or advanced 90 deg. In practice, this circuit may be unstable, since the offset error of the op amp is integrated, producing an output continuous drift. This may be overcome by a high-value resistor in parallel with the capacitor, at the cost of a low-frequency limit for the circuit operation as integrator.

Low pass filter. The circuit of Figure a.13.b is a second order low-pass filter. It uses a topology or configuration known as Sallen-Key, with a positive feedback (through C2) in addition of the gain-fixing negative feedback with R4/R5. It may be shown, using the techniques explained above and with a little algebra, that for $R_2=R_3$ and $C_2=C_3$ the response T(ω) of this circuit is (e.g. Millman, 1987)

$$T(\omega) \equiv \frac{v_{o2}}{v_2} = A_v \frac{\omega_0^2}{\omega_0^2 - \omega^2 + 2h\omega_0 \cdot i\omega} \qquad \text{(a.42)}$$

with $\omega_0 = \dfrac{1}{R_2 C_2}$ and $2h = 3 - A_v$ where A_v is the amplifier gain. In this case, since it uses a non-inverting configuration, it is $A_v=1+R_5/R_4$. With the component value shown, the corner frequency is $f_0=10.3$ Hz, $A_v=1.56$ and h=0.72, close to the nominal value of 0.707 for a second-order Butterworth filter. If the resistors R_2 and R_3 are interchanged with the capacitors C_2 and C_3, we have a high-pass filter, with the response

$$T'(\omega) \equiv \frac{v_{o2}}{v_2} = A_v \frac{-\omega^2}{\omega_0^2 - \omega^2 + 2h\omega_0 \cdot i\omega} \qquad \text{(a.43)}$$

This is formally identical to the response of a mechanical seismometer for ground displacement (2.33), including the sign for the polarity convention, except for the factor A_v. The low-pass filter has a response as a mechanical pendulum when used as accelerometer (2.35), except for a constant factor. These three responses are represented in Figure a.14.

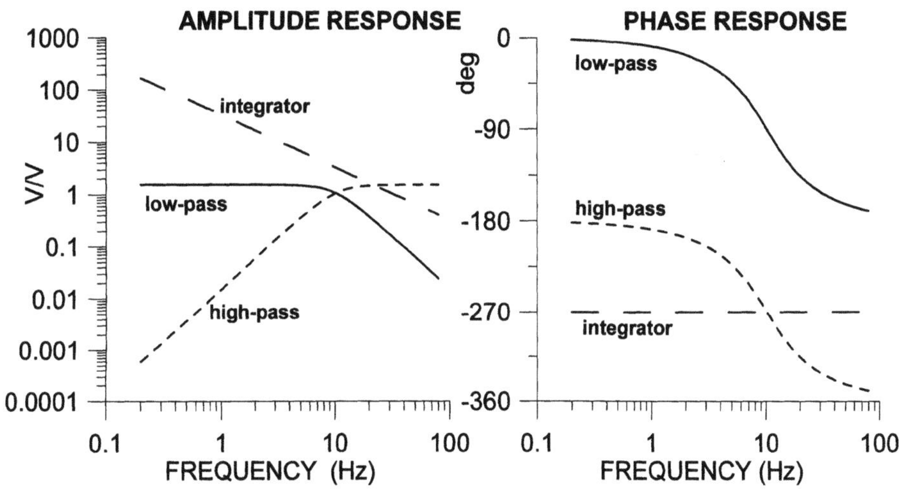

Figure a.14 Amplitude (left) and phase (right) response of the low-pass second order filter of Figure a.13b (solid), the integrator of Figure a.13a (long dashes) and a second order high pass filter (short dashes). Note the similarity of the latter with the mechanical seismometer response for ground displacement. The phase response of the integrator is –270 degrees instead of –90 because the sign inversion of the circuit in Figure a.13a.

A.7 Introduction to logical circuits

Logical or digital circuits operate with only two nominal levels of voltage or current that represent a *binary digit* or *bit*. Each family of digital circuits may use different definitions for the two levels, but the logical representation is the same. The two possible values are often represented as 0 and 1, TRUE and FALSE or HIGH and LOW. The mathematical basis to operate with binary variables is the Boolean algebra. The fundamental operations between these logical variables are AND, OR and NOT, defined by their *truth tables*:

x	y	x AND y
0	0	0
0	1	0
1	0	0
1	1	1

x	y	x OR y
0	0	0
0	1	1
1	0	1
1	1	1

NOT 0=1, NOT 1=0.

Abbreviated logical symbols for these operations are:

$x\ AND\ y \equiv x \cdot y$; $x\ OR\ y \equiv x + y$; $NOT\ x \equiv \bar{x}$

An auxiliary common operation is the exclusive OR (XOR), whose symbol is $\oplus$, which is true if ONLY one of the two variables is true. This may be expressed by the following relation: $x \oplus y = x \cdot \bar{y} + \bar{x} \cdot y$ (read “x and not y or not x and y”).

The symbols for the *logical gates* that implement these functions are represented in Figure a.15.

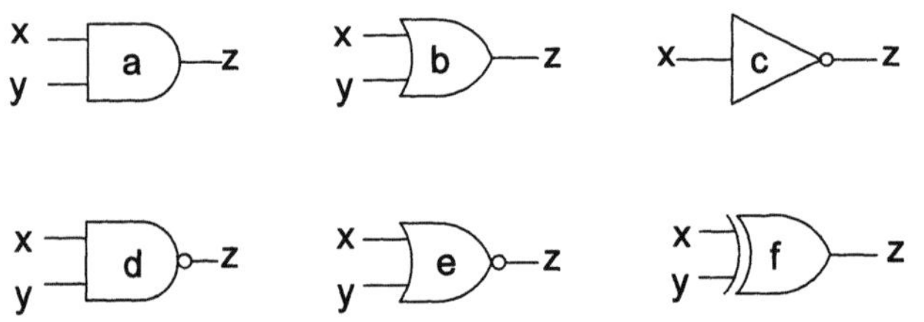

a) AND $z = x \cdot y$

b) OR $z = x + y$

c) NOT $z = \bar{x}$

d) NAND

$z = \overline{x \cdot y} = \bar{x} + \bar{y}$

e) NOR

$z = \overline{x + y} = \bar{x} \cdot \bar{y}$

f) XOR $z = x \oplus y$

Figure a.15 Left: The standard symbols for the main types of logic gates. Right: The logic functions that each of them performs.

These functions are implemented by specialized integrated circuits of different families, such as TTL (*transistor-transistor logic*), TTLLS (*TTL low power Schottky*), CMOS (*complementary metal-oxide semiconductor*), etc. Each family has a different range of power supply voltage, transition levels, input and output impedance, maximum clock frequency (how fast they can follow input state changes) and so on. Seismic equipment often requires low power consumption and does not need high speed. CMOS 4xxx series is suitable for many such applications. The power drained by these circuits is proportional to the transition frequency. Some seismic sensors include control circuits (e.g. for driving mass centering or calibration pulse sequences) with this type of gates.

The circuits with only logical gates perform *combinational logic*. Other kinds of logic functions require the use of circuits with *memory*, i.e. circuits whose output depends not only on the present input, but also on the past states. The basic building block of these *sequential circuits* is the flip-flop or one-bit-memory cell. There exist a number of implementations of such blocks. One of the simplest is the D-flip-flop. An example is the CD4013 chip, which has four inputs (*D, CK, CL* and *PR*) and two outputs (Q and $\overline{Q}$) (Figure a.16). $\overline{Q}$ is just the opposite of Q. The inputs *CL* (clear), *PR* (preset) and *D* (data) are normal logical inputs. *CK* stands for clock input and it is different from the other inputs in that it only reacts to the rising edge of a pulse, not the falling edge, so it can only register a change of state from 0 to 1, but not from 1 to 0. *CK* is usually used with a synchronous clock signal, while *CL* and *PR* are used with asynchronous signals and the output reacts immediately to a change in any of them, therefore they are labeled *asynchronous* inputs. If *PR* and *CL* are not active, the output Q follows the input *D* *synchronously* with the next rising edge of the clock input *CK*. Figure a.16 reproduces the complete truth table.

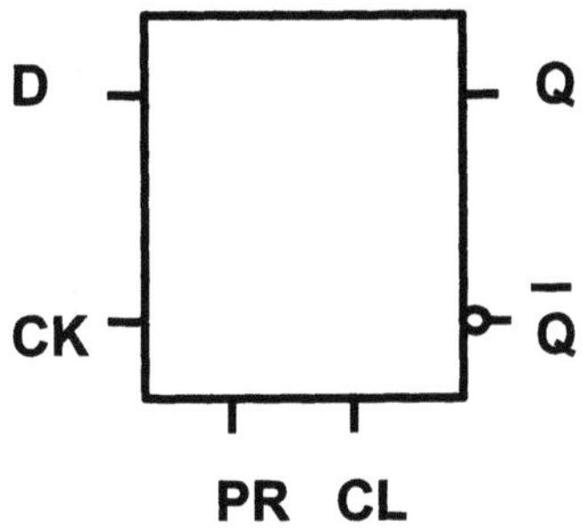

CK	D	CL	PR	Q_{n+1}	$\overline{Q}_{n+1}$
↑	0	0	0	0	1
↑	1	0	0	1	0
↓	x	0	0	Q_n	$\overline{Q}_n$
x	x	1	0	0	1
x	x	0	1	1	0
x	x	1	1	1	1

x: Don't care
↑ Clock rising edge

Figure a.16 Left, the symbol of a D flip-flop. Right, its truth table. Q_{n+1} represents the state of Q after the clock edge has passed and Q_n is the previous state of Q.

It is easy to implement a binary counter with flip-flops. For example, a D type may be configured so its output $\overline{Q}$ is connected to its input D. Then, with every rising edge of the input clock, the output state changes. This output may be used as clock input for another identical stage and so on. In this way, the output of each stage has half the frequency of the precedent one. Of course, there exist cheap integrated counters (or frequency dividers) of e.g. 4, 7, 20 or 24 stages (two counters of 24 and 14 stages are used in the circuit in Figure a.17 as an example).

Logic gates and flip-flops are the building blocks for every digital function. The microprocessors or the computer central processors, for instance, include thousands of such elements in a single integrated circuit. The ability to be programmed makes these systems versatile for implementing a variety of complex functions.

A.7.1 EXAMPLE: A DAILY CALIBRATION SEQUENCE GENERATOR FOR SEISMOMETERS.

A remote field station may suffer from changes in the constants, not so constant actually, or malfunction with time. A convenient way of monitoring the correct working and even of keeping a record of the possible changes is to periodically generate calibration pulses, for instance by injecting a known current to the calibration coil.

Let us describe in detail a circuit (Figure a.17) for this purpose and how it works. This will illustrate several concepts discussed in the previous sections.

The circuit has two distinct main parts: A digital sequence generator (can make a defined sequence of pulses or logical states) and a linear current generator (gives an output current proportional to an input voltage).

The digital block is implemented with CMOS circuits, for low power drain. As the frequency is not high, the 4000 series performs well. U1 is an oscillator plus a 24-stage binary counter. With a quartz crystal of 2.097152 MHz (2^{21} Hz), we obtain at the output of its last stage, Q24, a frequency of $2^{21}/2^{24} = 2^{-3}$ Hz, or a period of 8 s. This clock signal is then fed to a 14-stage counter, U2. If we want to get a calibration pulse sequence two times per day, to account for possible diurnal variations, we need a period of 12 h = 43200 s. Since our clock has a period of 8 s, we should divide its frequency by 43200/8 = 5400. Expressed in binary form, this is 5400 (decimal)=1010100011000 (binary). Then we must reset the counter when it reaches this final count (FC). This condition becomes true when "Q4 and Q5 and Q9 and Q11 and Q13" are 1 (there would be many other states that meet this condition, but the first one to occur is the count 5400). Its Boolean expression is

$$FC=Q4\cdot Q5\cdot Q9\cdot Q11\cdot Q13 \qquad (a.44)$$

This logic function is implemented in this case with four two-input AND gates (U3) as:

$$FC=((Q11\cdot Q13)\cdot(Q5\cdot Q9))\cdot Q4 \qquad (a.45)$$

This signal resets both counters, thus making the condition false immediately. Therefore it is high just for the time needed to propagate the change through the circuits, somewhat less than 1 μs. This pulse occurs every twelve hours.

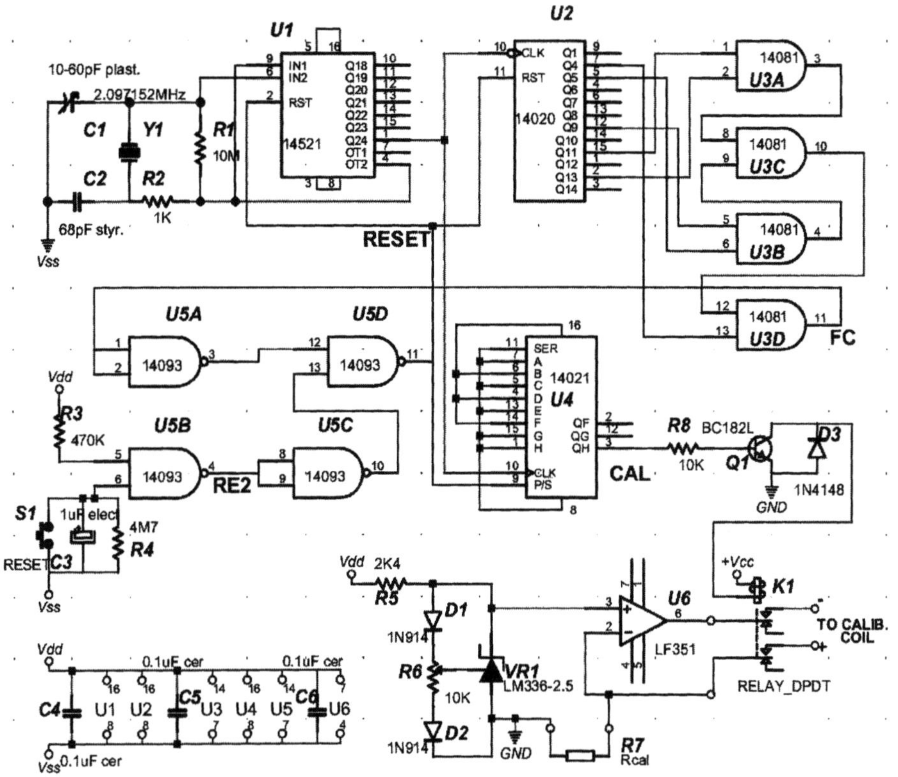

Figure a.17 A circuit to generate a sequence of three calibration pulses every twelve hours. U1 has a crystal oscillator and a 24-stage counter, U2 divides the frequency further and together with the logic about U3 and U5, provides two pulses per day. U4 generates then the three pulse sequence. The voltage reference VR1 and the opamp U6 form a current source connected to the sensor coil by K1, controlled by the pulses. See text.

To assure that the circuit starts up at a known state, the NAND gates of U5 provide a power-on reset. U5B acts as an inverter with hysteresis, that is, the input threshold level at rising edge is higher than the input threshold level at falling edge, just to prevent multiple resets due to possible noise. At power on, C3 is initially discharged and starts to charge through R3. The same effect is obtained when the manual reset button is pushed and left, discharging C3 for a while. The output of U5B, RE2, is 1 until the capacitor voltage reaches the input threshold, then it switches back to 0. U5C inverts again this pulse. The output of U5A is the signal FC inverted. The output of U5D, RESET, is thus:

$$RESET = \overline{(\overline{FC} \cdot \overline{RE2})} = FC + RE2 \tag{a.46}$$

In this way, the counters are RESET with the power-on OR the manual reset pushbutton OR *the final count.*

This same RESET pulse is fed to the *shift register* U4. This circuit charges asynchronously its eight flip-flops in parallel with the present inputs values (hardwired through inputs A to H) and then shifts them serially to the output synchronously with the next clock pulses. The effect is to generate a sequence 00101010, that is, three calibration pulses, each eight seconds high and eight low (since the input clock frequency is 1/8 Hz on CLK). Figure a.18 shows the waveforms at several points of the circuit at times around the final count.

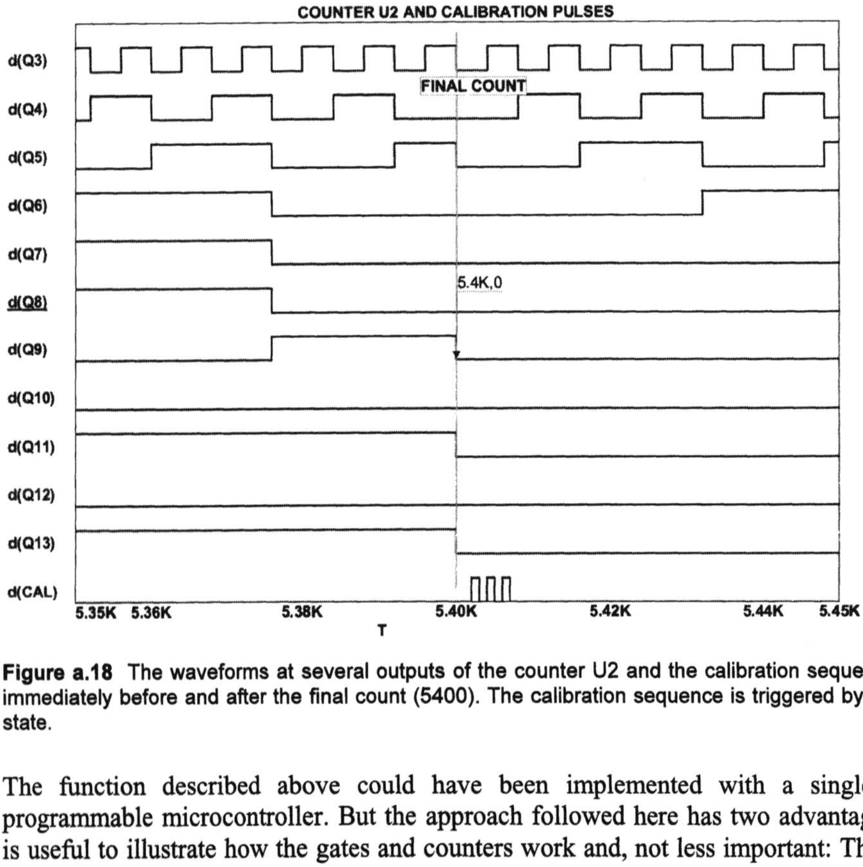

Figure a.18 The waveforms at several outputs of the counter U2 and the calibration sequence, immediately before and after the final count (5400). The calibration sequence is triggered by this state.

The function described above could have been implemented with a single-chip programmable microcontroller. But the approach followed here has two advantages: It is useful to illustrate how the gates and counters work and, not less important: There is no warranty that a program running on a microcontroller does not get out of control (techniques to prevent it do exist, such as *watch-dog* circuits), which is undesirable if the calibration unit is far away at a remote station or on the sea bottom! The simple design described has been working without a single fault for twenty years, as part of several stations of a radio-telemetered local network.

The analog part includes a voltage reference and a current source. The LM336 is a precision voltage reference of 2.5 or 5V. In this case we use the former. The diodes D1 and D2 and the adjustable resistor R6 is recommended by the manufacturer for maximum temperature stability. The voltage output is tied to the non-inverting input (pin 3) of an opamp with very high input impedance to avoid input current errors. Because of the *virtual ground* (see a.6.1), the voltages at pin 2 and pin 3 have to be the same when the circuit of the coil is closed. Therefore, since the opamp input current is negligible, the current I_C on the charge coil and the resistor R7 will be 2.5V/R7. For example if R7=1 Mohm, I_C=2.5 μA.

The sensor coil is connected to this current source by a relay switch K1, activated by the three pulse sequence every 12 hours. A transistor Q1 is used to provide enough current for the relay to be switched on. D3 protects the transistor from the inverse extra-voltage generated on the relay coil at disconnection.

The whole circuit is powered by a dual supply at ±5 V. Vdd is the positive voltage and Vss the negative one. The capacitors C4-C6, each mounted close to the integrated circuit, have the function of "decoupling" the power line, i.e., to avoid that the little current pulses drained by the CMOS circuits propagate as voltage noise on the power line within the circuit board. The power voltage may be up to ±7.5 V, for the CMOS circuits to be safe, but in this case the value of R5 should be changed to keep the current around 1 mA (Vdd-2.5 V≅R5·1 mA): This minimizes the temperature coefficient of VR1.

References

Aki, K. and P. G. Richards (1980). *Quantitative seismology – theory and methods*. W. H. Freeman and Company, San Francisco, Vol 1, Chapter 10: principles of seismometry, 477-524.

Alguacil, G., F. J. Almendros, E. Del Pezzo, A. García, J. M. Ibáñez, M La Rocca, J. Morales and R. Ortiz (1999): Observations of volcanic earthquakes and tremor at Deception Island – Antarctica. *Annali di Geofisica*, 42, 417-436.

Allen, R. V. (1978). Automatic earthquake recognition and timing from single traces. *Bull. Seism. Soc. Am.* 68,1521-1532.

Al-Shukri, H., G. L. Pavlis, F. L. Vernon III (1995). Site Effect Observations from Broadband Arrays. *Bull. Seism. Soc. Am.* 85, 1758-1769.

Almendros, J., J. M. Ibáñez, G. Alguacil, E. Del Pezzo and R. Ortiz (1997). Array tracking of the volcanic tremor source at Deception Island, Antarctica. *Geophys. Res. Lett.* 24, 3069-3072.

Almendros, J., J. Ibáñez, G. Alguacil and E. Del Pezzo (1999). Array analysis using circular wavefront geometry: An application to locate the nearby seismo-volcanic source. *Geophys. J. Int.* 136, 159-170.

Anderson, K. R. (1978). *Automatic Processing of Local Earthquake Data.* Ph. D. Thesis. Massachussetts Institute of Technology.

Aoi, S., T. Iwata, H. Fujiwara, K. Irikura (1997). Boundary Shape Waveform Inversion for Two-Dimensional Basin Structure Using Three-Component Array Data of Plane Incident Wave with an Arbitrary Azimuth. *Bull. Seism. Soc. Am.* 87, 222-233.

Asten, M.W. (1977). Theory and practice of geophone calibration in situ using a modified step method. *IEEE Trans. Geosc. Electr.*, GE-15, 208-214.

Asten, M. W., J. D. Henstridge (1984). Array estimators and the use of microseism for reconnaisse of sedimentary basins. *Geophysics* 49, 1828-1837.

Aziz, P.M., H.V. Sorensen and J. Van der Spiegel (1996): An overview of sigma-delta converters. *IEEE Signal Processing Magazine*, 13, No. 1, 61-84.

Barker, J. S., M. Campillo, F. J. Sánchez-Sesma, D. Jongmans, S. K. Singh (1996). Analysis of Wave Propagation in the Valley of Mexico from a Dense Array of Seismometers. *Bull. Seism. Soc. Am.* 86, 1667-1680.

Bartal, Y., Z. Somer, G. Leonhard, D.M. Steinberg and Y.B. Horin (2000). Optimal seismic networks in Israel in the context of the Comprehensive Test Ban Treaty. *Bull. Seism. Soc. Am.* 90, 1, pp. 151-165

Barzilai, A (2000). Improving a geophone to produce an affordable, broadband seismometer. Ph. D, thesis, Stanford University, 2000.

Barzilai, A., T. V. Zandt and T. Kerry (1998). Technique for measurement of the noise of a sensor in the presence of large background signals. *Review of Scientific Instruments*, 69, 2767-2772.

Berger, J., D. C. Agnew, R. L. Parker and W. E. Farrell (1979). Seismic system calibration: 2. Cross-spectral calibration using random binary signals. *Bull. Seism. Soc. Am.*, 69, 271-288.

Blackman, R.B. and J.W. Tukey (1966): *The measurement of Power Spectra.* Dover Publication Inc. New York, 190pp.

Bodin, P., J. Gomberg, S. K. Singh, M. Santoyo (1997). Dynamic Deformations of Shallow Sediments in the Valley of Mexico, Part I: Three-Dimensional Strains and Rotations Recorded on a Seismic Array. *Bull. Seism. Soc. Am.,* 87, 528-539.

Bormann, P. (editor) (2002). IASPEI *New manual of seismological observatory practice (NMSOP)*, GeoForchungsZentrum Potsdam.

Brune, J. N and J. Oliver (1959). The seismic noise at the earth's surface. *Bull. Seism. Soc. Am.*, 49, 349-353.

Bungum, H., E. S. Husebye and F. Ringdal (1971). The NORSAR array and preliminary results of data analysis. *Geophys. J.* 25, 115-126.

Bungum, H., E. S. Husebye (1974). Analysis of the operational capabilities for detection and location of seismic events at NORSAR. *Bull. Seism. Soc. Am.* 64, 637 – 656.

Byrne, C.J. (1961): Instrument noise in seismometers. *Bull. Seism. Soc. Am.* 51, 69-84.

Chiu, J.M.; Steiner, G.; Smalley, R.; Johnston, A.C. (1991): PANDA: A simple, portable seismic array for local- to regional-scale seismic experiments. *Bull. Seism. Soc. Am.* 81, 1000-1014.

Chouet, B., G. De Luca, G. Milana, P. Dawson, M. Martini, R. Scarpa (1998). Shallow Velocity Structure of Stromboli Volcano, Italy Derived from Small-Aperture Array Measurements of Strombolian Tremor. *Bull. Seism. Soc. Am.* 88, 653-666.

Camelbeck, T, J. Flick and B. Ducarme, editors (1990) *Seismic networks and rapid digital data transmission and exchange.* Cahiers du Centre Européen de Géodynamique et de Séismologie, Luxembourg, 1.

Capon, J. (1970). Analysis of Rayleigh-wave multipath propagation at LASA. *Bull. Seism. Soc. Am.* 60, 1701-1731.

Dainty, A.M., M. N. Toksoz, (1990). Array analysis of seismic scattering. *Bull. Seism. Soc. Am.* 80, 2242-2260.

Del Pezzo, E., M. La Rocca, J. M. Ibañez (1997). Observations of High-Frequency Scattered Waves Using Dense Arrays at Teide Volcano. *Bull. Seism. Soc. Am.* 87, 1637-1647.

Ferrazzini, V., K. Aki, B. A. Chouet (1991). Characteristics of seismic waves composing Hawaiian volcanic tremor and gas-piston events observed by a near-source array. *J. Geophys. Res.* 96, 6199-6209.

Frankel, A. (1994). Dense Array Recordings in the San Bernardino Valley of Landers-Big Bear Aftershocks: Basin Surface Waves, Moho Reflections, and Three-Dimensional Simulations. *Bull. Seism. Soc. Am.* 84, 613-624.

GSETT-3 (1997) Provisional GSE 2.1, Message formats & protocols, Operations Annex 3.

Goldstein, P. (1999). *SAC's user's manual.* Lawrence-Livermore Laboratory, University of California.

Green, P.E., R. A. Frosch and C. F. Romney (1965). Principles of an experimental large aperture seismic array (LASA). *Proc. IEEE* 53, 1821-1833.

Gupta, I. N, C. S. Lynnes, T. W. McElfresh, R. A. Wagner (1990). F-K analysis of NORESS Array and single-station data to identify sources of near-receiver and near-source scattering. *Bull. Seism. Soc. Am.,* 80, 2227-2241.

Hanka, W and R. Kind (1994). The GEOFON program. *Annali di Geofisica*, 37, 1060-1065.

Hardt, M. and F. Scherbaum (1994). The design of optimum networks for aftershock recordings. *Geophys. J. Int.* 117, 716-726.

Havskov, J., L.B. Kvamme, R.A. Hansen, H. Bungum and C.D. Lindholm (1992). The Northern Norway seismic Network: Design, Operation and Results. *Bull. Seism. Soc. Am.* 82, 481-496.

Havskov, J. and L. Ottemöller (1999). SEISAN earthquake analysis software. *Seismological Research letters*, 70, 532-534.

Holcomp, G. L. (1989). *A direct method for calculating instrument noise levels in side-by-side seismometer evaluations. Open-file Report 89-214*, U. S. Geol. Survey.

Holcomp, G. L (1990). *A numerical study of some potential sources of error in side-by-side seismometer evaluations. Open-file Report 92-302*, U. S. Geol. Survey.

Holland, A. (2003): Earthquake Data Recorded by the MEMS Accelerometer: Field Testing in Idaho. *Seismological Research Letters, 74*, 20-26.

Hutton, L. K. and D. Boore (1987). The Ml scale in Southern California. *Bull. Seism. Soc. Am.* 77, 2074-2094.

Iida, M., T. Miyatake, K. Shimazaki (1990). Relationship between strong-motion array parameters and the accuracy of source inversion, and physical waves. *Bull. Seism. Soc. Am.* 80, 1533-1552.

IRIS (1993). *Standard for exchange of earthquake data, manual*, published by "Incorporated Research Institutions in Seismology", printed by US. Geological Survey, 205 pp.

Kanasewich, E. R. (1973). *Time sequence analysis in geophysics.* The University of Alberta Press, 352 pp.

Kijko, A. (1977). An algorithm for the optimum distribution of a regional seismic network - I. *Pageoph* 115, 999-1009.

Kradolfer, U (1996). AutoDRM The first five years, *Seismological Research letters*, 67, 30-33.

Kromer, R. P. , and T. S McDonald (1999). *Report on the test evaluation of the Kinemetrics/Quanterra Q73B borehole digitizer.* Sandia National Laboratories report SAND99-2709, 36 pp

LaCoste, L. J. B. (1934). A new type long period seismograph, *Physics*, 5, 178-180.

Lay, T and T. C. Wallace (1995). *Modern global seismology*, Academic Press, 521 pp.

Lee, W. H. K. and S. W. Stewart (1981). *Principles and applications of microearthquake networks*, Academic Press.

Lee, W. H. K. (1989). *Toolbox for seismic data acquisition, processing and analysis. IASPEI software library, Vol. 1.* Published by IASPEI in collaboration with Seismological Society of America.

Lee, Y.W. (1967). *Statistical Theory of Communication.* John Wiley & Sons, New York, 507 pp.

Li, B.; D. Lu and W. Wang (2001): Micromachined Accelerometer with area-changed capacitance. *Mechatronics, 11,* 811-819.

Liu, C. H. and T. W. Kenny (2001): A High-Precision, Wide Bandwidth Micromachined Tunneling Accelerometer. *Journal of Electromechanical Systems, 10*, 425-433.

MacArthur, A. (1985). Geophone frequency calibration and laser verification. *Geophysics,* 50, 49-55.

MacWilliams, F. J. and N. J. Sloan (1976). Pseudo-random sequences and arrays. *Proc. IEEE* 64, 1715-1729.

Malone, S. (1996). "Near real time seismology", *Seismological Research Letters*, 67, 52-54.

Maxwell, P.; J. Tessman and B. Reichert (2001): Design through to production of a MEMS digital accelerometer for seismic acquisition. *First Break, 19*, 141-143.

Millman, J. (1987): *Microelectronics: Digital & Analog Circuits & Systems* McGraw-Hill Education, New York, 996 pp.

Mitronovas, W and E. Wielandt (1975). High-precision phase calibration of long period electromagnetic seismographs, *Bull. Seism. Soc Am.*, 60, 411-424.

Niazi, M. and Y. Bozorgnia (1991). Behavior of near-source peak horizontal and vertical ground motions over SMART-1 Array, Taiwan. *Bull. Seism. Soc. Am.* 81, 715-732.

Oppenheim, A. V. and R. W. Shafer (1989). *Discrete time signal processing.* Prentice Hall.

Ottemöller, L and J. Havskov (1999). SEISNET: A general purpose virtual seismic network, *Seismological Research Letters*, 70, 522-528

Pavlis, G.L. and F.L. Vernon (1994). Calibration of Seismometers Using Ground Noise. *Bull. Seism. Soc. Am.* 84, 1243-1255.

Peterson, J. (1993). *Observations and modeling of seismic background noise.* U. S. Geol. Survey Open-File Report 93-322, 95 pp

Press, W. H., S. A. Teukolsky, W. T. Vetterling and B. P. Flannery (1995): *Numerical Recipes in C: The Art of Scientific Computing.* Cambridge University Press, 994pp.

Proakis, J. G and D. G. Manolakis (1992). *Digital signal processing principles, algorithms and applications*. Macmillan Publishing Company.

Rabinowitz, N. and D. M. Steinberg (1990). Optimal configuration of a seismographic network: A statistical approach. *Bull. Seism. Soc. Am.* 80, 1, 187-196.

Riedesel, M., R. D. Moore and J. A. Orcutt (1990). Limits of sensitivity of inertial seismometers with velocity transducers and electronic amplifiers. *Bull. Seism. Soc. Am.* 80, 1725-1752.

Rodgers, P.W. (1992) a. Frequency limits for seismometers as determined from signal-to-noise ratios. Part 1. The electromagnetic seismometer. *Bull. Seism. Soc. Am.* 82, 1071-1098.

Rodgers, P.W. (1992) b. Frequency limits for seismometers as determined from signal-to-noise ratios. Part 2. Feedback seismometers. *Bull. Seism. Soc. Am.* 82, 1071-1098.

Shearer, P. M. (1999). *Introduction to seismology*. Cambridge University Press.

Scherbaum, F. (2001). Of poles and zeros, fundamentals of digital seismology, second edition. Kluwer Academic Publishers.

Silicon Designs, Inc. (2003): Model 2220 Analog Accelerometer Module (pdf). *www.silicondesigns.com*

Steck, L. and Prothero, W.A. (1989): Seismic calibration using the simplex algorithm. *Bull. Seism. Soc. Am. 79*, 1618-1628.

Steim, J.M. and Wielandt, E. (1985). *Report on the Very Broad Band Seismograph.* Harvard University, Cambridge, Massachusetts. 34 pp.

Stein, S. and M. Wysession (2003). *Introduction to seismology, earthquakes and earth structure.* Blackwell Publishing.

Steinberg, D. M., N. Rabinowitz, Y. Shimshoni, and D. Mizrachi (1995). Configuring a seismograph network for optimal monitoring of fault lines and multiple sources. *Bull. Seism. Soc. Am.* 85, 6, 1847-1857.

Stich, D.; G. Alguacil and J. Morales (2001): The relative location of multiplets in the vicinity of the Western Almería (Southern Spain) earthquake series of 1993-1994. *Geophys. J. Int. 146*, 801-812.

Trnkoczy, A. and Zivcic, M. (1992) Design of Local Seismic Network for Nuclear Power Plant Krsko, *Cahiers du Centre Européen de Géodynamique et de Séismologie*, Luxembourg, **5**, 31-41.

Usher, M.J.; Buckner, I.W. and Burch, R.F. (1977): A miniature wideband horizontal-component feedback seismometer. *Journal of Physics E: Scientific Instruments* 10, 1253-1260.

Usher, M.J., Burch, R.F. and Guralp, C. (1979). Wide-band feedback seismometers. *Physics of the Earth and Planetary Interiors*, 18, 38-50.

Utheim, T, J. Havskov and Y. Natvik (2001). Seislog data acquisition systems, *Seismological Research Letters*, 72, 77-79.

Ward, P., 1989. SUDS - Seismic Unified Data System. USGS Open-File Report 89-188,Denver, Colorado.

Wielandt, E. and Streckeisen, G. (1982): The leaf-spring seismometer: design and performance. *Bull. Seism. Soc. Am.,* **72,** 2349-2367.

Wilmore, P. L. editor (1979). *Manual of seismological observatory practice*, Report SE-20, World Data Center A for Solid Earth Geophysics, US Dep. of Commerce, NOAA.

COMPANY REFERENCES

Adaptive Broadband
75 Science Parkway, Rochester, New York 14620, USA
Tel: 716 242-9699, Fax: 716 242-9620
www.adaptivebroadband.com
webmaster@adaptivebroadband.com

AGECODAGIS SARL
16 rue d'Auriac, 31310 Rieux Volvestre, France
Tel: 33 5 61 97 84 52, Fax: 33 3 61 97 84 51
www.ageco.fr.fm
agecodagis@libertysurf.fr

Akashi Corporation
3286Lkebe-cho,Tsuzki-ku,Yokhama-shi, Kanagawa, 224-0053, Tokyo, Japan
Tel:45 932 6138, Fax: 45 932 5688
www.akashi-grp.co.jp
seismograph@akashi-grp.co.jp

Data-Linc
2635 151st Place N.E. Redmond, WA 98052, USA
Tel: 1 425-882-2206, Fax: 1 425-867-0865
www.data-linc.com
info@data-linc.com

DataRadio
5500 Royalmount Avenue, Suite 200, TMR, Montreal, Quebec, Canada H4P 1H7
Tel: 1 514-737-0020, Fax: 1 514-737-7883
www.dataradio.data.com
info@dataradio.com

Earth data
Kenda Electronic Systems, Nutsey Lane, Totton, Southampton, UK, SO40 3NB

Tel: 44 2380 869922 , Fax : 44 2380 86 0800
www.earthdata.co.uk
sales@earthdata.co.uk

Electronic Systems Technology, ESTEEM
415 North Quay Street, Kennewick, WA 99336, USA
Tel: 1 509 735 909, Fax: 1 509 783 5475
www.esteem.com
infor@esteem.com

Eentec, PMD Scientific
625 N. Euclid Ave., Ste. 404, St. Louis, MO 63108, USA
Tel: 1 314 454-9977, Fax: 1 314 454-9979
www.eentec.com
sales@eentec.com

ES&S, Environmental System and Services
8 River Street, Richmond, Victoria, 3121 Australia
Tel: 61 3 8420 8999, Fax: 61 3 8420 8900
www.esands.com
info@esands.com

FreeWave Technologies
1880 S. Flatiron Court, Suite F,Boulder, CO 80301, USA
Tel: 1 303 444-3862, Fax:1 303 786-9948
www.freewave.com
moreinfo@freewave.com

Geodevice
63 Fu Xing Avenue, Beijing, China, Post Box POB 166, Beijing, China
Tel: 86 10-68212359, Fax: 86 10-68162055
info@mail.geodevice.com.cn
http://www.geodevice.cn

Geospace
7334 N. Gessner, Houston, Texas 77040, USA
Tel: 1 713 939-7093 , Fax: 1 713 937-8012
www.geospacelp.com
sales@geospace.com

GeoSIG Ltd.
Europastrasse 11,8152, Glattbrugg, Switzerland
Tel.: 41 1 810 21 50, Fax: 41 1 810 23 50
www.geosig.com
info@geosig.com

Geotech Instruments, LLC
10755 Sanden Drive, Dallas, TX 75238-1336, USA

Tel: 1 214 221 0000, Fax: 1 214 3434400
www.geoinstr.com
info@geoinstr.com

Güralp
3 Midas House, Calleva Park, Aldermaston, Reading, Berks, RG7 8EA, UK.
Tel: 44 118 9819056 Fax:44 118 9819943
www.guralp.com
sales@guralp.com

Hakusan Corporation
J Tower10F, 1-1 Nikko Chou, Fuchu City, Tokyo Zip Code 183-0044
Tel: 81 42 333 080, Fax: 81 42 333 0096
www.datamark.co.jp
support@datamark.co.jp

Input/Output, Inc.
12300 Parc Crest Drive, Stafford, TX 77477
Tel: 1 281 933 3339, Fax: 1 281 879 3626
www.i-o.com
info@I-o.com

Kinemetrics Inc.
222 Vista Avenue, Pasadena, CA 91107, USA
Tel: 1 626 795 2220, Fax: 1 626 795 0868
www.kinemetrics.com
sales@kmi.com

LEAS SA
ZA de la Batie, route de Lancey BP 38, 38332 SAINT ISMIER CEDEX – FRANCE
Tel: 33 4 76 52 13 30, Fax: 33 4 76 52 18 60
www.lab-leas.fr
info@www.lab-leas.fr

Lennartz electronic
Bismarchstrasse 136, D-72072 Tübingen, Germany
Tel 49 7071 93550, Fax 49 7071 935530
www.lennartz-electronic.de
info@lennartz-electronic.de

Nanometrics
250 Hertzberg Road, Kanata, Ontario, Canada K2K2A1
Tel 1 613 592 6776 , Fax 1 613 592 5929
www.nanometrics.ca,
info@nanometrics.ca

RefTek
2626 Lombardy Ln, #105 Dallas, TX 75220, USA
Tel: 1 214 353 0609, Fax: 1 214 353 9659
www.reftek.com
info@reftek.com

SARA
Via A.Mercuri 4 - 06129 - Perugia - Italy
Tel: 39 075 5051014, Fax: 39 075 5006315
www.sarar.pg.it
info@sara.pg.it

Sercel
SERCEL Inc.,17200 Park Row, Houston, Texas 77084, USA
Tel: 1 281 492 66 88, Fax: 1 281 579 75 05
www.sercel.com
Sales.hou@sercelus.com

Sprengnether Instruments, Inc.
53 Millwell Drive, Maryland Heights, MO 63043, USA
Tel: 1 314 209 1682, Fax: 1 314 209 0026
www.sprengnether.com
info@sprengnether.com

Symmetric Research
9805 NE 116th St #7407, Kirkland WA 98034
Tel: 1 702 341-9325, Fax: 1 702 341-9326 24
www.symres.com
info@symres.com

Streckeisen AG Messgeräte
Dätlikonerstrasse 5, 8422 Pfungen, Switzerland
Tel: 41-52-3152161, Fax: 41-315 2710
gstr@swissonline.ch

TAIDE Enterprise Co
China
Tel: 86 756-2521081, Fax: 86 756-2521061
www.tai-de.com
market@tai-de.com

WiLan
2891 Sunridge Way N.E. Calgary, Alberta,Canada T1Y 7K7
Tel: (403) 273-9133, Fax: (403) 273-5100
www.wi-lan.com
info@wi-lan.com

Index

MODERN APPROACHES IN GEOPHYSICS

1. E.I. Galperin: *Vertical Seismic Profiling and Its Exploration Potential.* 1985 ISBN 90-277-1450-9
2. E.I. Galperin, I.L. Nersesov and R.M. Galperina: *Borehole Seismology.* 1986 ISBN 90-277-1967-5
3. Jean-Pierre Cordier: *Velocities in Reflection Seismology.* 1985 ISBN 90-277-2024-X
4. Gregg Parkes and Les Hatton: *The Marine Seismic Source.* 1986 ISBN 90-277-2228-5
5. Guust Nolet (ed.): *Seismic Tomography.* 1987 Hb: ISBN 90-277-2521-7; Pb: ISBN 90-277-2583-7
6. N.J. Vlaar, G. Nolet, M.J.R. Wortel and S.A.P.L. Cloetingh (eds.): *Mathematical Geophysics.* 1988 ISBN 90-277-2620-5
7. J. Bonnin, M. Cara, A. Cisternas and R. Fantechi (eds.): *Seismic Hazard in Mediterranean Regions.* 1988 ISBN 90-277-2779-1
8. Paul L. Stoffa (ed.): *Tau-p: A Plane Wave Approach to the Analysis of Seismic Data.* 1989 ISBN 0-7923-0038-6
9. V.I. Keilis-Borok (ed.): *Seismic Surface Waves in a Laterally Inhomogeneous Earth.* 1989 ISBN 0-7923-0044-0
10. V. Babuska and M. Cara: *Seismic Anisotropy in the Earth.* 1991 ISBN 0-7923-1321-6
11. A.I. Shemenda: *Subduction.* Insights from Physical Modeling. 1994 ISBN 0-7923-3042-0
12. O. Diachok, A. Caiti, P. Gerstoft and H. Schmidt (eds.): *Full Field Inversion Methods in Ocean and Seismo-Acoustics.* 1995 ISBN 0-7923-3459-0
13. M. Kelbert and I. Sazonov: *Pulses and Other Wave Processes in Fluids.* An Asymptotical Approach to Initial Problems. 1996 ISBN 0-7923-3928-2
14. B.E. Khesin, V.V. Alexeyev and L.V. Eppelbaum: *Interpretation of Geophysical Fields in Complicated Environments.* 1996 ISBN 0-7923-3964-9
15. F. Scherbaum: *Of Poles and Zeros.* Fundamentals of Digital Seismology. 1996 Hb: ISBN 0-7923-4012-4; Pb: ISBN 0-7923-4013-2
    2nd Edition. Hb: ISBN 0-7923-6834-7; Pb: ISBN 0-7923-6835-5
16. J. Koyama: *The Complex Faulting Process of Earthquakes.* 1997 ISBN 0-7923-4499-5
17. L. Tauxe: *Paleomagnetic Principles and Practice.* 1998 ISBN 0-7923-5258-0
18. C.H. Thurber and N. Rabinowitz (eds.): *Advances in Seismic Event Location.* 2000 ISBN 0-7923-6392-2
19. A.P. Nickolaenko and M. Hayakawa: *Resonances in the Earth-Ionosphere Cavity.* 2002 ISBN 1-4020-0754-X
20. R. Sabadini and B. Vermeersen: *Global Dynamics of the Earth.* Applications of Normal Mode Relaxation Theory to Solid-Earth Geophysics. 2004 Hb: ISBN 1-4020-1267-5: Pb: ISBN 1-4020-1268-3

21. W. Sandham and M. Leggett (eds.): *Geophysical Applications of Artificial Neural Networks and Fuzzy Logic*. 2003. ISBN 1-4020-1729-4
22. J. Havskov and G. Alguacil: *Instrumentation in Earthquake Seismology*. 2004. ISBN 1-4020-2968-3

MIX
Papier aus verantwortungsvollen Quellen
Paper from responsible sources
FSC® C105338

If you have any concerns about our products,
you can contact us on
ProductSafety@springernature.com

In case Publisher is established outside the EU,
the EU authorized representative is:
Springer Nature Customer Service Center GmbH
Europaplatz 3, 69115 Heidelberg, Germany

Printed by Libri Plureos GmbH
in Hamburg, Germany